Charged Particles

Edited by Malek Maaza
and Mahmoud Izerrouken

Published in London, United Kingdom

IntechOpen

Supporting open minds since 2005

Charged Particles
http://dx.doi.org/10.5772/intechopen.73999
Edited by Malek Maaza and Mahmoud Izerrouken

Contributors
Ian Lazarus, Takemi Otsuki, Suni Lee, Kei Yoshitome, Nagisa Sada, Megumi Maeda, Hidenori Matsuzaki, Naoko Kumagai-Takei, Yasumitsu Nishimura, Thomas Taylor, Jaydip Singh, Srishti Nagu, Jyotsna Singh, Takaaki Nomura, Ishaq Ahmad

First published in London, United Kingdom, 2019 by IntechOpen
IntechOpen is the global imprint of INTECHOPEN LIMITED, registered in England and Wales, registration number: 11086078, The Shard, 25th floor, 32 London Bridge Street
London, SE19SG – United Kingdom
Printed in Croatia

British Library Cataloguing-in-Publication Data
A catalogue record for this book is available from the British Library

Additional hard copies can be obtained from orders@intechopen.com

Charged Particles
Edited by Malek Maaza and Mahmoud Izerrouken
p. cm.
Print ISBN 978-1-78985-395-7
Online ISBN 978-1-78985-396-4

For EU product safety concerns:
IN TECH d.o.o., Prolaz Marije Krucifikse Kozulić 3, 51000 Rijeka, Croatia, info@intechopen.com or visit our website at intechopen.com.

Meet the editors

Prof. M. Maaza holds a PhD in Quantum Neutron Optics from the University of Pierre-Marie Curie and the Commissariat a lEnergie Atomique, respectively. He is currently a permanent joint staff member of the National Research Foundation of South Africa, in charge of the Africa-International relations desk at iThemba LABS and the University of South Africa. Prof. Maaza is the chair of reputable international organizations and commissions. His expertise is in the multidisciplinary field of nanosciences, photonics, and solar energy. Within the STI and RDI landscapes, with a Hirsch index of 33, Prof. Maaza has published more than 300 ISI international publications and supervised several postgraduates from all over Africa and the south.

Mahmoud Izerrouken joined the Nuclear Research Centre of Draria in 1996 and currently holds the position of Head of the Nuclear Techniques Department, Nuclear Research Center of Draria, Algiers, Algeria. He has coauthored over 30 publications in SCI journals and supervised several Master's and PhD students. His research interests are focused on ion- and neutron-induced damage in materials. He obtained his PhD at Ferhat Abbas University, Faculty of Sciences, Setif, Algeria.

Contents

Preface

Particle physicists believe that the creation of the universe and matter was started as a result of a very big explosion known as the big bang. At this time, matter and antimatter were created in equal amounts. Just after the big bang at $t \sim 10^{-35}$ s came the era of grand-unification transition, i.e. inflation, baryogenesis, monopoles, cosmic strings, etc. Sometime later, electro-weak phase transition occurred, i.e. electromagnetic and weak nuclear forces became differentiated. At $t \sim 10^{-6}$ protons and neutrons formed. Then nucleo-synthesis occurred, i.e. light elements were created, e.g. D, He, Li, etc. With the creation of light elements the amount of positive and negative charges was balanced inside an atom. There are electrons carrying a negative charge and protons carrying a positive charge inside atoms. This book briefly looks at high-energy charged particle physics and gives a flavor of the application of charged particles.

The first part of the book covers the cutting-edge and important research topic of flavor physics to search for new areas of physics via charged particles that appear in different extensions of the standard model. The latest research on the analysis of ultra-high energy muons using the pair-meter technique is also presented in a Geant4 simulation study for iron plates. In this study, the feasibility of detecting high-energy muons (1–1000 TeV) in the underground iron calorimeter detector is demonstrated. The idea of the Eloisatron to Pevatron is also included in this book.

The second part of this book covers the important research field of electrostatic waves in magnetized electron/positron plasmas where the behavior of the arbitrary amplitude of electrostatic waves propagation in an electron/positron plasma is discussed in detail. Well-known fluid and kinetic approaches have been used to describe linear waves, whereas nonlinear analysis of electrostatic waves is done via fluid modeling.

The final part of the book addresses the most important applications of charged bodies. Apart from high-energy particle physics, the application of charged bodies is also included in this book, such as immune effects of negatively charged particles dominating indoor air conditions and many others.

The authors gratefully acknowledge all those who contributed to this book : Prof. Dr. Ishaq Ahmad, Prof. Dr. Jaydip Singh, Prof. Dr. Srishti Nagu, Prof. Dr. Jyotsna Singh, Prof. Dr. Thomas Taylor, Prof. Dr. Takaaki Nomura, Prof. Dr. Ian Joseph Lazarus, Prof. Dr. Suni Lee, Prof. Dr. Yasumitsu Nishimura, Prof. Dr. Naoko Kumagai-Takei,

Prof. Dr. Hidenori Matsuzaki, Prof. Dr. Megumi Maeda, Prof. Dr. Nagisa Sada, Prof. Dr. Kei Yoshitome, Prof. Dr. Takemi Otsuki, and Ms. Marijana Francetic, Author Service Manager.

Prof. Dr. Maaza Malek
iThemba LABS-National Research Foundation, Somerset West, South Africa,
UNESCO-UNISA Africa Chair in Nanosciences/Nanotechnology,
University of South Africa

Prof. Dr. Mahmoud Izerrouken
Nuclear Research Center Draria, Algiers, Algeria

Section 1

Introductory Chapter

Chapter 1

Introductory Chapter: Charged Particles

Mahmoud Izerrouken and Ishaq Ahmad

1. Overview

Bringing history back in August 1912, Austrian physicist Victor Hess discovered cosmic rays coming from outer space. These cosmic rays consist of high-energy particles, entering from outer space, such as mainly protons, helium, and heavier nuclei up to uranium. When these cosmic rays come to earth to interact with upper atmosphere, they collide with the nuclei of atoms, creating more high-energy particles such as pions. The charged pions can quickly decay into two particles, a muon and a muon neutrino or antineutrino. Several high-energy particles were also discovered, which is long list. Studies of cosmic rays opened the door to a world class of particles.

It is concluded that charged particle is a particle that carries an electric charge. In atomic levels, the atom consists of nucleus around which the electrons turn. The nucleus is formed by proton and neutron and thus carries a positive charge (the proton charge is 1.602×10^{-19} Coulombs). The electron carries a negative charge (-1.602×10^{-19} Coulombs). An atom is called neutral if the number of protons equals the number of electrons. Thus, an atom can be positive, negative, or neutral. The charged particle is negative when it gains electron from another atom. It is positively charged if it loses electron from it. Applications of charged particles are subjected to control their motion and energy through electric field and magnetic field. Therefore, motion of charged particle in electric and magnetic fields is discussed in order to understand the beam of charged particles and their applications.

2. Charged particles motion in an electromagnetic field ($\vec{E}$, $\vec{B}$)

The motion of charged particle of mass m and charge q with a velocity $\vec{v}$ in uniform magnetic field $\vec{B} \neq \vec{0}$ and uniform electric field $\vec{E}$ is subjected to an electromagnetic force called the Lorentz force given by:

$$\vec{F} = q\vec{E} + q\vec{v} \times \vec{B} \tag{1}$$

From Newton second law, the particle's equation of motion is written as:

$$m\frac{d\vec{v}}{dt} = q\left(\vec{E} + \vec{v} \times \vec{B}\right) \tag{2}$$

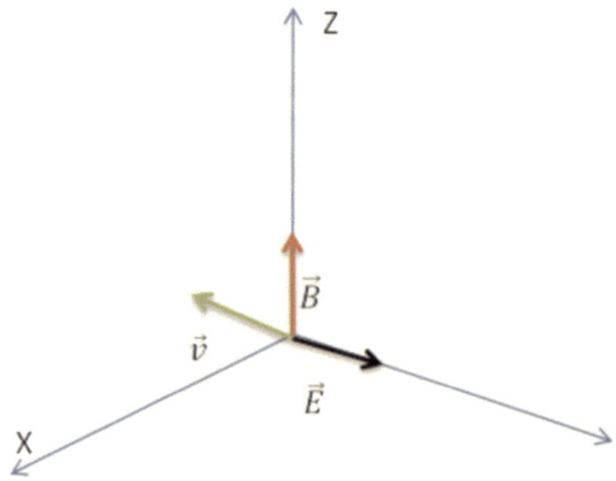

Let us assume that the magnetic field is applied in the direction Oz and electric field $\vec{E}$ is applied in the direction Oy.

1. In the case of motion of charged particle through a stationary electric field $\vec{E} = \vec{E_0}$ and $\vec{B} = \vec{0}$, the equation of motion is:

$$m\frac{d\vec{v}}{dt} = q\vec{E} \Leftrightarrow \vec{v} = \frac{q\vec{E}}{m}t + \vec{v_0} \tag{3}$$

The projection of (3) on the axes gives

$$\begin{aligned} m\frac{dv_x}{dt} &= 0 \\ m\frac{dv_y}{dt} &= qE \\ m\frac{dv_z}{dt} &= 0 \end{aligned} \tag{4}$$

The integration of (4) with the initial conditions
$x(0) = 0, y(0) = 0,$ and $z(0) = 0; vx(0) = v_{x0}, vy(0) = v_{yo},$ and $vz(0) = v_{z0}$ gives:

$$\begin{aligned} x &= v_{x0}t \\ y &= \frac{qE}{2m}t^2 + v_{y0}t \\ z &= v_{0z}t \end{aligned} \tag{5}$$

If $\vec{E}$ and $\vec{v}$ are collinear, the motion is rectilinear and uniformly accelerated.

2. In the case of motion of charged particle in uniform magnetic field $\vec{B} \neq \vec{0}$, the projection of (2) on the axes gives:

$$\begin{aligned} m\frac{dV_x}{dt} &= qBV_y \Leftrightarrow \frac{dV_x}{dt} = \frac{qB}{m}V_y \\ m\frac{dV_y}{dt} &= -qBV_x + qE \Leftrightarrow \frac{dV_y}{dt} = -\frac{qB}{m}V_x + \frac{q}{m}E \\ m\frac{dV_z}{dt} &= 0 \end{aligned} \tag{6}$$

We note $\Omega = \frac{qB}{m}$ the so-called cyclotron frequency.

The solution of (6) is:

$$x = \frac{E}{B}t + \frac{1}{\Omega}\left(v_{x0} - \frac{t}{B}\right)\sin(\Omega t) + \frac{v_{y0}}{\Omega}(1 - \cos(\Omega t))$$

$$y = \frac{1}{\Omega}\left(v_{x0} - \frac{t}{B}\right).(\cos(\Omega t) - 1) + \frac{v_{y0}}{\Omega}\sin(\Omega t) \quad (7)$$

$$z = v_{z0}\, t$$

According to the above equations, we can see that the charged particle can have different trajectories. Depending on the initial conditions, the trajectory could be straight, parabolas, circles, cycloids, spirals, etc. Lorentz force is then a base of charged particle acceleration and beam guidance.

3. Charged particle accelerators

After understanding the concept of controlling and generating the charged particles, different machines were developed to deflect or accelerate the charged particles through electromagnetic fields. The machines that generate and push charged particles to very high speed and energies and contain them in well-defined beams are called charged particle accelerators. A large variety of accelerators were developed since that fabricated by Cockcroft and Walton in 1932. Using such accelerator, the authors achieved the first nuclear reaction using artificially accelerated particle:

$$p + Li \rightarrow 2He$$

Since then, more and more successful accelerators appeared according to the progress in particle acceleration techniques. Depending on the accelerated particle trajectory, we can distinguish linear accelerators and circular accelerators. The accelerator during the last century can be found chronologically in [1]. Currently available electrostatic accelerators and cyclotrons over the world can produce and accelerate intense and stable charged particle beams with energy varying between few keV and few TeV. Charged particle accelerators are classified as per their applications:

4. Classification of charged particles

Charged particle accelerators are classified mainly into electrostatic and electromagnetic.

4.1 Electrostatic charged particle accelerators

In electrostatic accelerators, the static high voltage was generated and then applied across the ion source. The charged particles are accelerated through static electric field generated from static high voltage due to the electrostatic force. These types of accelerators are suitable to accelerate light and heavy ions from keV to few MeV energies. These ion beams, charged particle beams of various energies, are standard research tools in many areas of sciences and engineering having many applications in nuclear physics, atomic physics, medicine, materials science, agriculture, industry, and so on [2–6]. It is an advanced and versatile tool frequently applied across a broad range of discipline and fields.

4.2 Electromagnetic charged particle accelerators

Electrostatic charged particle accelerators have limitations on its beam energy due to high electrical voltage discharge. To avoid electrical discharge and increase charged particle energies, techniques involved electromagnetic fields instead of electrostatic fields. Electromagnetic acceleration is possible from two mechanisms either nonresonant magnetic induction or resonant circuits or cavities excited by oscillating radio frequency. High-energy charged particle colliders are installed around the world for forefront of scientific discoveries. These colliders are based on electrostatic charged particle accelerators. Through high-energy colliders, standard models are verified experimentally. Moreover, the cutting-edge and important research topic of flavor (particle physics) to search for new physics via charged particles that appears in the different extension of standard model is presented in this book. The latest research on analysis of ultrahigh-energy muon using pair-meter technique is also presented in Geant4 simulation study for iron plates. In this study, the feasibility for detection of high-energy muons at the underground iron calorimeter detector is demonstrated. The basic aim of this study is to detect high-energy muons (1–1000 TeV). The idea of the Eloisatron to Pevatron is also included in this book.

5. Charged particles applications

Charged particles interact with electrons and atom nuclei via Coulomb force, also called electrostatic force. When two charges are placed near to each other, they will be repulsed if they have the same charge or attract each other if they are of opposite charges. Each particle exerts a force on each other given by Coulomb's law expressed as:

$$F = k\frac{Q1Q2}{r^2}$$

where Q1, Q2, and r are the charge of the two particles (1 and 2) in Coulomb and the distance between the charges and k is the proportionality constant:

$$k = \frac{1}{4\pi\varepsilon}$$

Thus, when accelerated charged particle moves in materials, it interacts with orbital electron and nuclei via Coulomb interaction depending on its energy. At low energy (<0.01 MeV/u), the interaction is with nuclei, known as elastic interaction. This interaction leads to atomic displacement. At high energy (>0.01 MeV/u), the interaction is mainly with orbital electron known as inelastic interaction and leads to ionization and excitation. At very high energy, nuclear reactions can be produced and give rise to new particles (neutron, proton alpha, gamma rays). The basic interaction process of charged particle with matter is well known, and much performed detectors are now available. So ion beam is actually used in several applications. Electrostatic waves in magnetized electron-positron plasmas are covered in this book where the behavior of arbitrary amplitude of electrostatic wave propagation in electron-positron plasma is discussed. The well-known fluid and kinetic approaches have been used to describe linear waves, whereas the nonlinear analysis of ESW is done via fluid modeling. Apart from the high-energy particle physics, charged bodies are also included in this book such as immune effects of negative charged particles dominated by indoor air conditions and many others.

6. Charged particle beams for materials analysis

Several techniques of ion beam analysis (IBA) are being used for the study of the chemical composition and structure of surfaces, interface, and thin layers and are explained as below.

6.1 Rutherford backscattering spectroscopy

The accelerated charged particle with energy E_0 and mass M_1 scatters from the analyzed surface containing the particle M_2 with energy E_1 and scattering angle θ. From the conservation laws of energy and momentum and the known Rutherford cross section, it is possible to deduce the mass M_2 and estimate its abundance [7, 8].

6.2 Nuclear reaction analysis (NRA)

The NRA technique is very useful as a tool for the detection and profiling of light elements. The fast charged particle (few MeV) initiates a nuclear reaction with target atom. The reaction products are characteristic for this reaction and can be used to identify the target atom and its concentration. For example, determination of hydrogen content in material: for this purpose, $H(^{15}N, \alpha\gamma)^{16}O$ nuclear reaction is used. This reaction produces alpha particle and excited ^{12}C isotope. The disintegration of the excited ^{12}C to ground state emits a gamma photon with a well-defined energy of $E\gamma$ = 4.43 MeV, which identifies hydrogen content in material.

6.3 Elastic recoil detection analysis (ERDA)

ERDA technique is a unique method to measure the H and D content in thin films. When He ion (alpha particle) interacts with material containing hydrogen (H) and deuterium (D), the H and D will be scattered in the forward direction. From the detection of the forwarded H and D, one can measure the quantitative depth profiling of these elements in the material. Similar experiments can be performed using heavy ion beam to study light element profiling.

6.4 Particle-induced X-ray emission (PIXE)

Ion beam of energy typically 1–2 MeV induces ionization of the target atom. If the ejected electron belongs to K-shell, an X-ray characteristic of the irradiated element is emitted. Using this technique, qualitative and quantitative analysis can be used where the trace element of about 1 ppm can be achieved [9].

Author details

Mahmoud Izerrouken[1*] and Ishaq Ahmad[2]

1 Nuclear Research Center Draria, Algiers, Algeria

2 National Centre for Physics, Islamabad, Pakistan

*Address all correspondence to: m-izerrouken@crnd.dz

References

[1] Brissaud I, Baron E. The Race of Particle Accelerators to High Energy and Log Periodicity, Cybergeo: European Journal of Geography [Online], Debates, Theory of Relativity Scale and Log Periodicity. Online: December 11, 2007. [Accessed Dec 14, 2018]. http://journals.openedition.org/cybergeo/14173

[2] Awais A, Hussain J, Usman M, Akram W, Shahzad K, Ali T, et al. The charge state distribution of B, C, Si, Ni, Cu and Au ions on 5 MV Pelletron accelerator. Nuclear Science and Techniques. 2017;**28**(5):64

[3] Ahmad I, Madhuku M, Sadaf A, Khan S, Hussain J, Ali A, et al. Tailoring the structural and optical characteristics of InGaN/GaN multilayer thin films by 12 MeV Si ion irradiation. Materials Science in Semiconductor Processing. 2017;**64**:95-100

[4] Ni Z, Ishaq A, Yan L, Gong J, Zhu D. Enhanced electron field emission of carbon nanotubes by Si ion beam irradiation. Journal of Physics D: Applied Physics. 2009;**42**:075408

[5] Ahmad I, Long Y, Suixia H, Dezhang Z, Xingtai Z. Optical absorption of ion irradiated multi-walled carbon nanotube sheets in the visible to terahertz ranges. Nuclear Science and Techniques. 2009;**20**:197-201

[6] Ishaq A, Yan L, Gong J, Zhu D. Graphite-to-amorphous structural transformation of multiwalled carbon nanotubes under proton beam irradiation. Materials Letters. 2009;**63**:1505-1507

[7] Husnain G, Shu-De Y, Ahmad I, Lin L. Role of substrate temperature on structure and magnetization of Cr-implanted GaN thin film. Applied Physics A: Materials Science & Processing. 2014;**117**:2275-2279

[8] Husnain G, Ahmad I, Rafique HM, Akrajas AU, Dee CF. Depth-dependent tetragonal distortion study of AlGaN epilayer thin film using RBS and channeling technique. Modern Physics Letters B. 2012;**26**(14):1250086

[9] Akram W, Madhuku M, Shehzad K, Awais A, Ahmad I, Ahmad I, et al. Roadside dust contamination with toxic metals along Islamabad industrial area. Nuclear Science and Techniques. 2014;**25**(3):030201

Section 2

High Energy Particle Physics

Chapter 2

Analysis of Ultra-High Energy Muons at INO-ICAL Using Pair Meter Technique

Jaydip Singh, Srishti Nagu and Jyotsna Singh

Abstract

The proposed magnetized Iron CALorimeter (ICAL) detector at India-based Neutrino Observatory (INO) is a large-sized underground detector. ICAL is designed to reconstruct muon momentum using magnetic spectrometers as detectors. Muon energy measurements using magnets fail for high energy muons (TeV range), since the angular deflection of the muon in the magnetic field is negligible and the muon tracks become nearly straight. A new technique for measuring the energy of muons in the TeV range, used by the CCFR neutrino detector is known as the pair meter technique. This technique estimates muon energy by measuring the energy deposited by the muon in several layers of an iron calorimeter through e^+ and e^- pair production. In this work we have performed Geant4-based preliminary analysis for iron plates and have demonstrated the feasibility to detect very high energy muons (1–1000 TeV) at the underground ICAL detector operating as a pair meter. This wide range of energy spectrum will not only be helpful for studying the cosmic rays in the Knee region which is around 5 PeV in the cosmic ray spectra but also useful for understanding the atmospheric neutrino flux for the running and upcoming ultra-high energy atmospheric neutrino experiments.

Keywords: pair meter techniques, cosmic rays, iron calorimeter

1. Introduction

ICAL at INO is a 52 ktons detector [1] proposed to be built at Theni district of Tamil Nadu in Southern India. It is designed to study the flavor oscillations of atmospheric neutrinos. The main goal of the ICAL detector is to precisely measure the neutrino oscillation parameters and to determine the neutrino mass hierarchy [1]. At a depth of around 1.2 km underground, the INO-ICAL detector will be the world's biggest magnetized detector to measure cosmic ray muon flux with the capability to distinguish $\mu+$ from $\mu-$. The existing direct and indirect methods of muon spectrometry at accelerator-based and in cosmic rays (magnetic spectrometers and transition radiation detectors) experiments involve certain technical problems and limitations in the energy region $\geq 10^{13}$ eV. These disadvantages vanquish in this alternate method where the muon energy is estimated by measuring the energy of secondary cascades formed by muons losing their energy in thick layers of matter, mainly due to the process of direct pair production of e^+ and e^-. By using this technique, muon energy can be estimated from INO-ICAL detector operating

IntechOpen

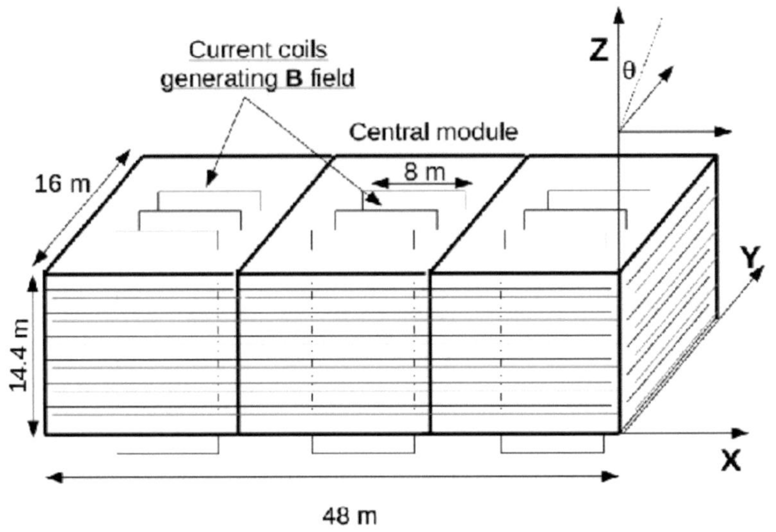

Figure 1.
Schematic view of three modules for the proposed INO-ICAL detector.

as a pair meter. The primary cosmic rays which are approximately in 50 TeV–50 PeV energy range correspond to this energy range [5].

This work presents a simulation based on the latest version of Geant4 [3] INO-ICAL code, developed by the INO collaboration for momentum reconstruction of muons in GeV energy range using INO-ICAL magnet. We have developed a separate Geant4 code for counting the muon bursts in iron plates for ultra-high energy muon analysis using the pair meter technique. The proposed detector will have a modular structure of total lateral size 48 m × 16 m, subdivided into three modules of size 16 m × 16 m. The height of the detector will be 14.5 m. It will consist of a stack of 151 horizontal layers of ~5.6 cm thick magnetized iron plates interleaved with 4 cm gaps to house the active detector layers. Detector geometry of ICAL magnet is presented in **Figure 1** and details of the components and dimensions are discussed in Ref. [1].

2. Momentum reconstruction analysis with ICAL magnet

In this section we have discussed the simulation for momentum reconstruction of muons in magnetic field. Details of the detector simulation for muons with energy of few 10's of GeV with older version of INO-ICAL code are already published in Ref. [4]. For simulating the response of high energy (100's of GeV) muons in the ICAL detector, 10,000 muons were propagated uniformly from a vertex randomly located inside 8 m × 8 m × 10 m volume. This is the central region of the central module where the magnetic field is uniform of 1.5 T. In our analysis we have considered only those events whose z coordinate of the input vertex lie within $z_{in} \leq 400$ cm which comprises the vertex to the central region. The input momentum and zenith angle are kept fixed in each case while the azimuthal angle is uniformly averaged over the entire range $-\pi \leq \phi \leq \pi$. In each case, we have studied the number of reconstructed tracks, the position resolution, including up/down discrimination and the zenith angle resolution. In this chapter, we have followed the same approach as in Ref. [4] for muon response analysis upto energy 500 GeV inside the detector [8]. Momentum reconstruction efficiency in the energy range of 1–400 GeV is shown in **Figure 2** and this energy range at the detector corresponds to the surface muon lying in the energy range 1600–2000 GeV from the top of the surface. Muon will lose around 1600 GeV in the rock overburden [8] to reach at the detector from the top surface. Energetic muons from other directions will also hit the detector since the rock cover in other directions of detector is very huge, so we have not incorporated it in our discussion.

The momentum reconstruction efficiency (ε_{recon}) is defined as the ratio of the number of reconstructed events, n rec, to the total number of generated events, N_{total}. We have

$$\varepsilon_{recon} = \frac{n_{rec}}{N_{total}} \tag{1}$$

with error, $\delta\varepsilon_{recon} = \sqrt{(\varepsilon_{recon}(1 - \varepsilon_{recon})/N_{total})}$.

Figure 2 shows the muon momentum reconstruction efficiency as a function of input momentum for different cosθ bins, here left and right panels demonstrate detector response for low and high energy muon momentum respectively. One can see that the momentum reconstruction efficiency depends on the incident particle momentum, the angle of propagation and the strength of the magnetic field. As the input momentum increases, the reconstruction efficiency increases for all angles because with increase in energy the particle crosses more number of layers producing more hits in the detector. But at sufficiently high energies, the reconstruction efficiency starts decreasing, since the muons travel nearly straight without being deflected in the magnetic field of the detector. Track reconstruction is done using Kalman Filter techniques [4], tracks for few typical energies are plotted in **Figure 3**, which shows the deflected and undeflected muon tracks depending on the energy of muons in a fixed magnetic field (1.5 T).

Pair meter techniques:

1. High energy muons produce secondary cascades mainly due to electron pair production process.

2. It is one of the most important processes for muon interaction at TeV energies, pair creation cross section exceeds those of other muon interaction processes in a wide range of energy transfer:

100 MeV $\leq E_0 \leq 0.1E_\mu$, E_0 is threshold energy.

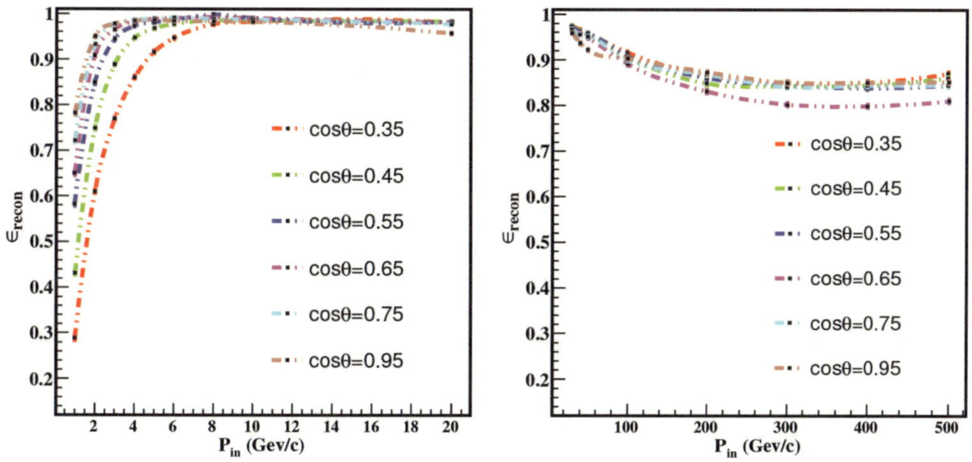

Figure 2.
Reconstructed momentum efficiency as a function of the input momentum for different cos θ values at low (1–20 GeV) and high energy (20–500 GeV).

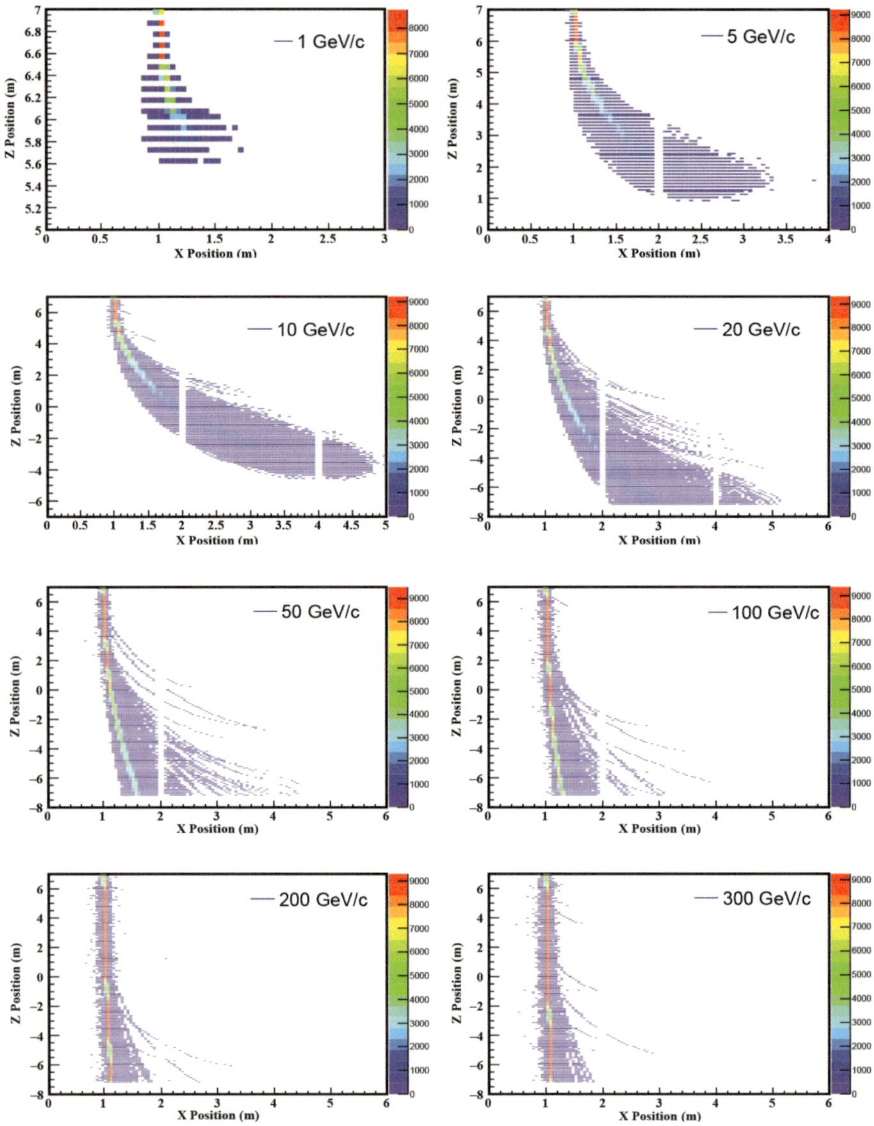

Figure 3.
Fixed energy muons track stored in X-Z plane of the detector going in the downward direction.

3. Average energy loss for pair production increases linearly with the increase in the muon energy and in the TeV region this process contributes more than 50% of the total energy loss rate.

4. Pair meter method for energy reconstruction of high energy muons has been used by the NuTeV/CCFR collaboration [6].

2.1 Average number of burst calculation

The e^+ and e^- pair production cross section by a muon of energy E_μ with energy transfer above a threshold E_0 increases as $\ln^2 (2m_e E_\mu/m_\mu E_0)$ where m_μ and m_e are the masses of the muon and the electron respectively [1]. Defining $v = E_0/E_\mu$, above $v^{-1} = 10$, this cross section dominates over other processes by which the muon loses its energy when it passes through dense matter, generating observable cascades.

- The pair production cross section depends upon E_μ/E_0 which allows one to estimate the energy of muon by counting the number of interaction cascades M in the detector with energies above a threshold E_0.

- Differential cross section for pair production is estimated in Ref. [1] and that expression is given by:

$$v\frac{d\sigma}{dv} \simeq \frac{14\alpha}{9\pi t_0} ln\left(\frac{km_e E_\mu}{E_0 m_\mu}\right) \tag{2}$$

where $\alpha = 1/137$, $k \simeq 1.8$ and t_0 is the radiation length(r.l.) of the material, for iron $t_0 = 13.75$ g/cm^2.

- The average number of interaction cascades M above a threshold E_0 is given by:

$$M(E_0, E_\mu) = Tt_0\sigma(E_0, E_\mu) \tag{3}$$

$$E_\mu = (E_0 m_\mu/km_e)\, exp\left(\sqrt{9\pi M/7\alpha T - A}\right) \tag{4}$$

where T is target thickness and σ (E_0,E_μ) is the integrated cross section(in unit of cm^2/g) and $A \simeq 1.4$.,

$$\sigma(E_0, E_\mu) \simeq \frac{7\alpha}{9\pi t_0}\left(ln^2\left(\frac{km_e E_\mu}{E_0 m_\mu}\right) + A\right) \tag{5}$$

3. Counting the burst using pair meter

A muon traversing vertically from the top will cover $151 \times 5.6 \simeq 845$ cm in iron plates, this is equivalent to a path-length of ≃480 r.l.(g/cm^2). The number of cascades produced by high energy muon for a path length of 450 r.l.(g/cm^2) can be calculated using Eq. (3). The number of cascades produced as a function of muon energy is shown in **Figure 4**, which increase with energy in the energy range 10^3–10^6 GeV (1–1000 TeV). One can interpret from **Figure 4** that the approximate number of interactions for a 1 TeV muon at threshold energy E_0 of 1 GeV is 4. Similarly for a 10 TeV muon, the number of interactions for $E_0 = 10$ GeV is 4 and for 1 GeV is 20. For a 100 TeV muon, the number of interactions for $E_0 = 100$ GeV is around 4, for $E_0 = 10$ GeV is close to 20 and for $E_0 = 1$ GeV is approximately 50.

3.1 Penetration depth of electron in the iron plates

The estimation of muon burst energy in iron plates is done by evaluating the energy of electron and positron pair, for that e^+ and e^- must hit the active elements of the detector, that is, the RPC which is a type of spark chamber with resistive electrodes placed parallel to each other. Measurement of penetration depth of the electron in iron gives us a handle to determine the energy loss of the electrons in the iron plates. The energy loss of electron in iron is given by: $E = E_0\, e^{-x/x_0}$, where x is distance traveled in the iron plate and x_0 is the radiation length. It is thus important to determine the range of electron in iron plates which is shown in **Figure 5**.

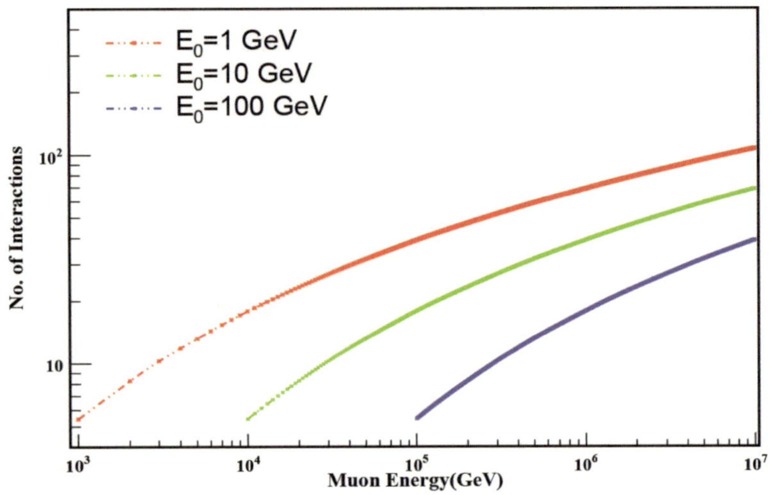

Figure 4.
Average number of bursts above a threshold E_0 versus muon energy for E_0 = 1, 10 and 100 GeV, with T fixed to 450 r.l.

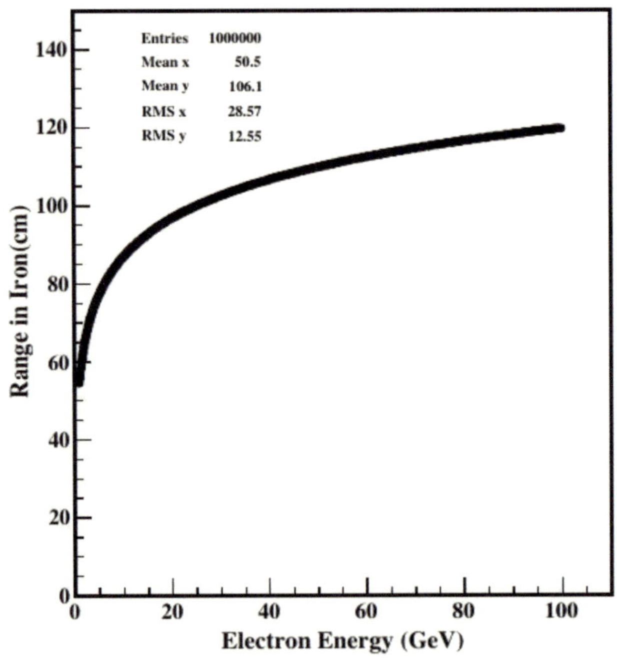

Figure 5.
Energy of the electron and its corresponding penetrating range in the iron plates.

3.1.1 Muon burst of a few typical energies in the iron plates

A Geant4-based code is developed for simulating the muons burst in iron plates, here horizontal axis is the z-axis of INO-ICAL detector, in which 152 layers of iron plates of width 5.6 cm are placed vertically, interleaved with 2.5 cm gaps for placing the RPC. Muons are propagated using Geant4 particle generator class and generated bursts in the iron plates are counted, muon bursts for a 10, 100, 500 and 1000 GeV are shown in **Figure 6(a)–(d)** respectively [7].

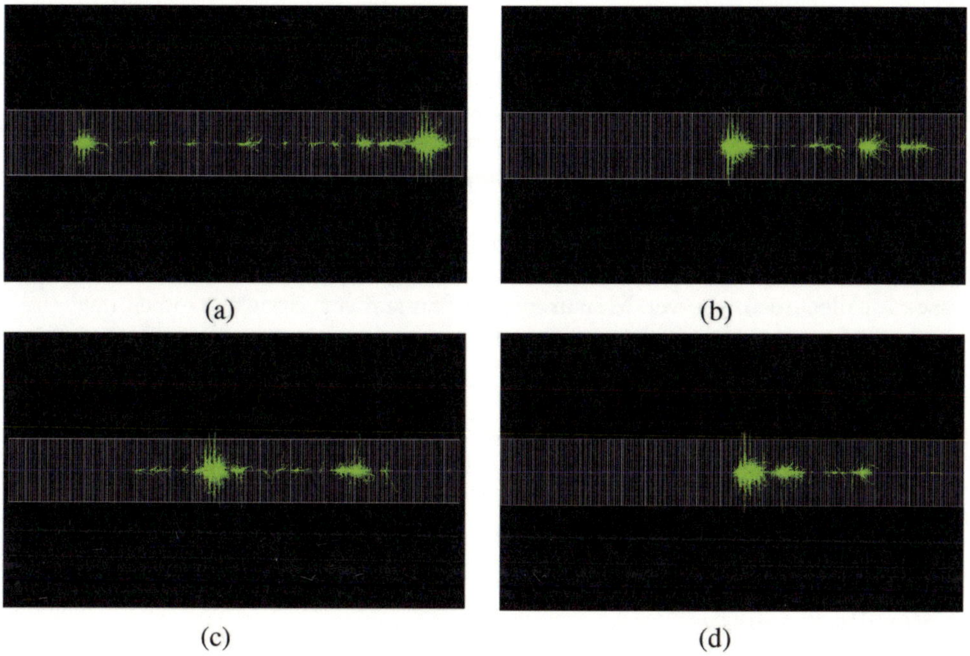

Figure 6.
Cascade generation for a few typical energies. Blue line (muon) represents z-axis of the detector and green line represents the electron-positron cascade in the x-y plane.

3.2 Operating ICAL using pair meter technique

Cosmic rays are composed of highly energetic particles mostly protons, alpha particles and a small fraction of heavier nuclei that reach the Earth from the outer space. Supernova bursts, quasars and other astronomical events are believed to be the sources of cosmic rays ranging over 10^{20} eV of energy [9]. The cosmic ray (CR) spectrum depicts a power law behavior over the entire spectral range but displaying two transition regions where the slope of the spectrum changes abruptly. The

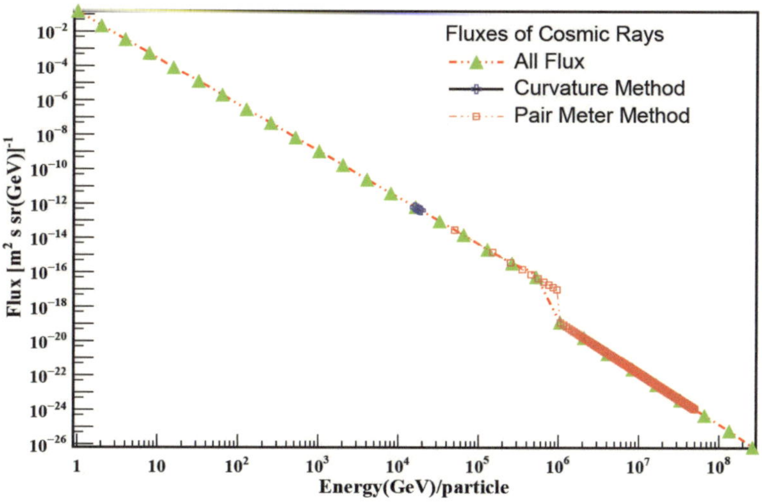

Figure 7.
Primary cosmic ray flux ($\phi \simeq KE^{-\alpha}$, where $\alpha \simeq 2.7$ and for KNEE (3PeV) α: $2.7 \rightarrow 3$) versus energy of primary particle, and limited range for ICAL to cover the spectrum using magnetic field and pair meter technique.

region around E ~ 5 PeV is where the CR spectrum becomes steeper, is known as the 'knee' region. At higher energy around E ~ 5000 PeV, the CR spectrum flattens at the 'ankle' where the sources of such high energies are believed to be of extra-galactic origin. The reason for these two appreciable 'breaks' in the CR spectrum is still unknown. Once accelerated at supernova shocks, cosmic rays have to propagate through the interstellar medium before they can be detected. CR muons are created when cosmic rays enter the Earth's atmosphere where they eventually collide with air molecules and decay into pions. The charged pions (π^+ and π^-) decay in flight. All these particles together create into muons (μ^+ and μ^-) and neutrinos (ν and ν^-). A cascade called an air shower. Measurement of air shower particles can be interpreted in terms of the energy spectrum and primary cosmic ray composition. To determine these measurements we require calculation of fluxes generated via CR nuclei of mass A, charge Z and energy E.

Figure 7 represents primary cosmic ray flux versus energy of the primary cosmic ray particle. From **Figure** 7, we can see the limit of the energy range upto which magnetic spectrometers work i.e. around E ~ 10^4 while the pair meter technique works well in a wide energy range from 5×10^4–5×10^7 GeV.

4. Results and discussion

In Section 2, we have discussed the limitation of magnetized ICAL detector to be used as a magnetic spectrometer, which limits the efficiency of the detector to discriminate between $\mu+$ and $\mu-$ at higher energies and reconstruct momentum. Variation in efficiency of muon momentum reconstruction as a function of input momentum is shown in **Figure 2**, which shows a clear fall in efficiency for energetic muons. In **Figure 3**, muon track could be seen, which is undeflected for highly energetic muons. Finally, it is concluded that with ICAL detector we can do analysis for muons in the energy range of 1–400GeV. This corresponds to surface muons in the energy range of 1600–2000 GeV, because muons lose around 1600 GeV energy [3] into the rock overburden to reach at the detector from the top of the INO-ICAL surface. ICAL cannot be used as a magnetic spectrometer for highly energetic cosmic ray muons. For energetic (TeV) muons, pair meter technique [2] can be used for momentum reconstruction as discussed in Section 3 [5]. This technique is tested by a few detectors, since INO-ICAL will be large in dimensions so it will be a perfect machine to test the capability of this technique. We have developed a separate Geant4 code for counting the bursts in the iron plates and also a technique to measure the energy of the bursts with the produced electron pairs into the iron plates. In **Figure 6**, we can see the burst of muons in iron plate. The variation of these burst number is shown in **Figure 4**, which is a function of the muon energy.

4.1 Summary and conclusions

- The pair meter technique can competently measure muon energy in the energy range of 1–1000 TeV at INO-ICAL detector operating as pair meter.
- One can probe very high energy muon fluxes and primary cosmic rays in the knee region which will aid in accurate background muon and neutrino flux measurement in the forthcoming detectors designed for ultra-high energy neutrino experiments.
- Our Geant4 analyses for central module of INO-ICAL detector are successfully performed and variation in the cascade number of varying energy is observed in the iron plates.

Acknowledgements

This work is partially supported by Department of Physics, Lucknow University, Department of Atomic Energy, Harish-Chandra Research Institute, Allahabad and INO collaboration. Financially it is supported by government of India, DST project no-SR/MF/PS02/2013, Department of Physics, Lucknow University. We thank Prof. Raj Gandhi for useful discussion and providing hospitality to work in HRI, Dr. Jyotsna Singh for her support and guidance in completing this work from Lucknow University.

Author details

Jaydip Singh*, Srishti Nagu and Jyotsna Singh
University of Lucknow, Lucknow, India

*Address all correspondence to: jaydip.singh@gmail.com

References

[1] Ahamed S, Sajjad Atahar M, et al. (INO Collab). Physics potential of the ICAL detector at the Indian based neutrino observatory. 9 May 2017; arXiv:1505.07380

[2] Kokoulin RP, Petrukhin AA. Theory of the pairmeter for high energy muon measurements. Nuclear Instruments and Methods in Physics Research. 1987;**A263**(1988):468-479

[3] Groom DE, Mokhov NV, Striganov S. Muon stopping power and range tables 10 MeV-100 TeV. Atomic Data and Nuclear Data Tables. 2001;**76**(2)

[4] Chatterjee A et al. (INO Collab). JINST 9(2014) PO 7001; July 2014

[5] Gandhi R, Panda S. Probing very high energy prompt muon and neutrino fluxes and the cosmic ray knee via underground muons. 31 August 2006; arXiv:hep-ph/0512179

[6] Chikkatur AP, Bugel L, et al. Test of a calorimetric technique for measuring the energy of cosmic ray muon in TeV energy range. Zeitschrift für Physik. 1997;**74**:279-289

[7] Agostinelli S et al. GEANT4: A simulation toolkit, GEANT4 collaboration. Nuclear Instruments and Methods in Physics Research Section A. 2003;**506**:250 http://geant4.cern.ch/

[8] Singh J, Singh J, et al. Atmospheric muons charge ratio analysis at the INO-ICAL detector. 29 Jun 2018; arXiv:1709.01064

[9] Ramesh N, Hawron M, Martin C, Bachri A. Flux variation of cosmic muons. Journal of the Arkansas Academy of Science. 2011;**65**:67-72

Chapter 3

From the Eloisatron to the Pevatron

Thomas Taylor

Abstract

In the late 1970s the experimental physics community was active in promoting the Large Electron Positron (LEP) collider and its associated experiments to study the Z- and W-bosons, and with the expectation that the tunnel could subsequently house a hadron collider (LHC), providing a center-of-mass energy for discoveries at the frontier of knowledge. At this time, Antonino Zichichi, who had chaired a Working Group in charge of promoting LEP among the community of experimental and accelerator physicists, realized that one should envisage building as large a ring as possible, for which LEP/LHC would be but a scale model, and it was thus the idea of the Eloisatron, or ELN, in a ring of about 300 km in circumference, was born. CERN and IHEP, China, are now engaged in studies for future colliders of 100 km in circumference, aiming to extend center-of-mass energy in hadron collisions to 100 TeV by using very high field magnets. The ELN idea lives on, but it is time to envision an update. A ring of diameter 300 km would make possible the installation of a sequence of increasingly complex accelerators culminating in one eventually capable of providing a center-of-mass energy of 1000 TeV, i.e. a peta-electron-volt or PeV.

Keywords: particle physics, standard model, CERN, colliders, ISR, LEP, LHC, FCC, Eloisatron, 100 TeV, 1 PeV, Pevatron

1. Introduction

Following the success of the first hadron collider, the Intersecting Storage Rings (ISR) [1] at CERN in the early 1970s, colliders have been the experimentalists' main tool for exploring, in laboratory conditions, particle physics at the frontier of knowledge. Enthusiasm for the large collider that was to become the Large Electron-Positron collider (LEP), was amplified in 1976 by the workings of the so-called LEP Working Group [2], which produced several visionary reports regarding the future possibilities of an accelerator of about 30 km in circumference [3]. Thanks to technological advances at the ISR, the Z- and W-bosons foreseen in the Standard Model were first observed (1984) in the proton-antiproton collider in the SPS tunnel at CERN [4], before LEP was completed. The lepton collider LEP, an accelerator of 27 km in circumference targeted detailed study of the intermediate bosons that had been previously discovered, providing unprecedented precision that helped to entrench the Standard Model of fundamental physics between 1989 and 2000 as being the best, albeit incomplete, description of the physical world of elementary particles at the present time [5]. In parallel, the feasibility of performing comparable experiments with leptons with linear colliders was ably demonstrated at SLAC, and a more powerful proton-antiproton collider was put into service at the Fermilab Tevatron [6]. Experimental particle physics had well and truly embraced the advantage of laboratory-based colliders for probing deeper into the unknown. In the quest for still higher center-of-mass energies, the 80 km Superconducting Super Collider (SSC) [7] in the USA was destined to provide conditions for experiments at center-of-mass energies of 25– 30 TeV, but due to funding problems this was not to be. It was therefore at the Large Hadron Collider [8], installed in the LEP tunnel, that events revealing the elusive Higgs boson were observed in 2012 [9]. Now the community is studying the possibility of providing the discovery potential of still higher center-of-mass energies, as well as precision measurements of the Higgs boson in a lepton collider in much the same way as was done at LEP for the Z- and W-bosons.

The idea of the Eloisatron, or ELN, to be installed in a tunnel of about 300 km of circumference, was born in 1979. Zichichi, an experimental physicist who had played an active role in the first g-2 experiment at CERN, and led experiments at the ISR, clearly understood the value of storage rings and colliders. He had been very active in promoting work on LEP and its associated experiments [2]—fully aware of the fact that the tunnel could be used subsequently to house a hadron collider. He realized the importance of equipping as large a ring as possible to enable to perform experiments at the highest possible center-of-mass energy, and for which LEP/LHC would be but a scale model. It was thus that he came up with the idea of the Eloisatron this being the largest that could be accommodated on the island of Sicily. Zichichi argued that such an instrument could be built for roughly the same cost as a bridge that was envisioned to join the island to the mainland, and this and several other sites in Italy were considered (perhaps the most appropriate being in the geologically stable island of Sardinia) [10]. Kjell Johnsen, who had previously led the highly innovative ISR project, was put in charge of the first studies for such an accelerator. However, this was to be overshadowed by work on the ill-fated 83 km Superconducting Synchrotron Collider (SSC) project in the USA, and on the LHC at CERN, only to be revived later in the ephemeral Very Large Hadron Collider (VLHC) studies for a 233 km long collider at Fermilab [11]—itself largely inspired by the Eloisatron concept. Nevertheless, it should be emphasized that it was due to the recognition of the importance of studies on detectors associated with the potential ELN (and *a fortiori* the LHC) that the LAA project was born. This auxiliary project provided the framework for vital work to be performed on detectors, enabling the adoption of techniques to address problems inherent with equipment required to observe very high energy collisions at the LHC [12].

Today we have the LHC, which will be upgraded to provide increased luminosity from 2026. Studies for a future generation of accelerator/colliders focus on either a linear collider (for leptons) aimed at detailed study of the Higgs (or Higgs-like) events seen at the LHC, or a large (100 km) circular collider. The idea is that this would first provide e^+e^- collisions (as a simpler alternative to the linear collider for a so-called Higgs factory), both in the European Future Circular Collider (FCC) and Chinese Circular Electron-Positron Collider (CEPC) versions, to be followed by installing a hadron collider with discovery potential [13, 14]. To achieve a sought-after 100 TeV center-of-mass energy, the new machine would require a large number of yet-to-be-developed, very high field superconducting magnets (16 T dipoles). The experimental physics community hopes that at least one of the large circular machines will be constructed, even if, as likely, it does not quite reach the presently advertised performance, it will allow groundbreaking studies both in particle and in accelerator physics and technology. However, the philosophy behind the idea of the ELN lives on. Closest to the ELN concept, there is a proposal [15] for a minimum cost route to a 100 TeV center-of-mass hadron collider: the idea being to finish the SCC tunnel in Texas, in which an injector synchrotron would be installed, and to bore a second tunnel, of length 270 km to house the main collider, using affordable 5 T superconducting dipoles. The SCC tunnel could also house an efficient circular e^+e^- collider to be used as a Higgs factory. The terrain in the vicinity of the SCC is propitious for tunneling, the per meter cost being less than that near CERN, at least in the sections of limestone [16] (but for a global costing this would have to be offset by the value of the existing laboratory infrastructure at CERN). However, unlike for LHC, where there was an identified goal (to reveal Higgs), indicating an appropriate energy, it is now desirable to foresee being able to access far higher collision energies, so the goal should perhaps evolve. It is thus proposed that one should consider a ring of about 300 km *in diameter*, enabling the installation of a sequence of accelerators culminating in one capable of achieving a hadronic center-of-mass energy of 1000 TeV, i.e. a PeV, or peta-electron-volt. In this chapter, a vision is presented of what would probably be the world's Ultimate Circular Collider, based on proven and foreseeable accelerator physics and engineering science: i.e. the UCC, or Pevatron.

2. The requirement

It is first assumed that a Higgs Factory, either in the form of a linear collider or phase 1 of an FCC or CEPC (or possibly "LEP3", an ultimate e^+e^- collider in the LEP/LHC tunnel) will have done its work before the first phase of experimentation starts at the UCC. While the tunnel could (and should, initially) undoubtedly be used to house an exciting e^+e^- collider, it is supposed to be principally an instrument for performing frontier physics using hadron collisions. The second assumption is that suitable stable sites can be identified for efficiently boring or excavating an approximately circular tunnel of about 300 km in diameter. For the first phase of operation of the hadron collider, the accelerator should be capable of delivering comfortably a center-of-mass energy of 100 TeV in at least two experiments using the simplest (and cheapest) possible guiding magnet system. This accelerator would serve as the injector for subsequent upgrades of the collision energy, culminating in 1000 TeV and (like the FCC) would depend on the development of affordable high performance superconducting material. The first phase would feature extremely simple magnets based on the use of existing low-cost superconductors. This, together with the fact that synchrotron radiation, and other effects associated with the deflection of stiff beams would not be a problem, could mean that the first phase of the proposed machine may even be cost-competitive with the hadron versions of FCC/CEPC. An added incentive to read on....

3. The site

There are undoubted advantages in using present laboratory infrastructure if at all possible, as has been done to render the sequence of accelerators at CERN both performant and affordable. The longstanding international nature of Geneva also facilitates the hosting of such a center. It is thus evident that one candidate should be CERN. However, the terrain is not ideal, as the tunnel would have to pass through many kilometers of mixed rock including limestone, which would complicate the process and have clear implications on the scale, cost and timescale of civil engineering. Other sites must therefore be considered for a larger circular accelerator than the FCC, and the cost analysis should consider what would be required to develop (or establish) suitable infrastructure for another, possibly entirely new, laboratory. As such an accelerator must be a fully international endeavor, and it will certainly take decades to achieve the ultimate goal, mechanisms will have to be enacted that render it attractive and guarantee its perennity. What springs to mind is a development of the successful CERN model, and CERN should most definitely be central to its establishment. Suitable geologically stable sites with good tunneling attributes certainly exist in China, Europe, Russia and the United States, but socio-political support will be as important as geographical location.

Considering the evident urbanization of the planet, this new collider would provide the incentive for a farseeing nation or region to combine the excitement of creating a laboratory to explore the very forefront of natural science, with that of establishing a cluster of cities, including some that are radically new. These should feature all the latest developments in sustainability and form a living exhibition of what can be done to enhance the quality of life and quest for perfection. Besides accelerator scientists and experimental physicists, architects, engineers, social scientists, artists and philosophers should all share the excitement of working together to create such a holistic ensemble showing the way for a harmonious future of the region of the world that is farsighted enough to seize the opportunity. We consider the establishment of at least four major agglomerations, or sub-cities, each housing at least 5 million inhabitants clustered around a circle defined by the accelerator, inter-linked by rapid train and highway systems. Ideally, to facilitate the setting up of the complex, at least one or two of the cities would be developments of existing conurbations. Each city would feature its own local subway system. The airport should be located approximately at the center of the circle with rapid local trains connecting to each of the mainline stations at the city nodes. Such an arrangement is shown schematically in **Figure 1**, but the actual layout would depend on local geography and a consensus based on overall requirements and planning, and responding to the constraints of sustainability, comfort and efficiency. Some of the glue holding the enterprise together would be the pride of hosting a forefront laboratory probing the mysteries of science using a unique instrument. It is confidently expected that such a complex of cities would be a breeding ground for experimentation in urban living as well as in physics research and associated technological developments and would pioneer advances in social well-being as well

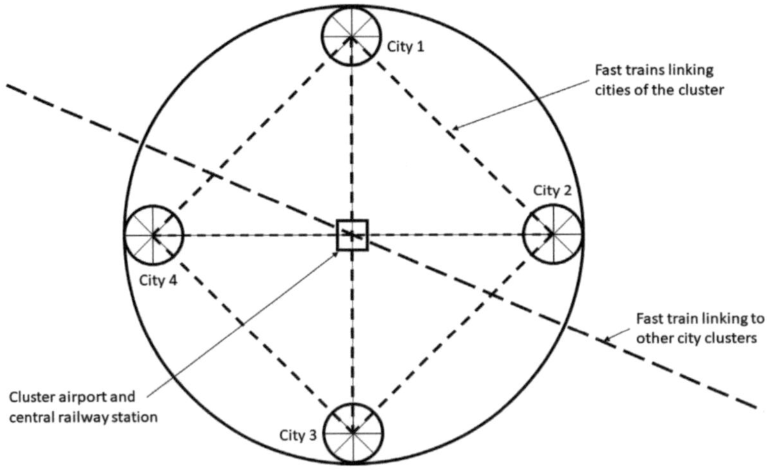

Figure 1.
Schematic layout of a cluster of cities hosting a very large particle accelerator/collider. The large circle represents the main ring. It is supposed that the injector synchrotron would be located at one of the cities, and the two major experiments at two of the other cities.

as science proper. It has been observed that clustering cities can lead to the creation of hotbeds of efficiency, creativity and innovation, and it is on this premise that China, for example, has already identified 19 regions that it intends to endow with the necessary infrastructure [17]. Other large nations are also considering moving in the same direction. It is opined that the fundamental research orientation of the complex suggested here would be more effective in providing the impetus for getting such a city cluster to work, than either bureaucratic edict or simply hoping that lavish infrastructure would somehow engender efficiency.

4. The accelerator/collider complex

It is foreseen that such a collider would be operated in three major phases, each phase taking its performance to new heights.

4.1 Phase 1

A large fraction of the experimental particle physics community is presently of the opinion that to provide worthwhile discovery potential the next hadron collider after the LHC should deliver a center-of-mass energy of about 100 TeV. This is the ultimate goal of the FCC as seen today, and it would be the target for the first phase of the UCC. For the 300 km diameter accelerator considered here, this would imply having a main dipole magnet field of 1.7 T. The transmission line magnet (pipetron), first developed for the VLHC study [11] provides such a field very efficiently. As shown in **Figure 2**, this device consists of field-shaping iron poles and yoke excited by a single superconducting cable carrying up to 100 kA and is extremely cost effective. Since the VLHC study, new cables based on magnesium diboride (MgB_2) material, which work comfortably at 20 K as opposed to 4.2 K for Nb-Ti, have been developed at CERN for interconnecting equipment required for the luminosity upgrade of the LHC [18]. As the specific heat of metals is proportional to the cube of the absolute temperature, such a cable is very stable, and, compared with the previous study for the VLHC, its use would also lead to a simplification of the associated cryogenic envelope and cooling system. To make best use of this technology the main magnet guiding and focusing system could be of the combined function type featured in the ISR. The estimated cost for such a

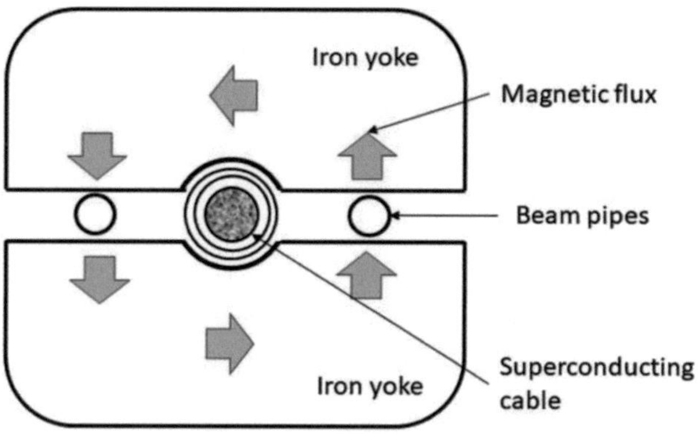

Figure 2.
Schematic cross-section of a transmission line magnet for bending the counter-rotating beams of elementary particles in a collider. The magnet is excited by a large current (about 100 kA) flowing in a superconducting cable. Such a device can deliver a field of up to about 1.7 T—sufficient for delivering a center-of-mass energy of 100 TeV to protons in a collider of 300 km in diameter.

magnet system for the UCC is of the order of twice the cost of the magnet system for the 8.3 T system for the (very much smaller) LHC, and probably more affordable (and certainly less risky) than that of the presently envisaged high field magnet system for the FCC. The accelerator would require an injector system to supply 4 TeV protons, consisting of a linear accelerator and two pulsed synchrotrons.

4.2 Phase 2

By adding a pair of rings, powered by coils wound from classical Nb-Ti conductor cooled to 4.2 T, delivering a dipole field of about 5 T, the second phase of the UCC would provide a center-of-mass energy of up to 330 TeV. By using pre-accelerated beams from the transmission line arrangement, the new magnet system would only be required to increase the momentum of circulating particles by a factor of 3.5 to 4, simplifying the process and enabling the inclusion of beam screen within a relatively small magnet aperture. It is reasonable to assume that, by careful design, the cost per meter of such a magnet system would be less than half that of the LHC system, and as the system is entirely classical, the technological risk would be low. The layout of the transmission line magnet in the tunnel will have to be such that the new magnet system can be installed during shutdowns without disrupting the Phase 1 operation.

4.3 Phase 3

It is confidently expected that during the construction and exploitation of phases 1 and 2 of the UCC, a vigorous R&D program on high field superconducting materials, and on their engineering into reliable, cost-effective cables, will have delivered conductors comparable to Nb-Ti today. That being the case, a third set of rings, with dipoles providing fields of up to 18 T, would allow us to envisage a center-of-mass energy of 1000 TeV, or 1 PeV, the ultimate goal of the complex. It is emphasized that such superconductors do not exist at present and that once they have been identified, it will also be necessary to identify a long-term use for the material that generates the need for its engineering development, just as MRI provided the "killer application" for Nb-Ti, and which led to it becoming an affordable commodity. The sole application to particle accelerators, being "one-off" in nature, is not sufficient. The push for increasingly high field magnets for NMR is an immediate application, but essentially small-scale; as seen today, the most likely long-term large-scale application would be for fusion containment—should that develop into a viable energy source. The technological development of such material, identification of applications, and the industrialization of the manufacture are examples of the high-tech activity that one could expect to flourish in the collider city complex.

5. The host laboratory

It is generally recognized that CERN provides a good example of how to organize a large international laboratory [19]. This is not an accident. The success of the laboratory is based on a combination of several important principles:

- A simple, well drawn-up convention
- Consistent rules-based funding and purchasing

- Accumulation of skills of the workforce
- Tight technical control of projects
- Skill in the use of infrastructure to build increasingly sophisticated accelerators
- Location in an internationally-oriented city

A new aspiring global laboratory should consider carefully the implication of each of these when planning how to undertake the work.

5.1 The convention

The CERN convention is a visionary 32-page document that spells out clearly the purpose of the Organization and the rules for its governance. The governing body is the CERN Council, made up of two delegates from each member state. The Council is assisted by the Finance Committee—which deals with material and personnel budgets, and the Scientific Policy Committee—which advises on the research agenda. The convention is drawn up in such a way as to vest the Council with significant autonomy and authority to negotiate and take decisions in the interest of the Organization, to empower its scientists, and to reduce to a strict minimum the bureaucracy.

5.2 Funding

Council allocates the annual budget, with funds provided by the member states in proportion to Net Nation Income. Any activity—in particular research and development—suffers from erratic funding. In the case of CERN this is avoided by a procedure of rolling forecasts: each year the budget for the following year is established, together with firm estimates for the next 2 years and provisional estimates for the following 2 years. This has provided the laboratory with a stable funding profile and enabled planning of both the day-to-day running and that of the medium and long-term scientific program. Purchasing of equipment is subject to strict rules that favor the lowest bid for a supply satisfying carefully drawn-up specifications. While there is not a policy of fair return ("juste retour"), some effort is put into distributing contracts fairly among member states.

5.3 Skills

Scientists, engineers and technicians are encouraged to hone their skills through their work on projects. This enables them to write comprehensive specifications for equipment that is available industrially, to follow up constructively the contracts, and to design and prototype special equipment that is not available on the market. The laboratory maintains well-equipped workshops for this purpose and for that of resolving technical problems which may occur due to accidents or malfunction of equipment.

5.4 Control

CERN maintains control of projects. The normal way of acquiring equipment is through buying from industry to a tight specification, written by staff competent in the field, and close technical follow-up. Cost is minimized by in-house design and prototyping, and by limiting the risk to manufacturers by confining the requirement to that of satisfying engineering standards: CERN specifically bears the technical risk for the correct functioning of complex equipment. While the LHC collider was mainly funded via CERN, that of the experiments was mainly financed via the participating institutes and universities, which led to frequent use of "in-kind" supplies. It was found to be necessary to ensure compliance of such equipment by tight control from the host laboratory. For this to work it is essential to have dedicated, competent, experienced and respected staff which is empowered with appropriate authority.

5.5 Infrastructure

The maintenance and development of infrastructure is of vital importance for a laboratory. As CERN has evolved, it capitalized on existing accelerators and associated equipment to build successive increasing complex and energetic accelerators and colliders at minimum cost. In parallel, there has been a continuing development in the technology of particle detectors required for the evolving experiments, and of course in that of the supporting informatics hardware and software. The maintenance of efficient and well-equipped workshops has also been of vital importance for the Laboratory. It is an understatement to say that the experience of laboratories established harboring the specific intention of excluding integrated workshops (i.e. relying exclusively on purchasing) has not been good.

5.6 Location

Geneva is a city with a long tradition of hosting international organizations. This activity is an important source of income for the city, and it makes a corresponding effort to simplify the bureaucratic problems that can occur with international staffing. Permanent staff is not subject to local income tax, and goods are not subject to value-added tax, which helps to keeps costs under control. The city also has a conveniently located international airport and an efficient public transport system that provides excellent access to the laboratory. The laboratory itself lies astride the frontier between Switzerland and France, so that both countries are in fact host states and provide facilities over and above their reglementary contributions to compensate for the advantages incurred. The country or region wishing to host a laboratory providing facilities for international big science projects is strongly advised to set up a framework of a similar nature.

6. City clusters and project funding

One of the reasons for proposing to associate the new collider with a cluster of cities is that it could facilitate the funding. If it is recognized that to host the collider has some value—be it for education, innovation, regional pride or some other factor—then investing in the success of the project could be rendered acceptable, compared with less well-focused requests to central government. Viewed alone, the cost of a very large particle accelerator/collider, just as an array of telescopes for astronomy, or a nuclear fusion device, is perceived to loom large in a national budget, even though much of the cost is simply pumping money around the economy. And to put the figures into perspective, the cost of running CERN, presently the largest collider facility in the world, is equivalent to that of a proverbial annual cup of coffee of the population of the member states—a small price to pay for motivational news generated directly by the laboratory, without considering the economic fall-out, and innovation derived from the activity [20, 21]. The complexity inherent in the forming of city clusters calls for the injection of capital investment on a huge scale, and the additional percentage cost of hosting a large research facility of the type discussed here would be small, whereas the publicity and enhancement of the attractiveness of the cluster would be considerable. It is anticipated that in the medium term, competition between the clusters will grow, and this will in turn accelerate the performance—and stimulate the desire to host big science projects.

7. Conclusion

While the FCC and the CEPC are thought of as being the next, and possibly final step in colliders, these do not reach the size dreamt of already 40 years ago—the Eloisatron, or ELN. The proposal for an accelerator in the USA, using the defunct SSC tunnel to house the injector gets close [15], but in the study presented here the possibility of going still further is addressed. It is suggested that the increasing desire of people to live in cities, and the expected increase in efficiency (and well-being) provided by setting up of clusters of cities, may provide an opportunity to consider associating such a city cluster with a collider that could ultimately deliver interactions at a center-of-mass energy of 1 PeV, the Pevatron.

Author details

Thomas Taylor[1,2]

1 AT Scientific LLC, Bernex, Switzerland

2 CERN, Geneva, Switzerland

*Address all correspondence to: tom.taylor@cern.ch

References

[1] Johnsen K. The ISR and accelerator physics. Particle Accelerators. 1986;**18**: 167-182

[2] Zichichi A, editor. ECFA-LEP Working Group 1979 Status Report, ECFA/79/39. 1980

[3] Camilleri L et al. Physics with very high energy e^+e^- colliding beams. CERN 76-18. 1976

[4] Gareyte J. The SPS p-pbar collider. CERN-SPS-84-03-DI-(MST). 1984

[5] Schopper H. LEP—The Lord of the Collider Rings at CERN 1980–2000. Berlin: Springer Verlag; 2009

[6] Wilson RR. The Tevatron. FERMILAB-TM-0753. 1978

[7] Jackson JD, editor. Conceptual design of the superconducting super collider. SSC-SR-2020. 1986

[8] Evans L, editor. The Large Hadron Collider. Lausanne: EPFL Press; 2009

[9] Randall L. Higgs discovery: the power of empty space. London: The Bodley Head; 2012

[10] Barletta WA, Leutz H, editors. Supercolliders and Super-Detectors. Singapore: World Scientific Publishing; 1993

[11] VLHC Design Study Group. Design study for a phased very large hadron collider. Fermilab-TM-2149. 2001

[12] Zichichi A et al. The LAA Project—Report No. 1. CERN/LAA. 1986

[13] Bogomyagkov AV et al. Projects for ultra-high-energy circular colliders at CERN. Physics of Particles and Nuclei Letters. 2016;**13**(7):870-875

[14] CEPC-SppC Preliminary conceptual design report, volume II: accelerator. Beijing: Institute of High Energy Physics, CAS; October 2015

[15] Assadi S, McIntyre P, et al. Higgs factory and 100 TeV hadron collider: Opportunity for a new world laboratory within a decade. 2014. Available from: https://arxiv.org/pdf/1402.5973

[16] Ballantyne B. Future proof. London: World Market Intelligence Ltd., Tunnels and Tunnelling; June 2016

[17] A tale of 19 megacities. London: The Economist magazine; June 23, 2018

[18] Ballarino A. Design of an MgB_2 feeder system to connect groups of superconducting magnets to remote power converters. Journal of Physics: Conference Series. 2010;**234**:032003

[19] Fabjan CF et al., editors. Technology Meets Research. Singapore: World Scientific Publishing; 2016

[20] Schmied H. A study of the economic utility resulting from CERN contracts. CERN-1975-005. 1975

[21] Barbalat O. Technology transfer from particle physics: The CERN experience 1974–1997. CERN-BLIT-97-1. 1997

Chapter 4

Flavor Physics and Charged Particle

Takaaki Nomura

Abstract

We have new charged particles in many scenarios of physics beyond the Standard Model where these particles are sometimes motivated to explain experimental anomalies. Furthermore, such new charged particles are important target at the collider experiments such as the Large Hadron Collider in searching for a signature of new physics. If these new particles interact with known particles in the Standard Model, they would induce interesting phenomenology of flavor physics in both lepton and quark sectors. Then, we review some candidate of new charged particles and its applications to flavor physics. In particular, vector-like lepton and leptoquarks are discussed for lepton flavor physics and B-meson physics.

Keywords: flavor physics, charged particle from beyond the standard model, B-meson decay, vector-like lepton/quark, leptoquarks, charged scalar boson

1. Introduction

Charged particles are often considered in the physics beyond the Standard Model (BSM) of particle physics as new heavy particles which are not observed at the experiments. Such charged particles can have rich phenomenology since it would interact with particles in the Standard Model (SM). Furthermore they are motivated to explain some experimental anomalies indicating deviation from predictions in the SM. For example, some charged particle interaction can accommodate with the anomalous magnetic moment of the muon, $(g-2)_\mu$, which shows a long-standing discrepancy between experimental observations [1, 2] and theoretical predictions [3–6],

$$\Delta a_\mu \equiv \Delta a_\mu^{\exp} - \Delta a_\mu^{\text{th}} = (28.8 \pm 8.0) \times 10^{-10}, \quad (1)$$

where $a_\mu = (g-2)_\mu/2$. This difference reaches to 3.6σ deviation from the prediction. In addition, new charged particles are introduced when we try to explain anomalies in B-meson decay like $B \to K^{(*)}\mu^+\mu^-$ and $B \to D^{(*)}\tau\nu$ [7–16].

In this chapter, we review some candidates of new charged particles from BSM physics. After listing some examples of them, the applications to some flavor physics will be discussed focusing on some specific cases. We find it interesting to consider new charged particles which are related to flavor physics in both lepton and quark sectors.

2. Some charged particles from beyond the standard model physics

In this section we review some examples of charged particles which are induced from BSM physics.

2.1 Charged scalar bosons

Singly charged scalar appears from two-Higgs doublet model (2HDM) [17, 18] in which two $SU(2)_L$ doublet Higgs fields are introduced:

$$H_1 = \begin{pmatrix} H_1^+ \\ (v_1 + \phi_1 + i\eta_1)/\sqrt{2} \end{pmatrix}, \quad H_2 = \begin{pmatrix} H_2^+ \\ (v_2 + \phi_2 + i\eta_2)/\sqrt{2} \end{pmatrix}, \tag{2}$$

where $v_{1,2}$ is the vacuum expectation values (VEVs) of Higgs fields. In general, one can write Yukawa interaction in terms of Higgs doublet fields as

$$\begin{aligned} -\mathcal{L}_Y = {} & \overline{Q}_L Y_1^d D_R H_1 + \overline{Q}_L Y_2^d D_R H_2 + \overline{Q}_L Y_1^u U_R \tilde{H}_1 + \overline{Q}_L Y_2^u U_R \tilde{H}_2 \\ & + \overline{L} Y_1^\ell \ell_R H_1 + \overline{L} Y_2^\ell \ell_R H_2 + H.c., \end{aligned} \tag{3}$$

where all flavor indices are hidden, $P_{R(L)} = (1 \pm \gamma_5)/2$; $Q_L^i = (u_L^i, d_L^i)$ and $L_L = (\nu_L^i, e_L^i)$ are the $SU(2)_L$ quark and lepton doublets with flavor index i, respectively; f_R ($f = U, D, \ell$) denotes the $SU(2)_L$ singlet fermion; $Y_{1,2}^f$ are the 3×3 Yukawa matrices; and $\tilde{H}_i = i\tau_2 H_i^*$ with τ_2 being the Pauli matrix. There are two CP-even scalars, one CP-odd pseudoscalar, and two charged Higgs particles in the 2HDM, and the relations between physical and weak eigenstates can be given by

$$\begin{aligned} h &= -s_\alpha \phi_1 + c_\alpha \phi_2, \\ H &= c_\alpha \phi_1 + s_\alpha \phi_2, \\ H^\pm(A) &= -s_\beta \phi_1^\pm(\eta_1) + c_\beta \phi_2^\pm(\eta_2), \end{aligned} \tag{4}$$

where $\phi_i(\eta_i)$ and $\eta_i^\pm$ denote the real (imaginary) parts of the neutral and charged components of H_i, respectively; $c_\alpha(s_\alpha) = \cos\alpha(\sin\alpha)$, $c_\beta = \cos\beta = v_1/v$, $s_\beta = \sin\beta = v_2/v$, and v_i are the vacuum expectation values (VEVs) of H_i and $v = \sqrt{v_1^2 + v_2^2} \approx 246$ GeV. In our notation, h is the SM-like Higgs, while H, A, and $H^\pm$ are new particles which appear in the 2HDM. In particular, Yukawa interactions with charged Higgs are given by

$$\begin{aligned} -\mathcal{L}_Y^{H^\pm} = {} & \sqrt{2}\overline{d}_L V^\dagger \left[-\frac{1}{v t_\beta} \mathrm{m_u} + \frac{\mathbf{X}^u}{s_\beta} \right] u_R H^- \\ & + \sqrt{2}\overline{u}_L V \left[-\frac{t_\beta}{v} \mathrm{m_d} + \frac{\mathbf{X}^d}{c_\beta} \right] d_R H^+ \\ & + \sqrt{2}\overline{\nu}_L \left[-\frac{t_\beta}{v} \mathrm{m}_\ell + \frac{\mathbf{X}^\ell}{c_\beta} \right] \ell_R H^+ + H.c., \end{aligned} \tag{5}$$

where V is the CKM matrix and the matrix $\mathbf{X}^f$ is defined by original Yukawa coupling and unitary matrix diagonalizing fermion mass

$$\mathbf{X}^{\mathrm{u}} = V_L^u \frac{Y_1^u}{\sqrt{2}} V_R^{u\dagger}, \quad \mathbf{X}^{\mathrm{d}} = V_L^d \frac{Y_2^d}{\sqrt{2}} V_R^{d\dagger}, \quad \mathbf{X}^{\ell} = V_L^{\ell} \frac{Y_2^{\ell}}{\sqrt{2}} V_R^{\ell\dagger}. \tag{6}$$

A doubly charged scalar boson also appears from $SU(2)_L$ triplet scalar field:

$$\Delta = \begin{pmatrix} \delta^+/\sqrt{2} & \delta^{++} \\ (v_\Delta + \delta^0 + i\eta^0)/\sqrt{2} & -\delta^+/\sqrt{2} \end{pmatrix}, \tag{7}$$

where v_Δ is the VEV of the triplet scalar. Such a triplet scalar is motivated to generate neutrino mass known as Higgs triplet model or type-II seesaw mechanism [19–26]. We can write Yukawa interaction of triplet scalar and lepton doublets by

$$L_Y = h_{ij} L_{L_i}^T C i\sigma_2 \Delta L_{L_j} + h.c. \ , \tag{8}$$

where $L_{L_i} = (\nu_i, \ell_i)_L^T$ with flavor index i and $C = i\gamma^2\gamma^0$ is the Dirac charge conjugation operator. In terms of the components, the Yukawa interaction can be expanded as

$$\begin{aligned} L_Y = h_{ij} &\left(-\frac{1}{\sqrt{2}} \ell_{iL}^T C \delta^+ \nu_{jL} - \ell_{iL}^T C \delta^{++} \ell_{jL} + \nu_{iL}^T C \delta^0 \nu_{jL} - \frac{1}{\sqrt{2}} \nu_{iL}^T C \delta^+ \ell_{jL} \right) \\ + h_{ji}^* &\left(\frac{1}{\sqrt{2}} \nu_{iL}^\dagger C \delta^- \ell_{jL}^* + \ell_{iL}^\dagger C \delta^{--} \ell_{jL}^* - \nu_{iL}^\dagger C \delta^0 \nu_{jL}^* + \frac{1}{\sqrt{2}} \ell_{iL}^\dagger C \delta^- \nu_{jL}^* \right) \end{aligned} \tag{9}$$

where $C^\dagger = -C$ is used. Another example of model including doubly charged scalar is Zee-Babu type model [27, 28] for neutrino mass generation at two-loop level. In such a type of model, one introduces singly and doubly charged scalars $(h^\pm, k^{\pm\pm})$ which are $SU(2)_L$ singlet. The Yukawa couplings associated with charged scalar fields are given by

$$L_Y = f_{ij} \overline{L}_{L_i}^c (i\sigma_2) L_{L_j} h^+ + g_{ee} \overline{e}_R^c e_R k^{++} + g_{ij} \overline{e}_{R_i}^c e_{R_j} k^{++} + h.c., \tag{10}$$

where f_{ij} is antisymmetric under flavor indices. These Yukawa interactions can be used to generate neutrino mass with the nontrivial interaction in scalar potential:

$$V \supset \mu k^{++} h^- h^- + c.c.. \tag{11}$$

Note that these charged scalars also contribute to lepton flavor violation processes.

2.2 Vector-like leptons

The vector-like leptons (VLLs) are discussed in Ref. [29]. They are new charged particles without conflict of gauge anomaly problem and induce rich lepton flavor physics. To obtain mixing with the SM leptons, the representations of VLL under $SU(2)_L \times U(1)_Y$ gauge symmetry can be singlet, doublet, and triplet under $SU(2)_L$. In order to avoid the stringent constraints from rare $Z \to \ell_i^\pm \ell_j^\mp$ decays, we here consider the triplet representations $(1, 3, -1)$ and $(1, 3, 0)$ with hypercharges $Y = -1$ and $Y = 0$, respectively. The new Yukawa couplings thus can be written such that

$$-\mathcal{L}_Y = \overline{L}\mathbf{Y}_1 \Psi_{1R} H + \overline{L}\mathbf{Y}_2 \Psi_{2R} \tilde{H} + m_{\Psi_1} Tr \overline{\Psi}_{1L} \Psi_{1R} + m_{\Psi_2} Tr \overline{\Psi}_{2L} \Psi_{2R} + H.c., \tag{12}$$

where we have suppressed the flavor indices; H is the SM Higgs doublet field, $\tilde{H} = i\tau_2 H^*$ and the neutral component of Higgs field is $H^0 = (v+h)/\sqrt{2}$. The representations of two VLLs are

$$\Psi_1 = \begin{pmatrix} \Psi_1^-/\sqrt{2} & \Psi_1^0 \\ \Psi_1^{--} & -\Psi_1^-/\sqrt{2} \end{pmatrix}, \quad \Psi_2 = \frac{1}{\sqrt{2}} \begin{pmatrix} \Psi_2^0/\sqrt{2} & \Psi_2^+ \\ \Psi_2^- & -\Psi_2^0/\sqrt{2} \end{pmatrix}, \tag{13}$$

with $\Psi_2^+ = C\overline{\Psi}_2^-$ and $\Psi_2^0 = C\overline{\Psi}_2^0$. Since Ψ_2 is a real representation of $SU(2)_L$, the factor of $1/\sqrt{2}$ in Ψ_2 is required to obtain the correct mass term for Majorana fermion Ψ_2^0. Due to the new Yukawa terms of $\mathbf{Y}_{1,2}$, the heavy neutral and charged leptons can mix with the SM leptons, after electroweak symmetry breaking (EWSB). Then the lepton mass matrices become 5×5 matrices and are expressed by

$$M_\ell = \begin{pmatrix} \mathbf{m}_\ell & \mathbf{Y}^\ell v \\ 0 & \mathbf{m}_\Psi \end{pmatrix}, \quad M_\nu = \begin{pmatrix} \mathbf{m}_\nu & \mathbf{Y}^\nu v \\ 0 & \mathbf{m}_\Psi \end{pmatrix}, \tag{14}$$

where the basis is chosen such that the SM lepton mass matrices are in diagonalized form, m_ℓ is the SM charged lepton mass matrix, $\mathbf{m}_\Psi = \mathrm{diag}(m_{\Psi_1}, m_{\Psi_2})$, and

$$\mathbf{Y}^\ell = \frac{1}{2} \begin{pmatrix} -Y_{11} & Y_{21} \\ -Y_{12} & Y_{22} \\ -Y_{13} & Y_{23} \end{pmatrix}, \quad \mathbf{Y}^\nu = \sqrt{2} \begin{pmatrix} Y_{11} & Y_{21}/2 \\ Y_{12} & Y_{22}/2 \\ Y_{13} & Y_{23}/2 \end{pmatrix}. \tag{15}$$

Note that the elements of $\mathbf{Y}^\chi$ should be read as $Y_{ij} = (\mathbf{Y}_i)_j$, where the index $i = 1, 2$ distinguishes the Yukawa couplings of the different VLLs and the index $j = 1, 2, 3$ stands for the flavors of the SM leptons.

To diagonalize M_ℓ and M_ν, the unitary matrices $V^\chi_{R,L}$ with $\chi = \ell, \nu$ so that $M^{\mathrm{dia}}_\chi = V^\chi_L M_\chi V^{\chi\dagger}_R$ are introduced. The information of V^χ_L and V^χ_R can be obtained from $M_\chi M_\chi^\dagger$ and $M_\chi^\dagger M_\chi$, respectively. According to Eq. (14), it can be found that the flavor mixings between heavy and light leptons in V^χ_R are proportional to the lepton masses. Since the neutrino masses are tiny, it is a good approximation to assume $V^\nu_R \approx 1$. If one further sets $m_e = m_\mu = 0$ in our phenomenological analysis, only τ-related processes have significant contributions among them. Unlike V^χ_R, the off-diagonal elements in flavor-mixing matrices V^χ_L are associated with $\mathbf{Y}_{1,2} v/\mathrm{m}_\Psi$. In principle, the mixing effects can be of the order of 0.1 without conflict. In our example later, we examine these effects on $h \to \tau\mu$. To be more specific, we choose parametrization that the unitary matrices in terms of $\mathbf{Y}_{1,2}$ as

$$V^\chi_L \approx \begin{pmatrix} 1_{3\times3} - \varepsilon^\chi_L \varepsilon^{\chi\dagger}_L/2 & -\varepsilon^\chi_L \\ \varepsilon^{\chi\dagger}_L & 1_{3\times3} - \varepsilon^{\chi\dagger}_L \varepsilon^\chi_L/2 \end{pmatrix}, \quad V^\ell_R \approx \begin{pmatrix} 1_{3\times3} & -\varepsilon^\ell_R \\ \varepsilon^{\ell\dagger}_R & 1_{3\times3} \end{pmatrix}, \tag{16}$$

where $V^\nu_R \approx 1$ is used in our approximation, $\varepsilon^\chi_L \approx v\mathbf{Y}^\chi/\mathbf{m}_\Psi$, and $\varepsilon^\ell_R \approx v\mathbf{m}^\dagger_\ell \mathbf{Y}^\ell/\mathbf{m}^2_\Psi$. Combining the SM Higgs couplings and new Yukawa couplings of Eq. (12), the Higgs couplings to all singly charged leptons are obtained such as

$$-\mathcal{L}_{h\ell'\ell'} = h\overline{\ell'}_L V^\ell_L \begin{pmatrix} \mathbf{m}_\ell/v & \mathbf{Y}^\ell \\ 0 & 0 \end{pmatrix} V^{\ell\dagger}_R \ell'_R + H.c., \tag{17}$$

where $\ell'^T = (e, \mu, \tau, \tau', \tau'')$ is the state of a physical charged lepton in lepton flavor space. We use the notations of τ' and τ'' to denote the heavy-charged VLLs in mass basis. Using the parametrization of Eq. (16), the Higgs couplings to the SM-charged leptons can be formulated by

$$-\mathcal{L}_{h\ell\ell} = C^h_{ij}\overline{\ell}_{iL}\ell_{jR}h + H.c.,$$
$$C^h_{ij} = \frac{m_{\ell j}}{v}\left[\delta_{ij} - \frac{3}{8}\left(\frac{v^2 Y_{1i}Y_{1j}}{m^2_{\Psi_1}} + \frac{v^2 Y_{2i}Y_{2j}}{m^2_{\Psi_2}}\right)\right]. \tag{18}$$

If one sets $m_e = m_\mu = 0$, it is clear that in addition to the coupling $h\tau\tau$ being modified, the tree-level flavor-changing couplings h- τ- μ and h- τ- e are also induced, and the couplings are proportional to $m_\tau/v \approx 7.2 \times 10^{-3}$. In order to study the VLL contributions to $h \to \gamma\gamma$, the couplings for $h\tau'\tau'$ and $h\tau''\tau''$ are expressed as

$$-\mathcal{L}_{h\Psi\Psi} = \frac{v\sum_i Y^2_{1i}}{2m_{\Psi_1}}h\tau'\tau' + \frac{v\sum_i Y^2_{2i}}{2m_{\Psi_2}}h\tau''\tau''. \tag{19}$$

2.3 Vector-like quarks

Here we consider vector-like triplet quarks (VLTQs) that are discussed in Ref. [30] The gauge invariant Yukawa couplings of VLTQs to the SM quarks, to the SM Higgs doublet and to the new Higgs singlet field are written as

$$\begin{aligned}-\mathcal{L}^Y_{VLTQ} &= \overline{Q}_L\mathbf{Y}_1F_{1R}\tilde{H} + \overline{Q}_L\mathbf{Y}_2F_{2R}H + \tilde{y}_1 Tr(\overline{F}_{1L}F_{1R})S + \tilde{y}_2 Tr(\overline{F}_{2L}F_{2R})S \\ &+ M_{F_1}Tr(\overline{F}_{1L}F_{1R}) + M_{F_2}Tr(\overline{F}_{2L}F_{2R}) + h.c.,\end{aligned} \tag{20}$$

where Q_L is the left-handed SM quark doublet and it could be regarded as mass eigenstate before VLTQs are introduced; here all flavor indices are hidden, $\tilde{H} = i\tau_2 H^*$, and $F_{1(2)}$ is the 2×2 VLTQ with hypercharge $2/3(-1/3)$. The representations of $F_{1,2}$ in $SU(2)_L$ are expressed in terms of their components as follows:

$$F_1 = \begin{pmatrix} U_1/\sqrt{2} & X \\ D_1 & -U_1/\sqrt{2}\end{pmatrix},\quad F_2 = \begin{pmatrix} D_2/\sqrt{2} & U_2 \\ Y & -D_2/\sqrt{2}\end{pmatrix}. \tag{21}$$

The electric charges of $U_{1,2}$, $D_{1,2}$, X, and Y are found to be $2/3$, $-1/3$, $5/3$, and $-4/3$, respectively. Therefore, $U_{1,2}(D_{1,2})$ could mix with up (down) type SM quarks. Here $M_{F_{1(2)}}$ is the mass of VLTQ, and due to the gauge symmetry, the VLTQs in the same multiplet state are degenerate. By the Yukawa couplings of Eq. (20), the 5×5 mass matrices for up and down type quarks are found by

$$M_u = \begin{pmatrix} (\mathbf{m}^{\text{dia}}_u)_{3\times 3} & | & v\mathbf{Y}_1/2 & v\mathbf{Y}_2/\sqrt{2} \\ ----- & | & ---- & ---- \\ 0_{2\times 3} & | & & (\mathbf{m}_F)_{2\times 2}\end{pmatrix},$$
$$M_d = \begin{pmatrix} (\mathbf{m}^{\text{dia}}_d)_{3\times 3} & | & v\mathbf{Y}_1/\sqrt{2} & -v\mathbf{Y}_2/2 \\ ----- & | & ---- & ---- \\ 0_{2\times 3} & | & & (\mathbf{m}_F)_{2\times 2}\end{pmatrix}, \tag{22}$$

where $(\mathbf{m}_u^{\rm dia})_{3\times3}$ and $(\mathbf{m}_d^{\rm dia})_{3\times3}$ denote the diagonal mass matrices of SM quarks and ${\rm dia}(m_F)_{2\times2} = (m_{F_1}, m_{F_2})$. Notice that a non-vanished v_s could shift the masses of VLTQs. Since $v_s \ll v$, we neglect the small effects hereafter. Due to the presence of $\mathbf{Y}_{1,2}$, the SM quarks, $U_{1,2}$, and $D_{1,2}$ are not physical states; thus one has to diagonalize M_u and M_d to get the mass eigenstates in general. If $vY^i_{1,2} \ll m_{F_{1,2}}$, we expect that the off-diagonal elements of unitary matrices for diagonalizing the mass matrices should be of order of $vY^i_{1,2}/m_{F_{1,2}}$. By adjusting $Y^i_{1,2}$, the off-diagonal effects could be enhanced and lead to interesting phenomena in collider physics.

2.4 Scalar leptoquarks

In this subsection we consider leptoquarks (LQs) which are discussed for example in Refs. [31, 32]. The three LQs are $\Phi_{7/6} = (2, 7/6)$, $\Delta_{1/3} = (3, 1/3)$, and $S^{1/3} = (1, 1/3)$ under $(SU(2)_L, U(1)_Y)$ SM gauge symmetry, where the doublet and triplet representations can be taken as

$$\Phi_{7/6} = \begin{pmatrix} \phi^{5/3} \\ \phi^{2/3} \end{pmatrix}, \quad \Delta_{1/3} = \begin{pmatrix} \delta^{1/3}/\sqrt{2} & \delta^{4/3} \\ \delta^{-2/3} & -\delta^{1/3}/\sqrt{2} \end{pmatrix}, \tag{23}$$

where the superscripts are the electric charges of the particles. Accordingly, the LQ Yukawa couplings to the SM fermions are expressed as

$$\begin{aligned} -L_{LQ} = & \left[\bar{u}\mathbf{V}\mathbf{k}P_R\ell\phi^{5/3} + \bar{d}\mathbf{k}P_R\ell\phi^{2/3}\right] + \left[-\bar{\ell}\tilde{\mathbf{k}}P_Ru\phi^{-5/3} + \bar{\nu}\tilde{\mathbf{k}}P_Ru\phi^{-2/3}\right] \\ & + \left[\overline{u^c}\,\mathbf{V}^*\mathbf{y}P_L\nu\delta^{-2/3} - \frac{1}{\sqrt{2}}\overline{u^c}\,\mathbf{V}^*\mathbf{y}P_L\ell\delta^{1/3} - \frac{1}{\sqrt{2}}\overline{d^c}\,\mathbf{y}P_L\nu\delta^{1/3} - \overline{d^c}\,\mathbf{y}P_L\ell\delta^{4/3}\right], \\ & + \left(\overline{u^c}\,\mathbf{V}^*\tilde{\mathbf{y}}P_L\ell - \overline{d^c}\,\tilde{\mathbf{y}}P_L\nu + \overline{u^c}\mathbf{w}P_R\ell\right)S^{1/3} + h.c., \end{aligned} \tag{24}$$

where the flavor indices are hidden, $\mathbf{V} \equiv U_L^u U_L^{d\dagger}$ denotes the Cabibbo-Kobayashi-Maskawa (CKM) matrix, $U_L^{u,d}$ are the unitary matrices used to diagonalize the quark mass matrices, and U_L^d and U_R^u have been absorbed into $\mathbf{k}$, $\tilde{\mathbf{k}}$, $\mathbf{y}$, $\tilde{\mathbf{y}}$, and $\mathbf{w}$. In this setup, we treat the neutrinos as massless particles and their flavor mixing effects are rotated away as an approximation. There is no evidence for any new CP violation, so in the following, we treat the Yukawa couplings as real numbers for simplicity.

The scalar LQs can also couple to the SM Higgs field via the scalar potential, and the cross section for the Higgs to diphoton can be modified in principle. However,

Particle type	$(SU(3), SU(2), U(1)_Y)$	Examples of application
Charged scalar	$(1,3,1)$, $(1,2,1/2)$	Neutrino mass, lepton flavor violation
Vector-like lepton	$(1,3,-1)$, $(1,3,0)$	Lepton flavor violation
Vector-like quark	$(3,3,2/3)$, $(3,3,-1/3)$	Quark flavor physics
Scalar leptoquark	$(3,2,7/6)$, $(3,3,1/3)$, $(3,1,1/3)$	Meson decay, lepton flavor violation

Table 1.
List of examples of charged particles from new physics discussed in this review showing $SU(3) \times SU(2) \times U(1)_Y$ representations and applications to phenomenology.

the couplings of the LQs to the Higgs are independent parameters and irrelevant to the flavors, so by taking proper values for the parameters, the signal strength parameter for the Higgs to diphoton can fit the LHC data. For detailed analysis see **Table 1** in Ref. [31].

3. Examples of applying charged particles to flavor physics

In this section, we review applications of charged particles to flavor physics by considering VLLs and LQs as examples.

3.1 Flavor physics from vector-like lepton

Introduction of VLLs contributes to lepton flavor physics via Yukawa interactions discussed in previous section. Here we review the leptonic decay of the SM Higgs and LFV decay of charged lepton as an illustration based on Ref. [29].

3.1.1 Modification to $h \to \tau^+\tau^-$ branching ratio

From Eq. (18), it can be seen that the modified Higgs couplings to the SM leptons are proportional lepton masses. By comparison with other lepton channels, it can be seen that the $\tau\tau$ mode is the most significant one, and thus we estimate the influence on $h \to \tau^+\tau^-$. Using the values that satisfy $BR(h \to \mu\tau) \approx 10^{-4}$, the deviation of $\Gamma(h \to \tau^+\tau^-)$ from the SM prediction can be obtained as

$$\kappa_{\tau\tau} \equiv \frac{\Gamma(h \to \tau^+\tau^-)}{\Gamma^{SM}(h \to \tau^+\tau^-)} = \left|1 - \frac{6v^2 Y_3^2}{8m_\Psi^2}\right|^2 \approx 0.88. \tag{25}$$

If the SM Higgs production cross section is not changed, the signal strength for $pp \to h \to \tau^+\tau^-$ in our estimation is $\mu_{\tau\tau} \approx 0.88$, where the measurements from ATLAS and CMS are $1.44^{+0.42}_{-0.37}$ [33] and 0.91 ± 0.27 [34], respectively. Although the current data errors for the $\tau\tau$ channel are still large, the precision measurement of $\mu_{\tau\tau}$ can test the effect or give strict limits on the parameters.

3.1.2 $\tau \to \mu\gamma$ process in vector-like lepton model

In the following, we investigate the contributions of new couplings in Eq. (18) to the rare tau decays and to the flavor-conserving muon anomalous magnetic moment. We first investigate the muon $g-2$, denoted by Δa_μ. The lepton flavor-changing coupling $h\mu\tau$ can provide contribution to Δa_μ through the Higgs-mediated loop diagrams. However, as shown in Eq. (18), the induced couplings are associated with $m_{\ell j}/v\overline{\ell}_{Li}\ell_{Rj}$; only the right-handed tau lepton has a significant contribution. The induced Δa_μ is thus suppressed by the factor of $m_\mu^2 m_\tau/(v m_h^2)$ so that the value of Δa_μ is two orders of magnitude smaller than current data $\Delta a_\mu = a_\mu^{exp} - a_\mu^{SM} = (28.8 \pm 8.0) \times 10^{-10}$ [2]. A similar situation happens in $\tau \to 3\mu$ decay also. Since the couplings are suppressed by m_τ/v and m_μ/v, the BR for $\tau \to 3\mu$ is of the order of 10^{-14}. We also estimate the process $\tau \to \mu\gamma$ via the h-mediation. The effective interaction for $\tau \to \mu\gamma$ is expressed by

$$\mathcal{L}_{\tau\to\mu\gamma} = \frac{e}{16\pi^2} m_\tau \overline{\mu}\sigma_{\mu\nu}(C_L P_L + C_R P_R)\tau F^{\mu\nu}, \tag{26}$$

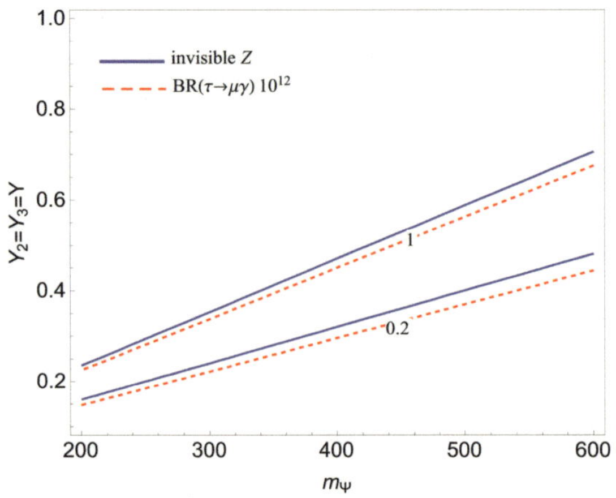

Figure 1.
Contours for $BR(\tau \to \mu\gamma)$ (dashed) as a function of Y and m_Ψ, where the constraint from Γ^Z_{inv} (solid) is included. (The plot is taken from ref. [29]).

where $C_L = 0$ and the Wilson coefficient C_R from the one loop is obtained as

$$C_R \approx \frac{C^h_{23} C^h_{33}}{2m_h^2} \left(\ln \frac{m_h^2}{m_\tau^2} - \frac{4}{3} \right). \tag{27}$$

Accordingly, the BR for $\tau \to \mu\gamma$ is expressed as

$$\frac{BR(\tau \to \mu\gamma)}{BR(\tau \to e\bar{\nu}_e \nu_\tau)} = \frac{3\alpha_e}{4\pi G_F^2} |C_R|^2. \tag{28}$$

We present the contours for $BR(\tau \to \mu\gamma)$ as a function of coupling Y and m_Ψ in **Figure 1**, where the numbers on the plots are in units of 10^{-12}. It can be seen that the resultant $BR(\tau \to \mu\gamma)$ can be only up to 10^{-12}, where the current experimental upper bound is $BR(\tau \to \mu\gamma) < 4.4 \times 10^{-8}$ [2].

3.2 *B*-meson flavor physics with leptoquarks

This section is based on Ref. [32]. Several interesting excesses in semileptonic B decays have been observed in experiments such as (i) the angular observable P_5' of $B \to K^* \mu^+ \mu^-$ [7], where a 3σ deviation due to the integrated luminosity of 3.0 fb $^{-1}$ was found at the LHCb [8, 9], and the same measurement with a 2.6σ deviation was also confirmed by Belle [10] and (ii) the branching fraction ratios R_{D,D^*}, which are defined and measured as follows:

$$R_D = \frac{\overline{B} \to D\tau\nu}{\overline{B} \to D\ell\nu} = \begin{cases} 0.375 \pm 0.064 \pm 0.026 & \text{Belle [11]}, \\ 0.440 \pm 0.058 \pm 0.042 & \text{BaBar [12, 13]}, \end{cases}$$
$$R_{D^*} = \frac{\overline{B} \to D^*\tau\nu}{\overline{B} \to D^*\ell\nu} = \begin{cases} 0.302 \pm 0.030 \pm 0.011 & \text{Belle [14]}, \\ 0.270 \pm 0.035 \pm^{+0.028}_{-0.025} & \text{Belle [15]}, \\ 0.332 \pm 0.024 \pm 0.018 & \text{BaBar [12, 13]}, \\ 0.336 \pm 0.027 \pm 0.030 & \text{LHCb [16]}, \end{cases} \tag{29}$$

where $\ell = (e, \mu)$, and these measurements can test the violation of lepton flavor universality. The averaged results from the heavy flavor averaging group are $R_D = 0.403 \pm 0.040 \pm 0.024$ and $R_{D^*} = 0.310 \pm 0.015 \pm 0.008$ [35], and the SM predictions are around $R_D \approx 0.3$ [36, 37] and $R_{D^*} \approx 0.25$, respectively. Further tests of lepton flavor universality can be made using the branching fraction ratios: $R_{K^{(*)}} = BR\left(B \to K^{(*)}\mu^+\mu^-\right)/BR\left(B \to K^{(*)}e^+e^-\right)$. The current LHCb measurements are $R_K = 0.745^{+0.090}_{-0.074} \pm 0.036$ [38] and $R_{K^*} = 0.69^{+0.11}_{-0.07} \pm 0.05$ [39], which indicate a more than 2.5σ deviation from the SM prediction. Furthermore, a known anomaly is the muon anomalous magnetic dipole moment (muon $g-2$), where its latest measurement is $\Delta a_\mu = a_\mu^{exp} - a_\mu^{SM} = (28.8 \pm 8.0) \times 10^{-10}$ [2]. These anomalies would be explained by introducing LQs and we review possible scenarios in the following.

3.2.1 Effective interactions for semileptonic B-decay

According to the interactions in Eq. (24), we first formulate the four-Fermi interactions for the $b \to c\ell'\overline{\nu}_{\ell'}$ and $b \to s\ell'^+\ell'^-$ decays. For the $b \to c\ell'\overline{\nu}_{\ell'}$ processes, the induced current-current interactions from $k_{3j}\tilde{k}_{i2}$ and $\tilde{y}_{3i}w_{2j}$ are $(S-P) \times (S-P)$, and those from $y_{3i}y_{2j}$ and $\tilde{y}_{3i}\tilde{y}_{2j}$ are $(S-P) \times (S+P)$, where S and P denote the scalar and pseudoscalar currents, respectively. Taking the Fierz transformations, the Hamiltonian for the $b \to c\ell'\overline{\nu}_{\ell'}$ decays can be expressed as follows:

$$\begin{aligned}\mathcal{H}_{b\to c} =& \left(-\frac{\tilde{y}_{3i}w_{2j}}{2m_S^2} + \frac{k_{3j}\tilde{k}_{i2}}{2m_\Phi^2}\right)\overline{c}P_L b\,\overline{\ell}_j P_L \nu_i + \left(\frac{\tilde{y}_{3i}w_{2j}}{2m_S^2} + \frac{k_{3j}\tilde{k}_{i2}}{2m_\Phi^2}\right)\frac{1}{4}\overline{c}\sigma_{\mu\nu}P_L b\,\overline{\ell}_j\sigma^{\mu\nu}P_L\nu_i \\ & - \sum_a V_{2a}\frac{y_{aj}y_{3i}}{4m_\Delta^2}\overline{c}\gamma_\mu P_L b\,\overline{\ell}_j\gamma^\mu P_L\nu_i + \sum_a V_{2a}\tilde{y}_{aj}\frac{\tilde{y}_{3i}}{2m_S^2}\overline{c}\gamma_\mu P_L b\,\overline{\ell}_j\gamma^\mu P_L\nu_i,\end{aligned} \quad (30)$$

where the indices i, j are the lepton flavors and the LQs in the same representation are taken as degenerate particles in mass. It can be seen that the interaction structure obtained from the triplet LQ is the same as that from the W-boson one. The doublet LQ generates an $(S-P) \times (S-P)$ structure as well as a tensor structure. However, the singlet LQ can produce currents of $(V-A) \times (V-A)$, $(S-P) \times (S-P)$, and tensor structures. Nevertheless, we show later that the singlet LQ makes the main contribution to the R_D and R_{D^*} excesses. Note that it is difficult to explain R_{D,D^*} by only using the doublet or/and triplet LQs when the R_K excess and other strict constraints are satisfied at the same time.

With the Yukawa couplings in Eq. (24), the effective Hamiltonian for the $b \to s\ell'^+\ell'^-$ decays mediated by $\phi^{2/3}$ and $\delta^{4/3}$ can be expressed as

$$\begin{aligned}\mathcal{H}_{b\to s} =& \frac{k_{3j}k_{2j}}{2m_\Phi^2}\left(\overline{s}\gamma^\mu P_L b\right)\left(\overline{\ell}_j\gamma_\mu P_R\ell_j\right), \\ & -\frac{y_{3j}y_{2j}}{2m_\Delta^2}\left(\overline{s}\gamma^\mu P_L b\right)\left(\overline{\ell}_j\gamma_\mu P_L\ell_j\right),\end{aligned} \quad (31)$$

where the Fierz transformations have been applied. By Eq. (31), it can be clearly seen that the quark currents from both the doublet and triplet LQs are left-handed; however, the lepton current from the doublet (triplet) LQ is right(left)-handed. When one includes Eq. (31) in the SM contributions, the effective Hamiltonian for the $b \to s\ell'^+\ell'^-$ decays is written as

$$\mathcal{H}_{b\to s} = \frac{G_F \alpha_{em} V_{tb} V_{ts}^*}{\sqrt{2}\pi} \left[H_{1\mu} L^{\mu} + H_{2\mu} L^{5\mu} \right], \tag{32}$$

where the leptonic currents are denoted by $L_{\mu}^{(5)} = \bar{\ell}\gamma_{\mu}(\gamma_5)\ell$, and the related hadronic currents are defined as

$$\begin{aligned} H_{1\mu} &= C_9^{\ell} \bar{s}\gamma_{\mu} P_L b - \frac{2m_b}{q^2} C_7 \bar{s} i \sigma_{\mu\nu} q^{\nu} P_R b, \\ H_{2\mu} &= C_{10}^{\ell} \bar{s}\gamma_{\mu} P_L b. \end{aligned} \tag{33}$$

The effective Wilson coefficients with LQ contributions are expressed as

$$\begin{aligned} C_{9(10)}^{\ell} &= C_{9(10)}^{\rm SM} + C_{9(10)}^{{\rm LQ},\ell'}, \\ C_9^{LQ,\ell_j} &= -\frac{1}{4c_{\rm SM}} \left(\frac{k_{3j}k_{2j}}{m_{\Phi}^2} - \frac{y_{3j}y_{2j}}{m_{\Delta}^2} \right), \\ C_{10}^{LQ,\ell_j} &= -\frac{1}{4c_{\rm SM}} \left(\frac{k_{3j}k_{2j}}{m_{\Phi}^2} + \frac{y_{3j}y_{2j}}{m_{\Delta}^2} \right), \end{aligned} \tag{34}$$

where $c_{\rm SM} = V_{tb} V_{ts}^* \alpha_{em} G_F / (\sqrt{2}\pi)$ and V_{ij} is the CKM matrix element. From Eq. (34), it can be seen that when the magnitude of C_{10}^{LQ,ℓ_j} is decreased, C_9^{LQ,ℓ_j} can be enhanced. That is, the synchrony of the increasing/decreasing Wilson coefficients of $C_9^{\rm NP}$ and $C_{10}^{\rm NP}$ from new physics is diminished in this model. In addition, the sign of $C_9^{LQ,\ell'}$ can be different from that of $C_{10}^{{\rm LQ},\ell'}$. As a result, when the constraint from $B_s \to \mu^+\mu^-$ decay is satisfied, we can have sizable values of $C_9^{LQ,\mu}$ to fit the anomalies of R_K and angular observable in $B \to K^*\mu^+\mu^-$. Although the LQs can contribute to the electromagnetic dipole operators, since the effects are through one-loop diagrams and are also small, the associated Wilson coefficient C_7 is mainly from the SM contributions.

3.2.2 Constraints from $\Delta F = 2$, radiative lepton flavor violating, $B^+ \to K^+\nu\bar{\nu}$, $B_s \to \mu^+\mu^-$, and $B_c \to \tau\nu$ processes

Before we analyze the muon $g-2$, $R_{D^{(*)}}$, and $R_{K^{(*)}}$ problems, we examine the possible constraints due to rare decay processes. Firstly, we discuss the strict constraints from the $\Delta F = 2$ processes, such as $F - \bar{F}$ oscillation, where F denotes the neutral pseudoscalar meson. Since $K - \bar{K}$, $D - \bar{D}$, and $B_d - \bar{B}_d$ mixings are involved, the first-generation quarks and the anomalies mentioned earlier are associated with the second- and third-generation quarks. Therefore, we can avoid the constraints by assuming that $k_{1\ell'} \approx \tilde{k}_{\ell'1} \approx y_{1\ell'} \approx \tilde{y}_{1\ell'} \approx w_{1i} \approx 0$ without affecting the analyses of $R_{D^{(*)}}$ and $R_{K^{(*)}}$. Thus, the relevant $\Delta F = 2$ process is $B_s - \bar{B}_s$ mixing, where $\Delta m_{B_s} = 2|\langle \bar{B}_s|\mathcal{H}|B_s\rangle|$ is induced from box diagrams, and the LQ contributions can be formulated as

$$\begin{aligned} \Delta m_{B_s} \approx{}& \frac{C_{\rm box}}{(4\pi)^2} \left[\frac{5}{4} \left(\frac{\sum_{i=1}^3 y_{3i} y_{2i}}{m_{\Delta}} \right)^2 + \left(\frac{\sum_{i=1}^3 k_{3i} k_{2i}}{m_{\Phi}} \right)^2 \right] \\ &+ \frac{C_{\rm box}}{(4\pi)^2} \left[\left(\frac{\sum_{i=1}^3 \tilde{y}_{3i} \tilde{y}_{2i}}{m_S} \right)^2 + 2 \frac{\left(\sum_{i=1}^3 y_{3i} \tilde{y}_{2i} \right) \left(\sum_{i=1}^3 \tilde{y}_{3i} y_{2i} \right)}{m_S^2 - m_{\Delta}^2} \ln \left[\frac{m_S}{m_{\Delta}} \right] \right], \end{aligned} \tag{35}$$

where $C_{\rm box} = m_{B_s} f^2_{B_s}/3, f_{B_s} \approx 0.224$ GeV is the decay constant of B_s-meson [40], and the current measurement is $\Delta m^{exp}_{B_s} = 1.17 \times 10^{-11}$ GeV [2]. To satisfy the $R_{K^{(*)}}$ excess, the rough magnitude of LQ couplings is $|y_{3i} y_{2i}| \sim |k_{3i} k_{2i}| \sim 5 \times 10^{-3}$. Using our parameter values, it can be shown that the resulting Δm_{B_s} agree with the current experimental data. However, Δm_{B_s} can indeed constrain the parameters involved in the $b \to c\ell' \bar{\nu}_{\ell'}$ decays.

In addition to the muon $g-2$, the introduced LQs can also contribute to the lepton flavor violating processes $\ell' \to \ell\gamma$, where the current upper bounds are $BR(\mu \to e\gamma) < 4.2 \times 10^{-13}$ and $BR(\tau \to e(\mu)\gamma) < 3.3(4.4) \times 10^{-8}$ [2], and they can strictly constrain the LQ couplings. To understand the constraints due to the $\ell' \to \ell\gamma$ decays, one expresses their branching ratios (BRs) such as

$$BR(\ell_b \to \ell_a \gamma) = \frac{48\pi^3 \alpha_{em} C_{ba}}{G_F^2 m^2_{\ell_b}} \left(\left|(a_R)_{ab}\right|^2 + \left|(a_L)_{ab}\right|^2 \right) \tag{36}$$

with $C_{\mu e} \approx 1$, $C_{\tau e} \approx 0.1784$, and $C_{\tau\mu} \approx 0.1736$. $(a_R)_{ab}$ is written as

$$(a_R)_{ab} \approx \frac{3}{(4\pi)^2} \int d[X]\, m_t \left(F_{k\tilde{k}} - F_{w\tilde{y}}\right)_{ab}, \tag{37}$$

where $\int [dX] \equiv \int dxdydz(1-x-y-z)$, $(a_L)_{ab}$ can be obtained from $(a_R)_{ab}$ by using $\left(F^\dagger_{\alpha\beta}\right)_{ab}$ instead of $(F_{\alpha\beta})_{ab}$, and the function $F_{k\tilde{k}}$ is given by

$$\begin{aligned} \left(F_{k\tilde{k}}\right)_{ab} &= (\mathbf{Vk})_{3b} \tilde{k}_{a3} \left(\frac{5}{3} \frac{x}{\Delta(m_t, m_\Phi)_{ab}} + \frac{2}{3} \frac{1-x}{\Delta(m_\Phi, m_t)_{ab}} \right), \\ \left(F_{w\tilde{y}}\right)_{ab} &= w_{3b} \left(\mathbf{V}\tilde{\mathbf{y}}\right)_{3a} \left(\frac{1}{3} \frac{x}{\Delta(m_t, m_S)_{ab}} + \frac{2}{3} \frac{1-x}{\Delta(m_S, m_t)_{ab}} \right), \\ \Delta(m_1, m_2)_{ab} &\approx x m_1^2 + (y+z) m_2^2. \end{aligned} \tag{38}$$

Note that $\mathbf{Vk}_{3b} \approx k_{3b}$ and $\mathbf{V}\tilde{\mathbf{y}}_{3a} \approx \tilde{y}_{3a}$ are due to $V_{ub,cb} \ll V_{tb} \approx 1$. From Eq. (24), we can see that the doublet and singlet LQs can simultaneously couple to both left- and right-handed charged leptons, and the results are enhanced by m_t. Other LQ contributions are suppressed by m_ℓ due to the chirality flip in the external lepton legs, and thus they are ignored. Based on Eq. (37), the muon $g-2$ can be obtained as

$$\Delta a_\mu \simeq -m_\mu (a_L + a_R)_{a=b=\mu}. \tag{39}$$

As mentioned earlier, the singlet LQ does not contribute to $b \to s\ell'^+ \ell^-$ at the tree level, but it can induce the $b \to s\nu\bar{\nu}$ process, where the current upper bound is $B^+ \to K^+ \nu\bar{\nu} < 1.6 \times 10^{-5}$, and the SM result is around 4×10^{-6}. Thus, $B^+ \to K^+ \nu\bar{\nu}$ can bound the parameters of $\tilde{y}_{3i}\tilde{y}_{2i}$. The four-Fermi interaction structure, which is induced by the LQ, is the same as that induced by the W-boson, so we can formulate the BR for $B^+ \to K^+ \nu\bar{\nu}$ as

$$BR(B^+ \to K^+ \nu\bar{\nu}) \approx \frac{1}{3} \left(\sum_{\ell'} |1 - r_{\ell'}|^2 \right) BR^{\rm SM}(B^+ \to K^+ \nu\bar{\nu}), \tag{40}$$

$$r_{\ell'} = \frac{1}{C^\nu_{SM}} \left(\frac{\tilde{y}_{3\ell'} \tilde{y}_{2\ell'}}{2m_S^2} + \frac{y_{3\ell'} y_{2\ell'}}{4m_\Delta^2} \right), \quad C^\nu_{SM} = \frac{G_F V_{tb} V^*_{ts}}{\sqrt{2}} \frac{\alpha_{em}}{2\pi \sin^2\theta_W} X(x_t), \tag{41}$$

where $x_t = m_t^2/m_W^2$ and $X(x_t)$ can be parameterized as $X(x_t) \approx 0.65 x_t^{0.575}$ [41]. According to Eq. (31), the LQs also contribute to $B_s \to \mu^+\mu^-$ process, where the BRs measured by LHCb [42] and prediction in the SM [43] are $BR(B_s \to \mu^+\mu^-)^{exp} = (3.0 \pm 0.6^{+0.3}_{-0.2}) \times 10^{-9}$ and $BR(B_s \to \mu^+\mu^-)^{SM} = (3.65 \pm 0.23) \times 10^{-9}$, respectively. The experimental data are consistent with the SM prediction, and in order to consider the constraint from $B_s \to \mu^+\mu^-$, we use the expression for the BR as [44].

$$\frac{\mathrm{BR}(B_s \to \mu^+\mu^-)}{\mathrm{BR}(B_s \to \mu^+\mu^-)^{SM}} = \left|1 - 0.24 C_{10}^{LQ,\mu}\right|^2. \tag{42}$$

In addition to the $B^- \to D^{(*)}\tau\overline{\nu}$ decay, the induced effective Hamiltonian in Eq. (30) also contributes to the $B_c \to \tau\overline{\nu}$ process, where the allowed upper limit is $BR(B_c^- \to \tau\overline{\nu}) < 30\%$ [45]. According to previous results given by [45], we express the BR for $B_c \to \tau\overline{\nu}$ as

$$BR(B_c \to \tau\overline{\nu}_\tau) = \tau_{B_c} \frac{m_{Bc} m_\tau^2 f_{B_c}^2 G_F^2 |V_{cb}|^2}{8\pi} \left(1 - \frac{m_\tau^2}{m_{B_c}^2}\right)^2 \left|1 + \varepsilon_L + \frac{m_{B_c}^2}{m_\tau (m_b + m_c)} \varepsilon_P\right|^2, \tag{43}$$

where f_{B_c} is the B_c decay constant, and the $\varepsilon_{L,P}$ in our model is given as

$$\varepsilon_L = \frac{\sqrt{2}}{4 G_F V_{cb}} \left[-\sum_a V_{2a} \frac{y_{a3} y_{33}}{4 m_\Delta^2} + \sum_a V_{2a} \frac{\tilde{y}_{a3} \tilde{y}_{33}}{2 m_S^2} \right],$$

$$\varepsilon_P = \frac{\sqrt{2}}{4 G_F V_{cb}} \left[\frac{\tilde{y}_{33} w_{23}}{2 m_S^2} - \frac{k_{33} \tilde{k}_{32}}{2 m_\Phi^2} \right].$$

Using $\tau_{B_c} \approx 0.507 \times 10^{-12}$ s, $m_{B_c} \approx 6.275$ GeV, $f_{B_c} \approx 0.434$ GeV [46], and $V_{cb} \approx 0.04$, the SM result is $BR^{SM}(B_c \to \tau\overline{\nu}_\tau) \approx 2.1\%$. One can see that the effects of the new physics can enhance the $B_c \to \tau\overline{\nu}_\tau$ decay by a few factors at most in our analysis.

3.2.3 Observables: $R_{D^{(*)}}$ and $R_{K^{(*)}}$

The observables of $R_{D^{(*)}}$ and $R_{K^{(*)}}$ are the branching fraction ratios that are insensitive to the hadronic effects giving clearer test of lepton universality in B-meson decay, but the associated BRs still depend on the transition form factors. In order to calculate the BR for each semileptonic decay process, we parameterize the transition form factors for $\overline{B} \to P$ by

$$\langle P(p_2)|q\gamma^\mu b|\overline{B}(p_1)\rangle = F_+(q^2)\left((p_1+p_2)^\mu - \frac{m_B^2 - m_P^2}{q^2} q^\mu\right) + \frac{m_B^2 - m_P^2}{q^2} q^\mu F_0(q^2),$$

$$\langle P(p_2)|q\sigma_{\mu\nu} b|\overline{B}(p_1)\rangle = -i\left(p_{1\mu} p_{2\nu} - p_{1\nu} p_{2\mu}\right) \frac{2F_T(q^2)}{m_B + m_P}, \tag{44}$$

where P can be the $D(q = c)$ or $K(q = s)$ meson and the momentum transfer is given by $q = p_1 - p_2$. For the $B \to V$ decay where V is a vector meson, the transition form factors associated with the weak currents are parameterized such that

$$\langle V(p_2,\varepsilon)|\bar{q}\gamma_\mu b|\overline{B}(p_1)\rangle = i\varepsilon_{\mu\nu\rho\sigma}\varepsilon^{\nu*}p_1^\rho p_2^\sigma \frac{2V(q^2)}{m_B+m_V},$$

$$\langle V(p_2,\varepsilon)|\bar{q}\gamma_\mu\gamma_5 b|\overline{B}(p_1)\rangle = 2m_V A_0(q^2)\frac{\varepsilon^*\cdot q}{q^2}q_\mu + (m_B+m_V)A_1(q^2)\left(\varepsilon_\mu^* - \frac{\varepsilon^*\cdot q}{q^2}q_\mu\right)$$
$$- A_2(q^2)\frac{\varepsilon^*\cdot q}{m_B+m_V}\left((p_1+p_2)_\mu - \frac{m_B^2-m_V^2}{q^2}q_\mu\right),$$

$$\langle V(p_2,\varepsilon)|\bar{q}\sigma_{\mu\nu} b|\overline{B}(p_1)\rangle = \varepsilon_{\mu\nu\rho\sigma}\left[\varepsilon^{\rho*}(p_1+p_2)^\sigma T_1(q^2) + \varepsilon^{\rho*}q^\sigma\frac{m_B^2-m_V^2}{q^2}\left(T_2(q^2)-T_1(q^2)\right)\right.$$
$$\left. +2\frac{\varepsilon^*\cdot q}{q^2}p_1^\rho p_2^\sigma\left(T_2(q^2)-T_1(q^2)+\frac{q^2}{m_B^2-m_V^2}T_3(q^2)\right)\right], \quad (45)$$

where $V = D^*(K^*)$ when $q = c(s)$, $\varepsilon^{0123} = 1$, $\sigma_{\mu\nu}\gamma_5 = (i/2)\varepsilon_{\mu\nu\rho\sigma}\sigma^{\rho\sigma}$, and ε^μ is the polarization vector of the vector meson. Here we note that the form factors associated with the weak scalar/pseudoscalar currents can be obtained through the equations of motion, i.e., $i\partial_\mu\bar{q}\gamma^\mu b = (m_b - m_q)\bar{q}b$ and $i\partial_\mu(\bar{q}\gamma^\mu\gamma_5 b) = -(m_b+m_q)\bar{q}\gamma_5 b$. For numerical estimations, the q^2-dependent form factors F_+, F_T, V, A^0, and T_1 are taken as [47]

$$f(q^2) = \frac{f(0)}{(1-q^2/M^2)(1-\sigma_1 q^2/M^2+\sigma_2 q^4/M^4)}, \quad (46)$$

and the other form factors are taken to be

$$f(q^2) = \frac{f(0)}{1-\sigma_1 q^2/M^2+\sigma_2 q^4/M^4}. \quad (47)$$

The values of$f(0)$, σ_1, and σ_2 for each form factor are summarized in **Table 2**. A detailed discussion of the form factors can be referred to [47]. The next-to-next-leading (NNL) effects obtained with the LCQCD Some Rule approach for the $B \to D$ form factors were described by [48].

According to the form factors in Eqs. (44) and (45), and the interactions in Eqs. (30) and (32), we briefly summarize the differential decay rates for the semileptonic B decay processes, which we use for estimating $R_{D^{(*)}}$ and R_K. For the $\overline{B} \to D\ell'\bar{\nu}_{\ell'}$ decay, the differential decay rate as a function of the invariant mass q^2 can be given by

	$B \to D$			$B \to D^*$						
	F_+	F_0	F_T	V	A_0	A_1	A_2	T_1	T_2	T_3
f(0)	0.67	0.67	0.69	0.76	0.69	0.66	0.62	0.68	0.68	0.33
σ_1	0.57	0.78	0.56	0.57	0.58	0.78	1.40	0.57	0.64	1.46
σ_2							0.41			
	$B \to K$			$B \to K^*$						
f(0)	0.36	0.36	0.35	0.44	0.45	0.36	0.32	0.39	0.39	0.27
σ_1	0.43	0.70	0.43	0.45	0.46	0.64	1.23	0.45	0.72	1.31
σ_2		0.27				0.36	0.38		0.62	0.41

Table 2.
$B \to P, V$ transition form factors, as parameterized in Eqs. (46) and (47).

$$\frac{d\Gamma_D^{\ell'}}{dq^2} = \frac{G_F^2|V_{cb}|^2\sqrt{\lambda_D}}{256\pi^3 m_B^3}\left(1-\frac{m_{\ell'}^2}{q^2}\right)^2\left[\frac{2}{3}\left(2+\frac{m_{\ell'}^2}{q^2}\right)\left|X_+^{\ell'}\right|^2+\frac{2m_{\ell'}^2}{q^2}\left|X_0^{\ell'}+\frac{\sqrt{q^2}}{m_{\ell'}}X_S^{\ell'}\right|^2\right.$$
$$\left.+16\left(\frac{2}{3}\left(1+\frac{2m_{\ell'}^2}{q^2}\right)\left|X_T^{\ell'}\right|^2-\frac{m_{\ell'}}{\sqrt{q^2}}X_T^{\ell'}X_0^{\ell'}\right)\right], \tag{48}$$

where the $\left\{X_\alpha^{\ell'}\right\}$ functions and LQ contributions are

$$\begin{aligned}
X_+^{\ell'} &=\sqrt{\lambda_D}\left(1+C_V^{\ell'}\right)F_+(q^2), \quad X_0^{\ell'} = (m_B^2-m_D^2)\left(1+C_V^{\ell'}\right)F_0(q^2)\\
X_S^{\ell'} &=\frac{m_B^2-m_D^2}{m_b-m_c}C_S^{\ell'}\sqrt{q^2}F_0(q^2), \quad X_T^{\ell'} = -\frac{\sqrt{q^2\lambda_D}}{m_B+m_D}C_T^{\ell'}F_T(q^2)\\
C_V^{\ell'} &=\frac{\sqrt{2}}{8G_FV_{cb}}\sum_a V_{2a}\left(\frac{\tilde{y}_{3\ell'}\tilde{y}_{a\ell'}}{m_S^2}-\frac{y_{3\ell'}y_{a\ell'}}{2m_\Delta^2}\right),\\
C_S^{\ell'} &= -\frac{\sqrt{2}}{4G_FV_{cb}}\left(\frac{\tilde{y}_{3\ell'}w_{2\ell'}}{2m_S^2}-\frac{k_{3\ell'}\tilde{k}_{\ell'2}}{2m_\Phi^2}\right), \quad C_T^{\ell'} = \frac{\sqrt{2}}{16G_FV_{cb}}\left(\frac{\tilde{y}_{3\ell'}w_{2\ell'}}{2m_S^2}+\frac{k_{3\ell'}\tilde{k}_{\ell'2}}{2m_\Phi^2}\right),\\
\lambda_H &=m_B^4+m_H^4+q^4-2(m_B^2m_H^2+m_H^2q^2+q^2m_B^2).
\end{aligned} \tag{49}$$

We note that the effective couplings $C_S^{\ell'}$ and $C_T^{\ell'}$ at the m_b scale can be obtained from the LQ mass scale via the renormalization group (RG) equation. Our numerical analysis considers the RG running effects with $\left(C_S^{\ell'}/C_T^{\ell'}\right)_{\mu=m_b}/\left(C_S^{\ell'}/C_T^{\ell'}\right)_{\mu=\mathcal{O}(\mathrm{TeV})} \sim 2.0$ at the m_b scale [49]. The $\overline{B}\to D^*\ell'\overline{\nu}_{\ell'}$ decays involve D^* polarizations and more complicated transition form factors, so the differential decay rate determined by summing all of the D^* helicities are

$$\frac{d\Gamma_{D^*}^{\ell'}}{dq^2} = \sum_{h=L,+,-}\frac{d\Gamma_{D^*}^{\ell'h}}{dq^2} = \frac{G_F^2|V_{cb}|^2\sqrt{\lambda_{D^*}}}{256\pi^3m_B^3}\left(1-\frac{m_{\ell'}^2}{q^2}\right)^2\sum_{h=L,+,-}V_{D^*}^{\ell'h}(q^2), \tag{50}$$

where λ_{D^*} is found in Eq. (53) and the detailed $\left\{V_{D^*}^{\ell'h}\right\}$ functions are shown in the appendix. According to Eqs. (48) and (50), R_M ($M=D, D^*$) can be calculated by

$$R_M = \frac{\int_{m_\tau^2}^{q_{\max}^2} dq^2\left(d\Gamma_M^\tau/dq^2\right)}{\int_{m_\ell^2}^{q_{\max}^2} dq^2\left(d\Gamma_M^\ell/dq^2\right)} \tag{51}$$

where $q_{\max}^2 = (m_B-m_M)^2$ and $\Gamma_M^\ell = \left(\Gamma_M^e+\Gamma_M^\mu\right)/2$. For the $B\to K\ell^+\ell^-$ decays, the differential decay rate can be expressed as [50].

$$\begin{aligned}
\frac{d\Gamma_{K\ell\ell}(q^2)}{dq^2} &\approx \frac{|c_{\mathrm{SM}}|^2m_B^3}{3\cdot 2^8\pi^3}\left(1-\frac{q^2}{m_B^2}\right)^{3/2}\\
&\times\left[\left|C_9^\ell F_+(q^2)+\tfrac{2m_bC_7}{m_B+m_K}F_T(q^2)\right|^2+\left|C_{10}^\ell F_+(q^2)\right|^2\right].
\end{aligned} \tag{52}$$

From Eq. (52), the measured ratio R_K in the range $q^2 = [q^2_{min}, q^2_{max}] = [1, 6]$ GeV^2 can be estimated by

$$R_K = \frac{\int_{q^2_{min}}^{q^2_{max}} dq^2 d\Gamma_{K\mu\mu}/dq^2}{\int_{q^2_{min}}^{q^2_{max}} dq^2 d\Gamma_{Kee}/dq^2}. \tag{53}$$

R_{K^*} is similar to R_K, and thus we only show the result for R_K.

3.2.4 Numerical analysis

After discussing the possible constraints and observables of interest, we now present the numerical analysis to determine the common parameter region where the $R_{D^{(*)}}$ and $R_{K^{(*)}}$ anomalies can fit the experimental data. Before presenting the numerical analysis, we summarize the relevant parameters, which are related to the specific measurements as follows:

$$\begin{aligned} &\text{muon } g-2 : k_{32}\tilde{k}_{23}, \tilde{y}_{32}w_{32}; \quad R_K : k_{3\ell}k_{2\ell}, y_{3\ell}y_{2\ell}; \\ &R_{D^{(*)}} : k_{3\ell'}\tilde{k}_{\ell' 2}, \sum_a V_{2a}\left(y_{3\ell'}y_{a\ell'}, \tilde{y}_{3\ell'}\tilde{y}_{a\ell'}\right), \tilde{y}_{3\ell'}w_{2\ell'}. \end{aligned} \tag{54}$$

The parameters related to the radiative LFV, $\Delta B = 2$, and $B^+ \to K^+ \nu\bar{\nu}$ processes are defined as

$$\begin{aligned} &\mu \to e\gamma : k_{32}\tilde{k}_{13}, \tilde{k}_{23}k_{31}, \tilde{y}_{32}w_{31}, w_{32}\tilde{y}_{31}; \\ &\tau \to \ell_a\gamma : k_{33}\tilde{k}_{a3}, \tilde{k}_{33}k_{3a}, \tilde{y}_{33}w_{3a}, w_{33}\tilde{y}_{3a}; \\ &B^+ \to K^+\nu\bar{\nu} : \tilde{y}_{3i}\tilde{y}_{2i}, y_{3i}y_{2i}; \quad B_s \to \mu^+\mu^- : k_{32}k_{22}, y_{32}y_{22}; \\ &\Delta m_{B_s} : \left(\sum_i z_{3i}z_{2i}\right)^2, \left(\sum_i y_{3i}\tilde{y}_{2i}\right)\left(\sum_i \tilde{y}_{3i}y_{2i}\right), \end{aligned} \tag{55}$$

where $z_{3i}z_{2i} = k_{3i}k_{2i}, y_{3i}y_{2i}, \tilde{y}_{3i}\tilde{y}_{2i}$. From Eqs. (54) and (55), we can see that in order to avoid the $\mu \to e\gamma$ and $\tau \to \ell\gamma$ constraints and obtain a sizable and positive Δa_μ, we can set $(\tilde{k}_{13,33}, k_{31,33}, w_{3i})$ as a small value. From the upper limit of $B^+ \to K^+\nu\bar{\nu}$, we obtain $\tilde{y}_{3i}\tilde{y}_{2i} < 0.03$, and thus the resulting Δm_{B_s} is smaller than the current data. In order to further suppress the number of free parameters and avoid large fine-tuning of couplings, we employ the scheme with $k_{ij} \approx \tilde{k}_{ji} \approx |y_{ij}|$, where the sign of y_{ij} can be selected to obtain the correct sign for C_9^{LQ,ℓ_j} and to decrease the value of $C_{10}^{LQ,\mu}$ so that $B_s \to \mu^+\mu^-$ can fit the experimental data. As mentioned above, to avoid the bounds from the K, B_d, and D systems, we also adopt $k_{1\ell'} \approx \tilde{k}_{\ell' 1} \approx y_{1i} \approx \tilde{y}_{1i} \approx w_{1i} \sim 0$. When we omit these small coupling constants, the correlations of the parameters in Eqs. (54) and (55) can be further simplified as

$$\begin{aligned} &\text{muon } g-2 : k_{32}\tilde{k}_{23}; \quad R_K : k_{32}k_{22}, \ y_{32}y_{22}; R_{D^{(*)}} : k_{32}k_{22}, y_{32}y_{22}, \tilde{y}_{3\ell'}w_{2\ell'}; \\ &B_s \to \mu^+\mu^- : k_{32}k_{22}, y_{32}y_{22}; \quad \Delta m_{B_s} : (k_{32}k_{22})^2, (y_{32}y_{22})^2, \end{aligned} \tag{56}$$

where $\tilde{y}_{3i}\tilde{y}_{2i}$ are ignored due to the constraint from $B^+ \to K^+\nu\bar{\nu}$. The typical values of these parameters for fitting the anomalies in the $b \to s\mu^+\mu^-$ decay are $y_{32}(k_{32}), y_{22}(k_{22}) \sim 0.07$, so the resulting Δm_{B_s} is smaller than the current data, but these parameters are too small to explain $R_{D^{(*)}}$. Thus, we must depend on the singlet LQ to resolve the R_D and R_{D^*} excesses, where the main free parameters are now $\tilde{y}_{3\ell'} w_{2\ell'}$.

After discussing the constraints and the correlations among various processes, we present the numerical analysis. There are several LQs in this scenario, but we use m_{LQ} to denote the mass of all LQs. From Eqs. (37), (39), and (56), we can see that the muon $g-2$ depends only on $k_{32}\tilde{k}_{23}$ and m_Φ. Here we illustrate Δa_μ as a function of $k_{32}\tilde{k}_{23}$ in **Figure 2(a)**, where the solid, dashed, and dotted lines denote the results for $m_\Phi = 1.5$, 5, and 10 TeV, respectively, and the band is the experimental value with 1σ errors. Due to the m_t enhancement, $k_{32}\tilde{k}_{23} \sim 0.05$ with $m_\Phi \sim 1$ TeV can explain the muon $g-2$ anomaly.

According to the relationships shown in Eq. (56), R_K, $B_s \to \mu^+\mu^-$, and Δm_{B_s} depend on the same parameters, i.e., $k_{32}k_{22}$ and $y_{32}y_{22}$. We show the contours for these observables as a function of $k_{32}k_{22}$ and $y_{32}y_{22}$ in **Figure 2(b)**, where the data with 1σ errors and $m_{LQ} = 1.5$ TeV are taken for all LQ masses. Based on these results, we see that $\Delta m_{B_s} < \Delta m_{B_s}^{exp}$ in the range of $|k_{32}k_{22}|$, $|y_{32}y_{22}| < 0.05$, where R_K and $BR(B_s \to \mu^+\mu^-)$ can both fit the experimental data simultaneously. In addition, we show $C_9^{LQ,\mu} = [-1.5, -0.5]$ in the same plot. We can see that $C_9^{LQ,\mu} \sim -1$, which is used to explain the angular observable P_5', can also be achieved in the same common region. According to **Figure 2(b)**, the preferred values of $k_{32}k_{22}$ and $y_{32}y_{22}$ where the observed R_K and $B_s \to \mu^+\mu^-$ and the $C_9^{LQ,\mu} = [-1.5, -0.5]$ overlap are around $(k_{32}k_{22}, y_{32}y_{22}) \sim (-0.001, 0.004)$ and $\sim (0.025, 0.03)$. The latter values are at the percentage level, but they are still not sufficiently large to explain the tree-dominated R_D and R_{D^*} anomalies.

After studying the muon $g-2$ and R_K anomalies, we numerically analyze the ratio of $BR(\bar{B} \to D^{(*)}\tau\bar{\nu}_\tau)$ to $BR(\bar{B} \to D^{(*)}\ell\bar{\nu}_\ell)$, i.e., $R_{D^{(*)}}$. The introduced doublet and triplet LQs cannot efficiently enhance $R_{D^{(*)}}$, so in the following estimations, we only focus on the singlet LQ contributions, where the four-Fermi interactions shown in Eq. (30) come mainly from the scalar- and tensor-type interaction structures. Based

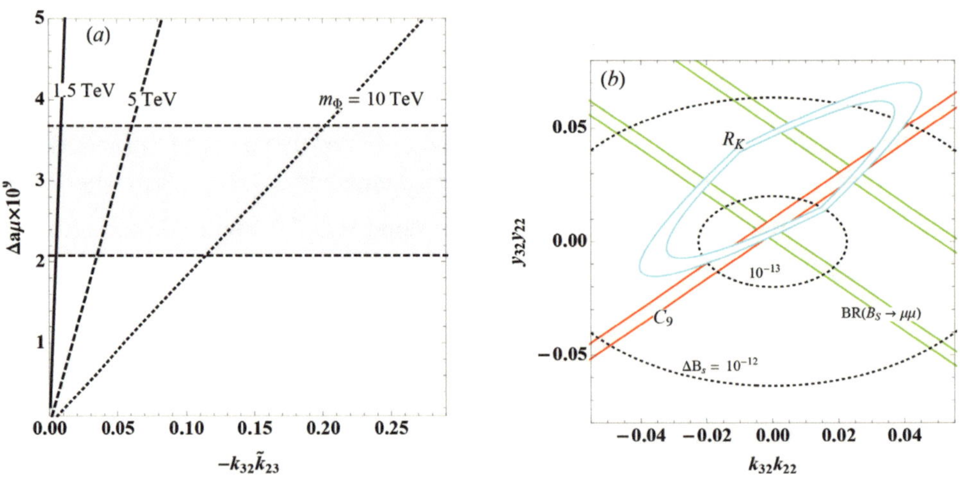

Figure 2.
(a) Δa_μ as a function of $k_{32}\tilde{k}_{23}$ with $m_\Phi = 1.5, 5, 10$ TeV, where the band denotes the experimental data with 1σ errors. (b) Contours for R_K, $B_s \to \mu^+\mu^-$, Δm_{B_s}, and $C_9^{LQ,\mu}$ as a function of $k_{32}k_{22}$ and $y_{32}y_{22}$, where the ranges of R_K and $B_s \to \mu^+\mu^-$ are the experimental values with 1σ errors and $m_{LQ} = 1.5$ TeV. For $C_9^{LQ,\mu}$, we show the range for $C^{LQ,\mu} = [-1.5, -0.5]$. (These plots are taken from Ref. [32]).

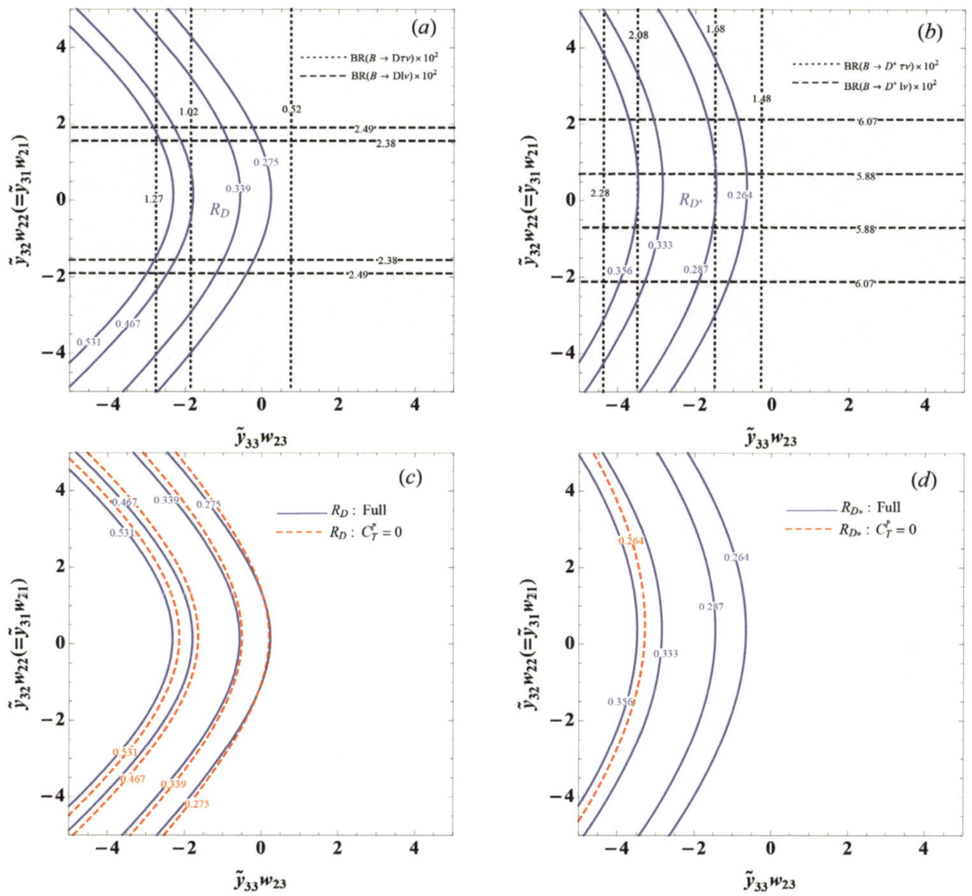

Figure 3.
Contours for (a) R_D and (b) R_{D^*}, where the solid lines denote the data with 1σ and 2σ errors, respectively. The horizontal dashed lines in both plots denote the $BR^{exp}(B^+ \to D^{(*)}\ell\nu_\ell)$, whereas the vertical dotted lines are the $BR^{exp}(B^+ \to D^{(*)}\tau\nu_\tau)$. Contours for (c) R_D and (d) R_{D^*}, where the solid and dashed lines denote the situations with and without tensor operator contributions, respectively. In this case, we take $m_{LQ} = 1.5$ TeV. (These plots are taken from ref. [32]).

on Eqs. (48), (50), and (51), we show the contours for R_D and R_{D^*} as a function of $\tilde{y}_{33}w_{23}$ and $\tilde{y}_{32}w_{22}(\tilde{y}_{31}w_{21})$ in **Figure 3(a)** and **(b)**, where the horizontal dashed and vertical dotted lines in both plots denote $BR^{exp}(B^- \to D[\ell\overline{\nu}_\ell, \tau\overline{\nu}_\tau]) = [2.27\pm 0.11, 0.77 \pm 0.25]\%$ and $BR^{exp}(B^- \to D^*[\ell\nu_\ell, \tau\overline{\nu}_\tau]) = [5.69 \pm 0.19, 1.88 \pm 0.20]\%$, respectively, and $m_{LQ} = 1.5$ TeV is used, and the data with 2σ errors are taken. For simplicity, we take $\tilde{y}_{31}w_{21} \approx \tilde{y}_{32}w_{22}$. When considering the limits from $BR(\overline{B} \to D^{(*)}\ell'\overline{\nu}_{\ell'})$, we obtain the limits $|\tilde{y}_{3\ell}w_{2\ell}|\leq 1.5$ and $\tilde{y}_{33}w_{23}>0$. In order to clearly demonstrate the influence of tensor-type interactions, we also calculate the situation by setting $C_T^{\ell'} = 0$. The contours obtained for R_D and R_{D^*} are shown in **Figure 3(c)** and **(d)**, where the solid and dashed lines denote the cases with and without $C_T^{\ell'}$, respectively. According to these plots, we can see that R_D and R_{D^*} have different responses to the tensor operators, where the latter is more sensitive to the tensor interactions. R_D and R_{D^*} can be explained simultaneously with the tensor couplings. In order to understand the correlation between $BR(B_c \to \tau\overline{\nu}_\tau)$ and $R_{D^{(*)}}$, we show the contours for $BR(B_c \to \tau\overline{\nu}_\tau)$ and $R_{D^{(*)}}$ as a function of $w_{23}\tilde{y}_{33}$ and m_S in **Figure 4**, where $\tilde{y}_{32}w_{22} \approx \tilde{y}_{31}w_{21} \approx 0$ are used, and the gray area is excluded by

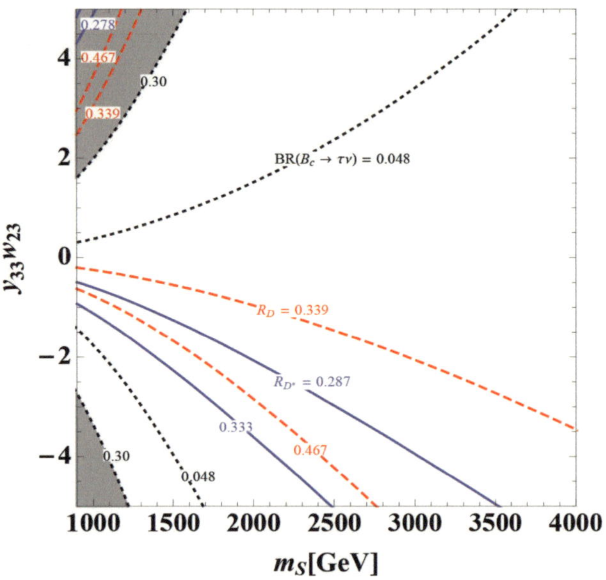

Figure 4.
Contours for $BR(B_c \to \tau\overline{\nu}_\tau)$ and $R_{D^{(*)}}$ as a function of $w_{23}\tilde{y}_{23}$ and m_S. (The plot is taken from ref. [32]).

$BR(B_c^- \to \tau\nu) < 0.3$. We can see that the predicted $BR(B_c \to \tau\overline{\nu}_\tau)$ is much smaller than the experimental bound.

Finally, we make some remarks regarding the constraint due to the LQ search at the LHC. Due to the flavor physics constraints, only the $S^{1/3}$ Yukawa couplings $\tilde{y}_{t\tau}$, $\tilde{y}_{b\nu_\tau}$, and $w_{c\tau}$ can be of $O(1)$. These couplings affect the $S^{1/3}$ decays but also their production. Therefore, in addition to the $S^{1/3}$-pair production, based on the $O(1)$ Yukawa couplings, the single $S^{1/3}$ production becomes interesting. In the pp collisions, the single $S^{1/3}$ production can be generated via the $gb \to S^{-1/3}\overline{\nu}_\tau$ and $gc \to S^{-1/3}\tau^+$ channels. Using CalcHEP 3.6 [51, 52] with the CTEQ6 parton distribution functions [53], their production cross sections with $|w_{23}| \sim |\tilde{y}_{b\nu_\tau}| \sim \sqrt{2}$ and $m_{LQ} = 1000$ GeV at $\sqrt{s} = 13$ TeV can be obtained as 3.9 fb and 2.9 fb, respectively, whereas the $S^{1/3}$-pair production cross section is $\sigma\left(pp \to S^{-1/3}S^{1/3}\right) \approx 2.4$ fb. If we assume that $S^{-1/3}$ predominantly decays into $t\tau$, $b\nu_\tau$, and $c\tau$ with similar BRs, i.e. $BR\left(S^{-1/3} \to f\right) \sim 1/3$, then the single $S^{1/3}$ production cross section $\sigma\left(S^{-1/3}X\right)$ times $BR\left(S^{-1/3} \to f\right)$ with X and f as the possible final states can be estimated as around 1 fb. The LQ coupling w_{23} involves different generations, so the constraints due to the collider measurements may not be applied directly. However, if we compare this with the CMS experiment [54] based on a single production of the second-generation scalar LQ, we find that the values of $\sigma \times BR$ at $m_{LQ} \sim 1000$ GeV are still lower than the CMS upper limit with few fb. The significance of this discovery depends on the kinematic cuts and event selection conditions, but this discussion is beyond the scope of this study, and we leave the detailed analysis for future research.

4. Conclusions

We have reviewed some charged particles which appear from physics beyond the Standard Model of particle physics. Some possible candidates of them are listed such as charged scalar boson, vector-like leptons, vector-like quarks, and leptoquarks. After showing some properties and interactions of these particles, we reviewed some applications to flavor physics in which lepton flavor physics with vector-like lepton and B-meson physics with leptoquarks are focused on as an illumination. We have seen rich phenomenology that would be induced from such new charged particles, and they will be also tested in the future experiments.

Author details

Takaaki Nomura
School of Physics, Korea Institute for Advanced Study, Seoul, Republic of Korea

*Address all correspondence to: nomura@kias.re.kr

References

[1] Bennett GW et al. [Muon g-2 Collaboration]. Final report of the muon E821 anomalous magnetic moment measurement at BNL. Physical Review D. 2006;**73**:072003. DOI: 10.1103/PhysRevD.73.072003. [hep-ex/0602035]

[2] Patrignani C et al. [Particle Data Group]. Review of particle physics. Chinese Physics C. 2016;**40**(10): 100001. DOI: 10.1088/1674-1137/40/10/100001

[3] Davier M, Hoecker A, Malaescu B, Zhang Z. Reevaluation of the hadronic contributions to the muon g-2 and to alpha (MZ). European Physical Journal C: Particles and Fields. 2011;**71**:1515. DOI: 10.1140/epjc/s10052-010-1515-z; Erratum: European Physical Journal C: Particles and Fields. 2012;**72**:1874. DOI: 10.1140/epjc/s10052-012-1874-8. [arXiv:1010.4180 [hep-ph]]

[4] Jegerlehner F, Szafron R. $\rho^0 - \gamma$ mixing in the neutral channel pion form factor F_π^e and its role in comparing e^+e^- with τ spectral functions. European Physical Journal C: Particles and Fields. 2011;**71**:1632. DOI: 10.1140/epjc/s10052-011-1632-3. [arXiv:1101.2872 [hep-ph]]

[5] Hagiwara K, Liao R, Martin AD, Nomura D, Teubner T. (g-2)μ and alpha (M_Z^2) re-evaluated using new precise data. Journal of Physics G. 2011;**38**: 085003. DOI: 10.1088/0954-3899/38/8/085003. [arXiv:1105.3149 [hep-ph]]

[6] Aoyama T, Hayakawa M, Kinoshita T, Nio M. Complete tenth-order QED contribution to the muon g-2. Physical Review Letters. 2012;**109**:111808. DOI: 10.1103/PhysRevLett.109.111808. [arXiv:1205.5370 [hep-ph]]

[7] Descotes-Genon S, Matias J, Ramon M, Virto J. Implications from clean observables for the binned analysis of $B->K*\mu^+\mu^-$ at large recoil. JHEP. 2013;**1301**:048. DOI: 10.1007/JHEP01(2013)048. [arXiv:1207.2753 [hep-ph]]

[8] Aaij R et al. [LHCb Collaboration]. Angular analysis of the $B^0 \to K^{*0}\mu^+\mu^-$ decay using 3 fb^{-1} of integrated luminosity. JHEP. 2016;**1602**:104. DOI: 10.1007/JHEP02(2016)104. [arXiv:1512.04442 [hep-ex]]

[9] Aaij R et al. [LHCb Collaboration]. Measurement of form-factor-independent observables in the decay $B^0 \to K^{*0}\mu^+\mu^-$. Physical Review Letters. 2013;**111**:191801. DOI: 10.1103/PhysRevLett.111.191801. [arXiv:1308.1707 [hep-ex]]

[10] Wehle S et al. [Belle Collaboration]. Lepton-Flavor-Dependent Angular Analysis of $B \to K^*\ell^+\ell^-$. Physical Review Letters. 2017;**118**(11):111801. DOI: 10.1103/PhysRevLett.118.111801 [arXiv:1612.05014 [hep-ex]]

[11] Huschle M et al. [Belle Collaboration]. Measurement of the branching ratio of $\overline{B} \to D^{(*)}\tau^-\overline{\nu}_\tau$ relative to $\overline{B} \to D^{(*)}\ell^-\overline{\nu}_\ell$ decays with hadronic tagging at Belle. Physical Review D. 2015;**92**(7):072014. DOI: 10.1103/PhysRevD.92.072014. [arXiv:1507.03233 [hep-ex]]

[12] Lees JP et al. [BaBar Collaboration]. Evidence for an excess of $\overline{B} \to D^{(*)}\tau^-\overline{\nu}_\tau$ decays. Physical Review Letters. 2012; **109**:101802. DOI: 10.1103/PhysRevLett.109.101802. [arXiv:1205.5442 [hep-ex]]

[13] Lees JP et al. [BaBar Collaboration]. Measurement of an excess of $\overline{B} \to D^{(*)}\tau^-\overline{\nu}_\tau$ decays and implications for charged Higgs bosons. Physical Review D. 2013;**88**(7):072012. DOI: 10.1103/PhysRevD.88.072012. [arXiv:1303.0571 [hep-ex]]

[14] Abdesselam A et al. [Belle Collaboration]. Measurement of the branching ratio of $\overline{B}^0 \to D^{*+}\tau^-\overline{\nu}_\tau$ relative to $\overline{B}^0 \to D^{*+}\ell^-\overline{\nu}_\ell$ decays with a semileptonic tagging method. arXiv: 1603.06711 [hep-ex]

[15] Hirose S et al. [Belle Collaboration]. Measurement of the τ lepton polarization and $R(D^*)$ in the decay $\overline{B} \to D^*\tau^-\overline{\nu}_\tau$. Physical Review Letters. 2017;**118**(21):211801. DOI: 10.1103/PhysRevLett.118.211801. [arXiv: 1612.00529 [hep-ex]]

[16] Aaij R et al. [LHCb Collaboration]. Measurement of the ratio of branching fractions $\mathcal{B}\left(\overline{B}^0 \to D^{*+}\tau^-\overline{\nu}_\tau\right)/\mathcal{B}\left(\overline{B}^0 \to D^{*+}\mu^-\overline{\nu}_\mu\right)$. Physical Review Letters. 2015;**115**(11):111803. DOI: 10.1103/PhysRevLett.115.111803; Addendum: Physical Review Letters. 2015;**115**(15):159901. DOI: 10.1103/PhysRevLett.115.159901. [arXiv: 1506.08614 [hep-ex]]

[17] Gunion JF, Haber HE, Kane GL, Dawson S. The Higgs hunter's guide front. Physiologie. 2000;**80**:1

[18] Branco GC, Ferreira PM, Lavoura L, Rebelo MN, Sher M, Silva JP. Theory and phenomenology of two-Higgs-doublet models. Physics Reports. 2012; **516**:1. DOI: 10.1016/j.physrep.2012.02.002. [arXiv:1106.0034 [hep-ph]]

[19] Magg M, Wetterich C. Neutrino mass problem and gauge hierarchy. Physics Letters B. 1980;**94**:61

[20] Lazarides G, Shafi Q, Wetterich C. Proton lifetime and fermion masses in an SO(10) model. Nuclear Physics B. 1981;**181**:287

[21] Mohapatra RN, Senjanovic G. Neutrino masses and mixings in gauge models with spontaneous parity violation. Physical Review D. 1981;**23**:165

[22] Ma E, Sarkar U. Neutrino masses and leptogenesis with heavy Higgs triplets. Physical Review Letters. 1998; **80**:5716. [hep-ph/9802445]

[23] Konetschny W, Kummer W. Nonconservation of total lepton number with scalar bosons. Physics Letters B. 1977;**70**:433

[24] Schechter J, Valle JWF. Neutrino masses in SU(2) × U(1) theories. Physical Review D. 1980;**22**:2227

[25] Cheng TP, Li L-F. Neutrino masses, mixings and oscillations in SU(2) × U(1) models of electroweak interactions. Physical Review D. 1980; **22**:2860

[26] Bilenky SM, Hosek J, Petcov ST. On oscillations of neutrinos with dirac and majorana masses. Physics Letters B. 1980;**94**:495

[27] Zee A. Quantum numbers of majorana neutrino masses. Nuclear Physics B. 1986;**264**:99

[28] Babu KS. Model of 'Calculable' majorana neutrino masses. Physics Letters B. 1988;**203**:132

[29] Chen CH, Nomura T. Bounds on LFV Higgs decays in a vector-like lepton model and searching for doubly charged leptons at the LHC. European Physical Journal C: Particles and Fields. 2016;**76**(6):353. DOI: 10.1140/epjc/s10052-016-4197-3. [arXiv:1602.07519 [hep-ph]]

[30] Chen CH, Nomura T. Single production of $X_{\pm 5/3}$ and $Y_{\mp 4/3}$ vector like quarks at the LHC. Physical Review D. 2016;**94**(3):035001. DOI: 10.1103/PhysRevD.94.035001. [arXiv: 1603.05837 [hep-ph]]

[31] Chen CH, Nomura T, Okada H. Explanation of $B \to K^{(*)}\ell^+\ell^-$ and muon $g-2$, and implications at the LHC. Physical Review D. 2016;**94**(11):115005.

DOI: 10.1103/PhysRevD.94.115005. [arXiv:1607.04857 [hep-ph]]

[32] Chen CH, Nomura T, Okada H. Excesses of muon $g-2$, $R_{D^{(*)}}$, and R_K in a leptoquark model. Physics Letters B. 2017;**774**:456. DOI: 10.1016/j.physletb.2017.10.005. [arXiv:1703.03251 [hep-ph]]

[33] Aad G et al. [ATLAS Collaboration]. Measurements of the Higgs boson production and decay rates and coupling strengths using pp collision data at $\sqrt{s}=7$ and 8 TeV in the ATLAS experiment. European Physical Journal C: Particles and Fields. 2016;**76**(1):6. DOI: 10.1140/epjc/s10052-015-3769-y. [arXiv:1507.04548 [hep-ex]]

[34] CMS Collaboration [CMS Collaboration]. Precise determination of the mass of the Higgs boson and studies of the compatibility of its couplings with the standard model. CMS-PAS-HIG-14-009

[35] Amhis Y et al. [HFLAV Collaboration]. Averages of b-hadron, c-hadron, and τ-lepton properties as of summer 2016. European Physical Journal C. 2017;**77**(12):895. DOI: 10.1140/epjc/s10052-017-5058-4 [arXiv:1612.07233 [hep-ex]]

[36] Bailey JA et al. [MILC Collaboration]. '$B \to D\tau\nu$ form factors at nonzero recoil and $|V_{cb}|$ from 2+1-flavor lattice QCD. Physical Review D. 2015;**92**(3):034506. DOI: 10.1103/PhysRevD.92.034506. [arXiv:1503.07237 [hep-lat]]

[37] Na H et al. [HPQCD Collaboration]. $B \to Dl\nu$ form factors at nonzero recoil and extraction of $|V_{cb}|$. Physical Review D. 2015;**92**(5):054510. DOI: 10.1103/PhysRevD.93.119906; Erratum: Physical Review D. 2016;**93**(11):119906. DOI: 10.1103/PhysRevD.92.054510 [arXiv:1505.03925 [hep-lat]]

[38] Aaij R et al. [LHCb Collaboration]. Test of lepton universality using $B^+ \to K^+\ell^+\ell^-$ decays. Physical Review Letters. 113, 151601 (2014) DOI: 10.1103/PhysRevLett.113.151601. [arXiv:1406.6482 [hep-ex]]

[39] Aaij R et al. [LHCb Collaboration]. Test of lepton universality with $B^0 \to K^{*0}\ell^+\ell^-$ decays. JHEP. 2017; **1708**:055. DOI: 10.1007/JHEP08(2017)055 [arXiv:1705.05802 [hep-ex]]

[40] Dowdall RJ et al. [HPQCD Collaboration]. B-Meson decay constants from improved lattice nonrelativistic QCD with physical u, d, s, and c quarks. Physical Review Letters. 2013;**110**(22):222003. DOI: 10.1103/PhysRevLett.110.222003. [arXiv:1302.2644 [hep-lat]]

[41] Buchalla G, Buras AJ, Lautenbacher ME. Weak decays beyond leading logarithms. Reviews of Modern Physics. 1996;**68**:1125. DOI: 10.1103/RevModPhys.68.1125. [hep-ph/9512380]

[42] Aaij R et al. [LHCb Collaboration]. Measurement of the $B_s^0 \to \mu^+\mu^-$ branching fraction and effective lifetime and search for $B^0 \to \mu^+\mu^-$ decays. Physical Review Letters. 2017;**118**(19):191801. DOI: 10.1103/PhysRevLett.118.191801. [arXiv:1703.05747 [hep-ex]]

[43] Bobeth C, Gorbahn M, Hermann T, Misiak M, Stamou E, Steinhauser M. $B_{s,d} \to l^+l^-$ in the standard model with reduced theoretical uncertainty. Physical Review Letters. 2014;**112**:101801. DOI:10.1103/PhysRevLett.112.101801 [arXiv:1311.0903 [hep-ph]]

[44] Hiller G, Schmaltz M. R_K and future $b \to s\ell\ell$ physics beyond the standard model opportunities. Physical Review D. 2014;**90**:054014. DOI: 10.1103/

PhysRevD.90.054014. [arXiv:1408.1627 [hep-ph]]

[45] Alonso R, Grinstein B, Martin Camalich J. Lifetime of B_c^- Constrains explanations for anomalies in $B \to D^{(*)}\tau\nu$. Physical Review Letters. 2017;**118**(8):081802. DOI: 10.1103/ PhysRevLett.118.081802. [arXiv: 1611.06676 [hep-ph]]

[46] Colquhoun B et al. [HPQCD Collaboration]. B-meson decay constants: A more complete picture from full lattice QCD. Physical Review D. 2015;**91**(11):114509. DOI: 10.1103/ PhysRevD.91.114509. [arXiv:1503.05762 [hep-lat]]

[47] Melikhov D, Stech B. Weak form-factors for heavy meson decays: An Update. Physical Review D. 2000;**62**: 014006. DOI: 10.1103/ PhysRevD.62.014006. [hep-ph/ 0001113]

[48] Wang YM, Wei YB, Shen YL, Lü CD. Perturbative corrections to B → D form factors in QCD. JHEP. 2017;**1706**: 062. DOI: 10.1007/JHEP06(2017)062. [arXiv:1701.06810 [hep-ph]]

[49] Sakaki Y, Tanaka M, Tayduganov A, Watanabe R. Testing leptoquark models in $\overline{B} \to D^{(*)}\tau\overline{\nu}$. Physical Review D. 2013; **88**(9):094012. DOI: 10.1103/ PhysRevD.88.094012. [arXiv:1309.0301 [hep-ph]]

[50] Chen CH, Geng CQ. Probing new physics in B —> K(*) l+ l– decays. Physical Review D. 2002;**66**:094018. DOI: 10.1103/PhysRevD.66.094018. [hep-ph/0209352]

[51] Belyaev A, Christensen ND, Pukhov A. CalcHEP 3.4 for collider physics within and beyond the Standard Model. Computer Physics Communications. 2013;**184**:1729. [arXiv:1207.6082 [hep-ph]]

[52] Belyaev A, Leroy C, Mehdiyev R, Pukhov A. Leptoquark single and pair production at LHC with CalcHEP/ CompHEP in the complete model. JHEP. 2005;**0509**:005. DOI: 10.1088/1126-6708/2005/09/005. [hep-ph/0502067]

[53] Nadolsky PM, Lai HL, Cao QH, Huston J, Pumplin J, Stump D, et al. Implications of CTEQ global analysis for collider observables. Physical Review D. 2008;**78**:013004. [arXiv:0802.0007 [hep-ph]]

[54] Khachatryan V et al. [CMS Collaboration]. Search for single production of scalar leptoquarks in proton-proton collisions at $\sqrt{s} = 8\ TeV$. Physical Review D. 2016;**93**(3):032005. DOI: 10.1103/PhysRevD.93.032005; Erratum: Physical Review D. 2017;**95** (3):039906. DOI: 10.1103/ PhysRevD.95.039906. [arXiv: 1509.03750 [hep-ex]]

Section 3

Electron Positron Plasma

Chapter 5

Electrostatic Waves in Magnetized Electron-Positron Plasmas

Ian Joseph Lazarus

Abstract

The behavior of arbitrary amplitude linear and nonlinear electrostatic waves that propagate in a magnetized four component, two-temperature, electron-positron plasma is presented. The characteristics of the dispersive properties of the associated linear modes using both fluid and kinetic theory are examined. The fluid theory analysis of the electrostatic linear waves shows the existence of electron acoustic, upper hybrid, electron plasma and electron cyclotron branches. A kinetic theory analysis is then used to study the acoustic mode, in particular the effect of Landau damping, which for the parameter regime considered is due to the cooler species. Consequently, it is found that a large enough drift velocity is required to produce wave growth. Nonlinear electrostatic solitary waves (ESWs), similar to those found in the broadband electrostatic noise observed in various regions of the earth's magnetosphere is further investigated. A set of nonlinear differential equations for the ESWs, which propagate obliquely to an external magnetic field is derived and numerically solved. The effect of various plasma parameters on the waves is explored and shows that as the electric driving force is increased, the electric field structure evolves from a sinusoidal wave to a spiky bipolar form. The results are relevant to both astrophysical environments and related laser-induced laboratory experiments.

Keywords: electrons, positrons, electrostatic waves, nonlinear waves

1. Introduction

Electron-positron plasmas play a significant role in the understanding of the early universe [1, 2], active galactic nuclei [3], gamma ray bursts (GRBs) [4], pulsar magnetospheres [5, 6] and the solar atmosphere [7]. These plasmas are also important in understanding extremely dense stars such as white dwarfs and pulsars, which are thought to be rotating neutron stars. The existence of these plasmas in neutron stars and in the pulsar magnetosphere is well documented [8]. The possibility for the co-existence of two types of cold and hot electron-positron populations in the pulsar magnetosphere has been suggested by [9] which was inspired by the pulsar model [10]. In their model, accelerated primary electrons moving on curved magnetic field lines emit curvature photons which produce electron-positron pairs. The secondary particles then produce curvature radiation, hence producing new electron-positron pairs, and so on. Therefore, both the electron and positron populations can be subdivided in two groups of distinct temperatures, one modeling the original plasma, and the second the higher-energy cascade-bred pairs. It is also

IntechOpen

known that in astrophysical and cosmic plasmas, a minority of cold electrons and heavy ions exist along with hot electron-positron pairs [11]. Hence, the formation of two temperature multispecies plasmas is possible due to the outflow of the electron-positron plasma from pulsars entering into an interstellar cold, low-density electron-ion plasma [12].

Investigations into electron-positron plasma behavior have focused primarily on the relativistic regime. It is however plausible that nonrelativistic astrophysical electron-positron plasmas may exist, given the effect of cooling by cyclotron emission [13]. The study of nonrelativistic astrophysical electron-positron plasmas therefore plays an important role in understanding wave fluctuations. Due to the equal charge to mass ratio for these oppositely charged species, only one frequency scale exists and due to this symmetry, there exists different physical phenomena to the conventional electron-ion plasmas. Further, the frequent instabilities that arise in space plasma and astrophysical environments (e.g., solar flames and auroras), involve the growth of electrostatic and electromagnetic waves which gives rise to a growing wave mode. In particular, the linear behavior of the electrostatic modes using fluid and kinetic theory approaches allows one to understand the effect of plasma parameters such as the propagation angle, cool to hot temperature ratios, density ratios and the magnetic field strength on the waves.

Investigations conducted have focused on modulational instabilities and wave localization [14], envelope solitons [15], multidimensional effects [16]. Large amplitude solitons and electrostatic nonlinear potential structures in electron-positron plasmas having equal hot and cold components of both species have been studied by a number of authors [17–19]. In one such study [20], using the two-fluid model with a single temperature they investigated linear and nonlinear longitudinal and transverse electrostatic and electromagnetic waves in a nonrelativistic electron-positron plasma in the absence and presence of an external magnetic field. They found that several of the modes present in electron-ion plasmas also existed in electron-positron plasmas, but in a modified form. Collective modes in nonrelativistic electron-positron plasmas using the kinetic approach was studied by [21]. The author found that the dispersion relations for the longitudinal modes in the electron-positron plasma for both unmagnetized and magnetized electron-positron plasmas were similar to the modes in one-component electron or electron-ion plasmas. Moreover, the hybrid resonances present in the former are not found in an electron-positron plasma.

The understanding of nonlinear wave structures which gives rise to electrostatic solitary wave (ESWs) in space is important since it is known that satellite measurements using high-time resolution equipment aboard spacecraft S3-3 [22], Viking [23], Geotail [24], Polar [25], and Fast [26] have indicated the presence of Broadband Electrostatic Noise (BEN) in the auroral magnetosphere at altitudes between 3000 km to 8000 km and beyond. These observations have shown the presence of electrostatic solitary waves (ESWs), which are characterized by their spiky bipolar pulses. Hence, the study of nonlinear wave behavior in electron-positron plasmas propagating at oblique angles to an ambient magnetic field is explored to understand electrostatic solitary waves in space. Specifically, the spiky nature of the electrostatic potential structures and the effects of the propagation angle, cold and hot drift velocities, cool to hot density and temperature ratios and Mach number on the ESWs are examined.

In this chapter a two-temperature magnetized four component electron-positron plasma model is used to study linear wave modes using both the fluid and kinetic approaches as well as the behavior of the nonlinear structures of these electrostatic solitary waves (ESWs) which plays an important role in space and astrophysical environments.

2. Linear waves in electron-positron plasmas: fluid theory approach

Let us consider a homogeneous magnetized, four component electron-positron plasma, consisting of cool electrons and cool positrons with equal temperatures and equilibrium densities denoted by T_c and n_{0c}, respectively, and hot electrons and hot positrons with equal temperatures and equilibrium densities denoted by T_h and n_{0h}, respectively. The temperatures are expressed in energy units and wave propagation is taken in the x-direction at an angle θ to the ambient magnetic field $\mathbf{B_0}$, which is assumed to be in the x–z plane.

Assuming that the hot isothermal species are described by the Boltzmann distribution, their densities are, respectively

$$n_{eh} = n_{0h} \exp\left(\frac{e\phi}{T_h}\right) \tag{1}$$

and

$$n_{ph} = n_{0h} \exp\left(\frac{-e\phi}{T_h}\right), \tag{2}$$

where n_{eh} (n_{ph}) is the density of the hot electrons (positrons) and ϕ is the electrostatic potential.

Using Boltzmann distribution of hot electrons and positrons is justified provided they have sufficiently high temperatures, much greater than that of cooler species such that their thermal velocities parallel to the magnetic field exceed the phase velocity of the modes so that they are able to establish the Boltzmann distribution. The magnetic field effects on hot species are not felt since the perturbation wavelengths are shorter than their gyroradii such that both hot electrons and positrons follow essentially straight line orbits across the magnetic field direction.

The dynamics of cooler isothermal species are governed by fluid equations, namely the continuity equations,

$$\frac{\partial n_{jc}}{\partial t} + \nabla.\left(n_{jc}\mathbf{v}_{jc}\right) = 0, \tag{3}$$

the equations of motion,

$$\frac{\partial \mathbf{v}_{jc}}{\partial t} + \mathbf{v}_{jc}.\nabla\mathbf{v}_{jc} = -\varepsilon_j \frac{e}{m}\nabla\phi + \varepsilon_j \frac{e}{m}\left(\mathbf{v}_{jc} \times B_0\right) - \frac{\gamma T_c}{n_{jc} m}\nabla n_{jc}, \tag{4}$$

where $\varepsilon_j = +1(-1)$ for positrons (electrons), $j = e(p)$ for the electrons (positrons). The system is closed by the Poisson equation

$$\varepsilon_0 \frac{\partial^2 \phi}{\partial x^2} = -e\left(n_{pc} - n_{ec} + n_{ph} - n_{eh}\right). \tag{5}$$

In the above equations, n_j and v_j are the number densities and fluid velocities respectively of the jth species. In order to derive the linear dispersion relation, equations (3)–(5) are linearized. For perturbations varying as $\exp\left(i(kx - \omega t)\right)$, $\partial/\partial t$ is replaced with $-i\omega$ and $\partial/\partial x$ with ik. Hence the perturbed densities for the electrons and positrons become

$$n_{ec} = -\left(\frac{n_{0c}ek^2\phi}{m}\right)\left(\frac{\omega^2 - \Omega^2\cos^2\theta}{\omega^4 - \omega^2(3k^2v_{tc}^2 + \Omega^2) + 3k^2v_{tc}^2\Omega^2\cos^2\theta}\right). \tag{6}$$

and

$$n_{pc} = \left(\frac{n_{0c}ek^2\phi}{m}\right)\left(\frac{\omega^2 - \Omega^2\cos^2\theta}{\omega^4 - \omega^2(3k^2v_{tc}^2 + \Omega^2) + 3k^2v_{tc}^2\Omega^2\cos^2\theta}\right). \tag{7}$$

From equations (1) and (2), the perturbed densities for the hot species are given by,

$$n_{eh} = n_{oh}\frac{e\phi}{T_h} \tag{8}$$

and

$$n_{ph} = -n_{oh}\frac{e\phi}{T_h}. \tag{9}$$

Substituting equations (6)–(9), into Poisson's equation (5), the general dispersion relation for the two temperature electron-positron plasma is found to be

$$\omega^2(\omega^2 - \Omega^2) - 3k^2v_{tc}^2(\omega^2 - \Omega^2\cos^2\theta) - \frac{k^2v_{ea}^2}{1 + \frac{1}{2}k^2\lambda_{Dh}^2}(\omega^2 - \Omega^2\cos^2\theta) = 0 \tag{10}$$

where $v_{ea} = (n_{0c}/n_{0h})^{1/2}v_{th}$ is the acoustic speed of the electron-positron plasma, analogous in form to the electron acoustic speed in an electron-ion plasma [27]. The thermal velocity of the cool species is $v_{tc} = (T_c/m)^{1/2}$, $\Omega_j = \Omega = q_jB_o/m$ is the gyrofrequency of the electrons and positrons and $\lambda_{dh} = (\varepsilon_0 T_h/n_{0h}e^2)^{1/2}$ is the Debye length of the hot species.

It is noted that the study of linear electrostatic waves using a simple fluid model cannot handle the possible Landau damping of the modes. Hence, Landau damping is not significant since phase velocities are far away from the thermal velocities of either the hot or cooler species, i.e., $v_{th} \gg v_\phi \gg v_{tc}$ with $T_h \gg T_c$. The effects of the temperature variation on the acoustic mode in terms of Landau damping using kinetic theory are discussed in the next section.

For a single species electron-positron plasma, with temperature T_c, equation (10) reduces to,

$$\omega^4 - \omega^2(\Omega^2 + 3k^2v_{tc}^2) + 3k^2v_{tc}^2\Omega^2\cos^2\theta = 0. \tag{11}$$

This is identical to the dispersion relation of [20] for their single temperature electron-positron model.

For wave frequencies much lower than the gyrofrequency and satisfying $\omega \ll \Omega\cos\theta$, the associated electron-acoustic (or positron-acoustic) mode is found to be,

$$\omega^2 = \frac{k^2v_{ea}^2\cos^2\theta}{1 + \frac{1}{2}k^2\lambda_{Dh}^2} + 3k^2v_{tc}^2\cos^2\theta. \tag{12}$$

Taking short wavelength limit ($k^2\lambda_{Dh}^2 \gg 1$), the dispersion relation equation (10) reduces to,

$$\omega^4 - \omega^2\left(3k^2v_{tc}^2 + \omega_{UH}^2\right) + \left(3k^2v_{tc}^2 + 2\omega_{pc}^2\right)\Omega^2\cos^2\theta = 0, \tag{13}$$

where

$$\omega_{UH}^2 = \Omega^2 + 2\omega_{pc}^2 \tag{14}$$

is the upper hybrid frequency associated with the cooler species [20], with $\omega_{pc} = \left(n_{oc}e^2/\varepsilon_0 m\right)^{1/2}$ as the plasma frequency of the cooler species. If one solves equation (13) in the limit $\left(3k^2v_{tc}^2 + \omega_{UH}^2\right)^2 \gg 4\left(3k^2v_{tc}^2\Omega^2\cos^2\theta + 2\omega_{pc}^2\Omega^2\cos^2\theta\right)$, one obtains for the upper hybrid mode,

$$\omega_+^2 = \left(3k^2v_{tc}^2 + \omega_{UH}^2\right) - \frac{\left(3k^2v_{tc}^2 + 2\omega_{pc}^2\right)\Omega^2\cos^2\theta}{3k^2v_{tc}^2 + \omega_{UH}^2}, \tag{15}$$

Taking the negative square root of equation (13) yields

$$\omega_-^2 = \frac{\left(3k^2v_{tc}^2 + 2\omega_{pc}^2\right)\Omega^2\cos^2\theta}{3k^2v_{tc}^2 + \omega_{UH}^2}, \tag{16}$$

In order to gain physical insight into the solution space of the dispersion relation, the two extreme limits of equation (10) will now be considered, viz. pure perpendicular and pure parallel propagations.

2.1 Case I: pure perpendicular propagation

Considering the pure perpendicular ($\theta = 90^o$) limit, the general dispersion relation (10), reduces to:

$$\omega^4 - \omega^2\left(\Omega^2 + 3k^2v_{tc}^2 + \frac{k^2v_{ea}^2}{1 + \frac{1}{2}k^2\lambda_{dh}^2}\right) = 0. \tag{17}$$

Hence the normal mode frequencies are, $\omega = 0$, which is a nonpropagating mode, and

$$\omega^2 = \Omega^2 + 3k^2v_{tc}^2 + \frac{k^2v_{ea}^2}{1 + \frac{1}{2}k^2\lambda_{dh}^2}. \tag{18}$$

Taking the short wavelength limit ($k^2\lambda_{dh}^2 \gg 1$) of the above relationship, one obtains,

$$\omega^2 = \omega_{UH}^2 + 3k^2v_{tc}^2. \tag{19}$$

showing that the behavior of the upper hybrid mode for the two temperature model is due to the cooler species, where $\omega_{UH}^2 = \Omega_p^2 + 2\omega_{pc}^2$.

Now taking the long wavelength limit ($k^2\lambda_{dh}^2 \ll 1$) of the dispersion relation for perpendicular propagation, equation (18) reduces to

$$\omega^2 = \Omega^2 + k^2\left(3v_{tc}^2 + v_{ea}^2\right). \tag{20}$$

This is the cyclotron mode for the electron-positron plasma with contributions from both the thermal motion of the adiabatic cooler species and the acoustic motion due to the two species of different temperatures. To try and understand the physical implications, the above expression for the dispersion relation can be written as,

$$\omega^2 = \Omega^2 + k^2 v_{ea}^2\left(1 + 3\frac{T_c}{T_h}\frac{n_{0h}}{n_{0c}}\right). \tag{21}$$

For $T_c/T_h \ll 1$, one requires $n_{0h} \gg n_{0c}$, i.e., a plasma dominated by the hot species, in order for the second term in brackets to affect the dispersive properties of the wave.

2.2 Case II: pure parallel propagation

Considering the limit of parallel propagation ($\theta = 0^o$), the general dispersion relation (10) reduces to,

$$\omega^4 - \omega^2\left(\Omega^2 + 3k^2v_{tc}^2 + \frac{k^2v_{ea}^2}{1 + \frac{1}{2}k^2\lambda_{dh}^2}\right) + \Omega^2\left(3k^2v_{tc}^2 + \frac{k^2v_{ea}^2}{1 + \frac{1}{2}k^2\lambda_{dh}^2}\right) = 0, \tag{22}$$

from which it can be shown

$$\omega^2 = \frac{1}{2}\left[\Omega^2 + 3k^2v_{tc}^2 + \frac{k^2v_{ea}^2}{1 + \frac{1}{2}k^2\lambda_{dh}^2} \pm \left(\Omega^2 - 3k^2v_{tc}^2 - \frac{k^2v_{ea}^2}{1 + \frac{1}{2}k^2\lambda_{dh}^2}\right)\right]. \tag{23}$$

There exist two possible solutions. Taking the positive sign of the relevant term in equation (23) as the first option yields,

$$\omega_+^2 = \Omega^2, \tag{24}$$

which is a constant frequency, nonpropagating cyclotron mode.

Now taking the negative sign of the term in equation (23) yields the normal mode frequency

$$\omega_-^2 = 3k^2v_{tc}^2 + \frac{k^2v_{ea}^2}{1 + \frac{1}{2}k^2\lambda_{dh}^2}, \tag{25}$$

which may be written for $k^2\lambda_{dh}^2 \ll 1$ as

$$\omega_-^2 = k^2v_{ea}^2\left(1 + 3\frac{T_c}{T_h}\frac{n_{0h}}{n_{0c}}\right), \tag{26}$$

which is identified fundamentally, as the electron-acoustic mode, with a correction term to its phase velocity due to the thermal motion of the cooler species.

In the limit $k^2\lambda_{dh}^2 \gg 1$, one obtains

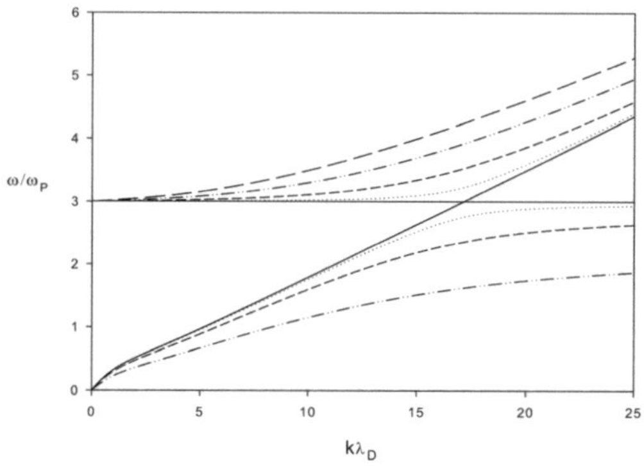

Figure 1.
Normalized real frequency as a function of the normalized wavenumber showing the acoustic and cyclotron branches for various angles of propagation θ = 0° (solid), 9° (dotted), 22.5° (broken), 45° (dashddot) and 90° (longbroken). The fixed plasma parameters are $R = 0.333$, $T_c/T_h = 0.01$ and $n_{0c}/n_{0h} = 0.11$.

$$\omega_-^2 = 3k^2v_{tc}^2 + 2\omega_{pc}^2 \tag{27}$$

Equating equations (24) and (27) in the limit $k^2\lambda_{dh}^2 \gg 1$, the critical k value for which the two modes may couple is determined to be,

$$(k\lambda_d)_{crit} = \left(\frac{T_h}{3T_c}\frac{n_{0c}}{n_0}\right)^{1/2}\left(\frac{n_0}{n_{0c}R^2} - 2\right)^{1/2}. \tag{28}$$

A numerical analysis of the general dispersion relation can be performed focusing on the effects of the density and temperature ratios of the hot and cool electrons and positrons. If one normalizes the fluid speeds by the thermal velocity v_{th} = $(T_h/m)^{1/2}$, the particle density by the total equilibrium plasma density $n_0 = n_{0c} + n_{0h}$, the temperatures by T_h, the spatial length by $\lambda_D = (\varepsilon_0 T_h/n_0e^2)^{1/2}$, and the time by $\omega_p^{-1} = (n_0e^2/\varepsilon_0 m)^{-1/2}$ in equation (10), you get the normalized general dispersion relation,

$$\omega'^4 - \omega'^2\left(\frac{1}{R^2} + 3k'^2\frac{T_c}{T_h} + \frac{k'^2 n'_{0c}}{n'_{0h} + \frac{1}{2}k'^2}\right) + \frac{\cos^2\theta}{R^2}\left(3k'^2\frac{T_c}{T_h} + \frac{k'^2 n'_{0c}}{n'_{0h} + \frac{1}{2}k'^2}\right) = 0, \tag{29}$$

where $\omega' = \omega/\omega_p$, $k' = k\lambda_D$, $n'_{0h} = n_{0h}/n_0$, $n'_{0c} = n_{0c}/n_0$ and $R = \omega_p/\Omega$ is a measure of the plasma densities and the strength of the magnetic field. A typical result can be seen in **Figure 1** [28] for the normalized real frequency as a function of the normalized wavenumber showing the acoustic and cyclotron branches for a range of propagation angles.

3. Linear waves in electron-positron plasmas: kinetic theory approach

In this section the kinetic theory approach is used to study the acoustic mode that was investigated in the previous section using fluid theory. The focus is on this mode since it is a micro-instability arising from resonances in velocity space. This instability is kinetic in nature and the growth rate of the wave is a function of the slope of the velocity distribution function. When the wave phase velocity along $\mathbf{B_0}$ sees a negative slope of the velocity distribution $\left(\partial f_0/\partial V_\parallel < 0\right)$, the particles on average will gain energy from the wave, consequently the wave losses energy and becomes damped, an effect known as Landau damping. The wave mode is hence subjected to Landau damping and wave enhancement. Therefore the focus in this section is primarily on the effect of the temperatures of the plasma species.

The same plasma model as in the previous section is considered, i.e., a four component magnetized electron-positron plasma, consisting of cool electrons and cool positrons with equal temperatures and equilibrium densities denoted by T_c and n_{0c} respectively, and hot electrons and hot positrons with equal temperatures and equilibrium densities denoted by T_h and n_{0h}, respectively.

We begin by deriving the general dispersion relation where each species j has an isotropic, drifting Maxwellian velocity distribution with temperatures T_j drifting parallel to the magnetic field $\mathbf{B_0} = B_0\hat{z}$, with drift velocities V_{oj}.

Hence, the equilibrium velocity distribution for the electron and positron species is chosen to be,

$$f_{\alpha 0} = \frac{n_{\alpha 0}}{\left(2\pi v_{tj}^2\right)^{\frac{3}{2}}} \exp\left\{\frac{-\left[V_x^2 + V_y^2 + \left(V_z - V_{oj}\right)^2\right]}{2v_{tj}^2}\right\}, \tag{30}$$

The Vlasov equations are,

$$\frac{\partial f_\alpha}{\partial t} + \mathbf{V}.\nabla f_\alpha + \frac{q_\alpha}{m}(\mathbf{E} + \mathbf{V} \times \mathbf{B}).\frac{\partial f_\alpha}{\partial \mathbf{V}} = 0, \tag{31}$$

and the equations of motion for the electrons and positrons is given by,

$$m\frac{d\mathbf{V}}{dt} = q_\alpha\{\mathbf{E} + \mathbf{V} \times \mathbf{B}\}, \tag{32}$$

where $j = c(h)$ for the cool (hot) species and $\alpha = ec, pc, eh$ and ph for the cool electrons, cool positrons, hot electrons and hot positrons respectively, and $v_{tj} = \left(T_j/m\right)^{1/2}$ is the thermal velocity of the jth species.

Following standard techniques for electron-ion plasmas [29], the general kinetic dispersion relation for the four component, two temperature electron-positron plasma is given by

$$\begin{aligned} k^2 &+ \frac{2}{\lambda_{Dc}^2}\left[1 + \frac{\omega - \mathbf{k}.\mathbf{V_{oc}}}{\sqrt{2}k_{\parallel}v_{tc}} \sum_{p=-\infty}^{\infty} Z(z_{pc})\Gamma_{pc}\right] \\ &+ \frac{2}{\lambda_{Dh}^2}\left[1 + \frac{\omega - \mathbf{k}.\mathbf{V_{oh}}}{\sqrt{2}k_{\parallel}v_{th}} \sum_{p=-\infty}^{\infty} Z(z_{ph})\Gamma_{ph}\right] = 0, \end{aligned} \tag{33}$$

where $\lambda_{Dc,h} = \left(\varepsilon_0 T_h/n_{0c,h}e^2\right)^{1/2}$ is the Debye length for the cool (hot) species and z_{pj} is the argument of the plasma dispersion function or Z-function [30] and is given by,

$$z_{pj} = \frac{\omega - \mathbf{k}.\mathbf{V_{oj}} - p\Omega_j}{\sqrt{2}k_{\parallel}v_{tj}}, \tag{34}$$

where,

$$\Gamma_{pj} = e^{-\alpha_j} I_p(\alpha_j), \tag{35}$$

and

$$\alpha_j = \frac{k_\perp^2 v_{tj}^2}{\Omega_j^2}, \tag{36}$$

where I_p is the modified Bessel function of order p. The components of $\mathbf{k}$ parallel (perpendicular) to $\mathbf{B_0}$ are given by $k_\parallel$ ($k_\perp$) respectively, while $\mathbf{V_{oc}}$ and $\mathbf{V_{oh}}$ are the drift velocities of the cool (hot) species, respectively.

3.1 Approximate solutions of the kinetic dispersion relation

The general dispersion relation (33) can be numerically solved without any approximations. However, to get some insight into the solutions, here, approximate expansions of the plasma dispersion function are used to obtain analytical expressions for the frequency and growth rate of the acoustic mode.

In proceeding, for the temperatures it is assumed that $T_h \gg T_c(\sim 0)$. In addition low frequency modes satisfying $|\omega| \ll \Omega$ are considered. The series expansion of the Z-function [30] is given by

$$Z(z) = i\sqrt{\pi}e^{-z^2} - 2z\left[1 - \frac{2z^2}{3} + \frac{4z^4}{15} - \dots\right] \text{ for } |z| \ll 1 \text{ and} \tag{37}$$

$$Z(z) = i\sqrt{\pi}\delta e^{-z^2} - \frac{1}{z}\left[1 + \frac{1}{2z^2} + \frac{3}{4z^4} + \dots\right] \text{ for } |z| \gg 1. \tag{38}$$

where for $|z| \gg 1$, $\delta = \begin{cases} 0, & \text{Im}(z)>0 \\ 1, & \text{Im}(z) = 0 \\ 2, & \text{Im}(z) < 0 \end{cases}$

Assuming the drift of the electrons and positrons to be weak (i.e., small V_{oc} and V_{oh}) [31] and $|\omega| \ll \Omega$,

$$z_{pc} = \frac{\omega - \mathbf{k}.\mathbf{V_{oc}} - p\Omega}{\sqrt{2}k_\parallel v_{tc}} \approx \frac{-p\Omega}{\sqrt{2}k_\parallel v_{tc}} \quad \text{for } p \neq 0 \tag{39}$$

and

$$z_{ph} = \frac{\omega - \mathbf{k}.\mathbf{V_{oh}} - p\Omega}{\sqrt{2}k_\parallel v_{th}} \approx \frac{-p\Omega}{\sqrt{2}k_\parallel v_{th}} \quad \text{for } p \neq 0. \tag{40}$$

Then for the cool species,

$$\sum_{p=-\infty}^{\infty} Z(z_{pc})\Gamma_{pc} \approx Z\left(\frac{\omega - \mathbf{k}.\mathbf{V_{oc}}}{\sqrt{2}k_\parallel v_{tc}}\right)\Gamma_{oc} + \sum_{p=1}^{\infty}\left\{Z\left(\frac{p\Omega}{\sqrt{2}k_\parallel v_{tc}}\right) + Z\left(\frac{-p\Omega}{\sqrt{2}k_\parallel v_{tc}}\right)\right\}\Gamma_{pc}. \tag{41}$$

From the definition of the Z-function, $Z(\xi) + Z(-\xi) = 0$, hence

$$\sum_{p=-\infty}^{\infty} Z(z_{pc})\Gamma_{pc} \approx Z(z_{oc})\Gamma_{oc}. \tag{42}$$

Taking the cooler species to be stationary, V_{oc} is therefore set to zero, allowing only the hot species to drift. Then,

$$z_{oc} = \frac{\omega}{\sqrt{2}k_{\parallel}v_{tc}}. \tag{43}$$

For modes satisfying $\omega/k_{\parallel} \gg v_{tc}$, one may assume $|z_{oc}| \gg 1$, i.e., the wave phase speed along $\mathbf{B_o}$ is much larger than the cool electron thermal speed. For instability (i.e., a growing wave with $\mathrm{Im}(z)>0$), δ is set equal to zero in equation (38). Hence using the series expansion equation (38), equation (41) becomes

$$\sum_{p=-\infty}^{\infty} Z(z_{pc})\Gamma_{pc} \approx \left[-\frac{1}{z_{oc}} - \frac{1}{2z_{oc}^3} - \frac{3}{4z_{oc}^5}\right]\Gamma_{oc}. \tag{44}$$

Similarly, using the series expansion equation (37) (where $e^{-z_{oh}^2} \approx 1$ for $|z_{oh}| \ll 1$), we have for the hot species,

$$\sum_{p=-\infty}^{\infty} Z(z_{ph})\Gamma_{ph} \approx \left(i\sqrt{\pi} - 2z_{oh} + \frac{4z_{oh}^3}{3}\right)\Gamma_{oh}. \tag{45}$$

It is noted that for relatively high temperature T_h, the thermal velocity of the hot species is much larger than the wave phase velocity. Hence, for large T_h, we have assumed that $|z_{oh}| \ll 1$.

Substituting (44) and (45), λ_D, λ_{Dc} and λ_{Dh}, into the dispersion relation (33), whereas before $\lambda_D = (\varepsilon_0 T_h/n_0 e^2)^{1/2}$, gives

$$k^2\lambda_D^2 + 2\frac{\frac{n_{0c}}{n_0}}{\frac{T_c}{T_h}}\left[i\sqrt{\pi}z_{oc}e^{-z_{oc}^2} - \frac{1}{2z_{oc}^2} - \frac{3}{4z_{oc}^4}\right] + 2\frac{n_{0h}}{n_0}\left[1 + i\sqrt{\pi}z_{oh}\Gamma_{oh}\right] = 0. \tag{46}$$

For the cool species we have assumed $|\alpha_c| = |k_\perp^2 v_{tc}^2/\Omega^2| = k^2\rho_c^2 \ll 1$ (where ρ_c is the gyroradius of the cool species), i.e., long wavelength fluctuations in comparison to ρ_c. Since in general for $|x| \ll 1$ we can write $\Gamma_p(x) = e^{-x}I_p(x) \approx (x/2)^p(1/p!)$ $(1-x)$, hence we have $\Gamma_{oc} \approx 1$.

Second and higher order terms in z_{oh} are also neglected since we have assumed $|z_{oh}| \ll 1$. Setting $\omega = \omega_r + i\gamma$ and assuming $\gamma/\omega_r \ll 1$ one may write

$$\frac{1}{\omega^2} \approx \frac{1}{\omega_r^2}\left(1 - \frac{2i\gamma}{\omega_r}\right). \tag{47}$$

Using the above manipulation the dispersion relation equation (46) becomes

$$\begin{aligned} k^2\lambda_D^2 &+ 2\frac{\frac{n_{0c}}{n_0}}{\frac{T_c}{T_h}}\left[i\sqrt{\pi}\left(\frac{\omega_r + i\gamma}{\sqrt{2}k_\parallel v_{tc}}\right)e^{-z_{oc}^2} - \frac{k_\parallel^2 v_{tc}^2}{\omega_r^2}\left(1 - \frac{2i\gamma}{\omega_r}\right) - \frac{3k_\parallel^4 v_{tc}^4}{\omega_r^4}\left(1 - \frac{2i\gamma}{\omega_r}\right)^2\right] \\ &+ 2\frac{n_{0h}}{n_0}\left[1 + i\sqrt{\pi}\left(\frac{\omega_r + i\gamma - \mathbf{k}.\mathbf{V_{oh}}}{\sqrt{2}k_\parallel v_{th}}\right)\Gamma_{oh}\right] = 0. \end{aligned} \tag{48}$$

Taking the real part of equation (48) with the charge neutrality condition $n_{oc} + n_{oh} = 1$, gives

$$\omega_r^2 = \frac{k^2 v_{ea}^2 \cos^2\theta}{1 + \frac{1}{2}k^2\lambda_{Dh}^2} + 3k^2 v_{tc}^2 \cos^2\theta, \tag{49}$$

where $\cos\theta = k_{\|}/k$ and $v_{ea} = (n_{0c}/n_{0h})^{1/2} v_{th}$ is the acoustic speed of the electron-positron plasma. It is noted that equation (49) is consistent with the expression (12) obtained from fluid theory.

The approximate solution of the growth rate is determined by taking the imaginary part of equation (48), and hence solving for γ, one finds

$$\gamma = \frac{\frac{\omega_r^4}{k_{\|}^3}\left(\frac{\pi}{8}\right)^{1/2}\left(\frac{m}{T_h}\right)^{3/2}\left[-\left(\frac{T_h}{T_c}\right)^{3/2} e^{-z_{oc}^2} + \left(\frac{n_{0h}}{n_{0c}}\right)\left(\frac{\mathbf{k}.\mathbf{V_{oh}}}{\omega_r} - 1\right)\Gamma_{oh}\right]}{\left[1 + \frac{6k_{\|}^2\frac{T_c}{m}}{\omega_r^2}\right]}. \tag{50}$$

We note that in equation (50), it is the cooler species that provides the Landau damping, i.e., the velocity distribution function sees a negative slope $\left(\partial f_0/\partial V_{\|} < 0\right)$. It is also seen from equation (50) that for an unstable mode (γ>0), it is necessary that $V_{0h} > \omega_r/k_{\|}$, i.e., the drift velocity parallel to $\mathbf{B_0}$ of the hot species has to be larger than the phase velocity to overcome the damping terms.

Normalizing the fluid speeds by the thermal velocity $v_{th} = (T_h/m)^{1/2}$, the particle density by the total equilibrium plasma density $n_0 = n_{0c} + n_{0h}$, the temperatures by T_h, the spatial length by $\lambda_{dj} = \left(\frac{\varepsilon_0 T_j}{n_{0j}e^2}\right)^{1/2}$, and the time by $\omega_p^{-1} = \left(\frac{n_0 e^2}{\varepsilon_0 m}\right)^{-1/2}$, one may write the normalized real frequency as,

$$\omega_r^2 = \frac{2n_{0c}k_{\|}^2\lambda_d^2}{2(1-n_{0c}) + k^2\lambda_d^2} + 3k_{\|}^2\lambda_d^2\frac{T_c}{T_h}, \tag{51}$$

and the approximate normalized growth rate as,

$$\gamma_r = \frac{\frac{\omega_r^4}{k_{\|}^3\lambda_d^3}\left(\frac{\pi}{8}\right)^{1/2}\left[\left(\frac{1-n_{0c}}{n_{0c}}\right)\left(\vec{k}\,.\frac{\vec{V}_{oh}}{\omega_r} - 1\right)\Gamma_{oh}\right]}{\left[1 + \frac{6k_{\|}^2\frac{T_c}{T_h}}{\omega_r^2}\right]}, \tag{52}$$

For a fixed value of $k\lambda_d$, the real frequency increases with an increase in the cool to hot temperature ratio. This can be seen from the approximate analytical expression (51). **Figure 2** displays the normalized growth rate as a function of the normalized wavenumber for varying cool to hot species temperature ratios T_c/T_h. It is noted that as the T_c/T_h decreases, the growth rate increases, implying that the

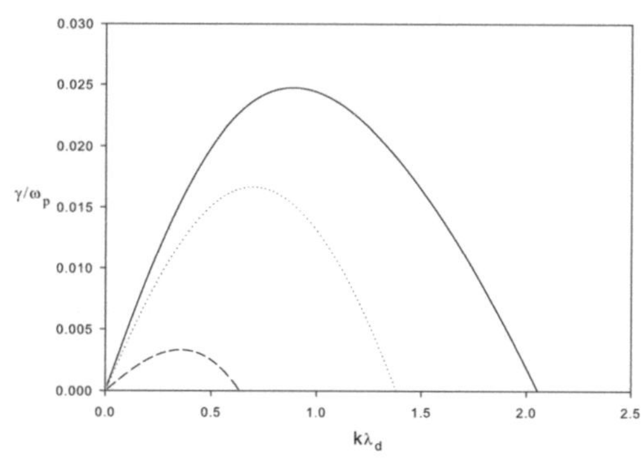

Figure 2.
Normalized growth rate as a function of the normalized wavenumber. The fixed parameters are $R = 0.333$, $V_{oh} = 0.5$, $n_{0c} = 0.1$ and $\theta = 45°$. The parameter labeling the curve is the cool to hot temperature ratio $T_c/T_h = 0.005$ (solid), 0.01 (dotted), and 0.02 (broken).

instability is more easily excited with lower temperature ratios. This may be explained as follows. As the temperature of the cooler species is increased, the associated Landau damping increases, resulting in a reduction of the overall growth rate. It is noted that a cutoff $k\lambda_d$ value is reached beyond which the mode is damped.

4. Nonlinear electrostatic solitary waves in electron-positron plasmas

The study of nonlinear effects in electron-positron plasmas is important since these plasmas exhibit different wave phenomena as compared to electron-ion plasmas. It is therefore important to understand the nonlinear structures, especially the solitary waves that exist in electron-positron plasmas. Satellite observations in the Earth's magnetosphere have shown the existence of electrostatic solitary waves which forms part of broadband electrostatic noise (BEN) and electrostatic solitary waves (ESWs) in various regions of the Earth's magnetosphere. The characteristic features of these ESWs are solitary bipolar pulses and consist of small scale, large amplitude parallel electric fields. These large amplitude spiky structures have been interpreted in terms of either solitons [32] or isolated electron holes in the phase space corresponding to positive electrostatic potential [33]. Given that electron-positron plasmas are increasingly observed in astrophysical environments, as well as in laboratory experiments [34], the above mentioned satellite observations also lead one to explore if such nonlinear structures are also possible in electron-positron plasmas. There is a distinct possibility that a pulsar magnetosphere can support coexistence of two types of cold and hot electron-positron populations [10, 35, 28]. In this section we investigate nonlinear electrostatic spiky structures in a magnetized four component two-temperature electron-positron plasma.

4.1 Basic equations

The model considered, as in the previous section is a homogeneous magnetized, four component, collisionless, electron-positron plasma, consisting of cool electrons (ec) and cool positrons (pc) with equal temperatures T_c and initial densities ($n_{ec0} = n_{pc0}$), and hot electrons (eh) and hot positrons (ph) with equal temperatures T_h and densities ($n_{eh0} = n_{ph0}$). Wave propagation is taken in the x-direction at an angle θ to the magnetic field $\mathbf{B_0}$, which is assumed to be in the x-z plane.

The continuity and momentum equations for the four species are given by

$$\frac{\partial n_j}{\partial t} + \frac{\partial (n_j v_{jx})}{\partial x} = 0 \tag{53}$$

$$\frac{\partial v_{jx}}{\partial t} + v_{jx}\frac{\partial v_{jx}}{\partial x} + \frac{1}{n_j m}\frac{\partial p_j}{\partial x} = -\frac{\varepsilon_j e}{m}\frac{\partial \phi}{\partial x} + \varepsilon_j \Omega v_{jy} \sin\theta \tag{54}$$

$$\frac{\partial v_{jy}}{\partial t} + v_{jx}\frac{\partial v_{jy}}{\partial x} = \varepsilon_j \Omega v_{jz} \cos\theta - \varepsilon_j \Omega v_{jx} \sin\theta \tag{55}$$

$$\frac{\partial v_{jz}}{\partial t} + v_{jx}\frac{\partial v_{jz}}{\partial x} = -\varepsilon_j \Omega v_{jy} \cos\theta, \tag{56}$$

where $\varepsilon_j = +1(-1)$ for positrons (electrons) and $j = ec, pc, eh, ph$ for the cool electrons, cool positrons, hot electrons, and the hot positrons, respectively.

The density of the cool electrons (positrons) is n_{ec} (n_{pc}), and that of the hot electrons (positrons) is n_{eh} (n_{ph}).

The general equation of state for the four species is given by

$$\frac{\partial p_j}{\partial t} + v_{jx}\frac{\partial p_j}{\partial x} + 3p_j\frac{\partial v_{jx}}{\partial x} = 0, \tag{57}$$

The system is closed by the Poisson equation

$$\varepsilon_0\frac{\partial^2\phi}{\partial x^2} = -e\left(n_{pc} - n_{ec} + n_{ph} - n_{eh}\right). \tag{58}$$

In the above equations, n_j, $\mathbf{v}_j$ and p_j are the densities, fluid velocities and pressures, respectively, of the jth species. $\Omega = \Omega_e = \Omega_p = eB_0/m$ is the cyclotron frequency. Here $m = m_e = m_p$ is the common mass of the electrons and the positrons. Adiabatic compression, $\gamma = (2+N)/N = 3$, is assumed, where $N = 1$ implies one degree of freedom.

Upon linearizing and combining equations (53)–(58) and taking the limit $v_{tc} \ll \omega/k \ll v_{th}$, where $v_{th} = (T_h/m)^{1/2}$ and $v_{tc} = (T_c/m)^{1/2}$ are the thermal velocities of the hot (cool) species, the dispersion relation equation for a magnetized two-temperature four component electron-positron plasma, where all species are governed by the fluid equations is,

$$\omega^4 - \omega^2\left(\Omega^2 + 2\omega_s^2 + 3k^2v_{tc}^2\right) + 2\omega_s^2\Omega^2\cos^2\theta = 0. \tag{59}$$

where $\omega_{pc,ph} = \left(n_{0c,h}e^2/\varepsilon_0 m\right)^{1/2}$ are the plasma frequencies of the cool and hot species respectively and $\omega_s = \omega_{pc}/\left(1 + 2/3k^2\lambda_{Dh}^2\right)^{1/2}$ and $\lambda_{Dh} = \left(\varepsilon_0 T_h/n_{oh}e^2\right)^{1/2}$.

Solving the above dispersion relation gives the cyclotron mode,

$$\omega_+^2 = \left(\Omega^2 + 2\omega_s^2 + 3k^2v_{tc}^2\right) - \frac{2\omega_s^2\Omega^2\cos^2\theta}{\Omega^2 + 2\omega_s^2 + 3k^2v_{tc}^2} \tag{60}$$

and the acoustic mode,

$$\omega_-^2 = \frac{2\omega_s^2\Omega^2\cos^2\theta}{\Omega^2 + 2\omega_s^2 + 3k^2v_{tc}^2}, \tag{61}$$

4.2 Nonlinear analysis

In the nonlinear regime, a transformation to a stationary frame $s = (x - Vt)(\Omega/V)$ is performed, and v, t, x and ϕ are normalized with respect to v_{th}, Ω^{-1}, $\rho = v_{th}/\Omega$, and T_h/e, respectively. V is the phase velocity of the wave. In equations (53)–(57), $\partial/\partial t$ is replaced by $-\Omega(\partial/\partial s)$ and $\partial/\partial x$ by $(\Omega/V)(\partial/\partial s)$, and the diving electric field amplitude is defined as $E = -(\partial\psi/\partial s)$, where $\psi = e\phi/T_h$.

Integrating equation (53) and using the initial conditions $n_{ec0} = n_0$ and $v_{ecx} = v_0$ at $s = 0$, yields the normalized velocity for the cool electrons in the x-direction.

$$v_{ecx} = -\left(\frac{n_{eco}}{n_{ec}}\right)(V - v_0) + V \tag{62}$$

Similarly the cool positrons, hot electrons and hot positrons velocities are determined. Substituting these into the normalized form of equations (53)–(57), gives the following set of nonlinear first-order differential equations for the cool electron species in the stationary frame.

$$\frac{\partial\psi}{\partial s} = -E \tag{63}$$

$$\frac{\partial E}{\partial s} = R^2M^2(n_{pcn} - n_{ecn} + n_{phn} - n_{ehn}) \tag{64}$$

$$\frac{\partial n_{ecn}}{\partial s} = \frac{n_{ecn}^3[E + M\sin\theta v_{ecyn}]}{\left(\frac{n_{ec0}}{n_0}\right)^2(M-\delta_c)^2 - 3\frac{T_c}{T_h}p_{ecn}n_{ecn}} \tag{65}$$

$$\frac{\partial v_{ecyn}}{\partial s} = \frac{Mn_{ecn}}{(M-\delta_c)}\left(\frac{n_0}{n_{ec0}}\right)\left[-\left(M - \frac{(M-\delta_c)}{n_{ecn}}\left(\frac{n_{ec0}}{n_0}\right)\right)\sin\theta + v_{eczn}\cos\theta\right] \tag{66}$$

$$\frac{\partial v_{eczn}}{\partial s} = -\left(\frac{n_0}{n_{ec0}}\right)\frac{n_{ecn}v_{ecyn}M\cos\theta}{(M-\delta_c)} \tag{67}$$

$$\frac{\partial p_{ecn}}{\partial s} = \frac{3p_{ecn}n_{ecn}^2[E + M\sin\theta v_{ecyn}]}{\left(\frac{n_{ec0}}{n_0}\right)^2(M-\delta_c)^2 - 3\frac{T_c}{T_h}p_{ecn}n_{ecn}} \tag{68}$$

The set of differential equations for the cool positrons are given by,

$$\frac{\partial n_{pcn}}{\partial s} = \frac{n_{pcn}^3}{(M-\delta_c)^2}\left(\frac{n_0}{n_{pc0}}\right)^2[-E - M\sin\theta v_{pcyn}] \tag{69}$$

$$\frac{\partial v_{pcyn}}{\partial s} = \frac{Mn_{pcn}}{(M-\delta_c)}\left(\frac{n_0}{n_{pc0}}\right)\left[\left(M - \frac{(M-\delta_c)}{n_{pcn}}\left(\frac{n_{pc0}}{n_0}\right)\right)\sin\theta - v_{pczn}\cos\theta\right] \tag{70}$$

$$\frac{\partial v_{pczn}}{\partial s} = \left(\frac{n_0}{n_{pc0}}\right)\frac{n_{pcn}v_{pcyn}M\cos\theta}{(M-\delta_c)} \tag{71}$$

$$\frac{\partial p_{pcn}}{\partial s} = \frac{3p_{pcn}n_{pcn}^2[-E - M\sin\theta v_{pcyn}]}{\left(\frac{n_{pc0}}{n_0}\right)^2(M-\delta_c)^2 - 3\frac{T_c}{T_h}p_{pcn}n_{pcn}} \tag{72}$$

Similar sets of differential equations can be derived for the hot electrons and hot positron species. The velocities are normalized with respect to the thermal velocity of the hot species $v_{th} = (T_h/m)^{1/2}$ and the densities with respect to the total density n_0. The equilibrium density of the cool (hot) electrons is n_{ec0} (n_{eh0}), and that of the cool (hot) positrons n_{pc0} (n_{ph0}), with $n_{ec0} + n_{eh0} = n_{pc0} + n_{ph0} = n_0$. $R = \omega_p/\Omega$, where $\omega_p = (n_0e^2/\varepsilon_0 m)^{1/2}$ is the total plasma frequency, $M = V/v_{th}$ is the Mach number and $\delta_{c,h} = v_{0c,0h}/v_{th}$ is the normalized drift velocity of cool (hot) species at s =0. The system of nonlinear first-order differential equations can now be solved numerically using the Runge-Kutta (RK4) technique [36]. The initial values can be determined self consistently where the actual normalized electric fields are given by $E_{norm} = -(1/M)(\partial\psi/\partial s)$ and wave propagation is taken almost parallel to the ambient magnetic field $\mathbf{B_0}$.

Numerical results to investigate the effect of parameters such as the electric driving force E_0, densities n_{ec0} and n_{ph0}, temperature ratio T_c/T_h, Mach number M, drift velocities $\delta_{c,h}$ and propagation angle θ on the wave can be explored. A typical numerical result is seen in **Figure 3a–d** [37] showing the evolution of the system for various driving electric field amplitudes E_0. It is seen that as E_0 increases, the electric field structure evolves from a sinusoidal wave to a sawtooth structure. For a

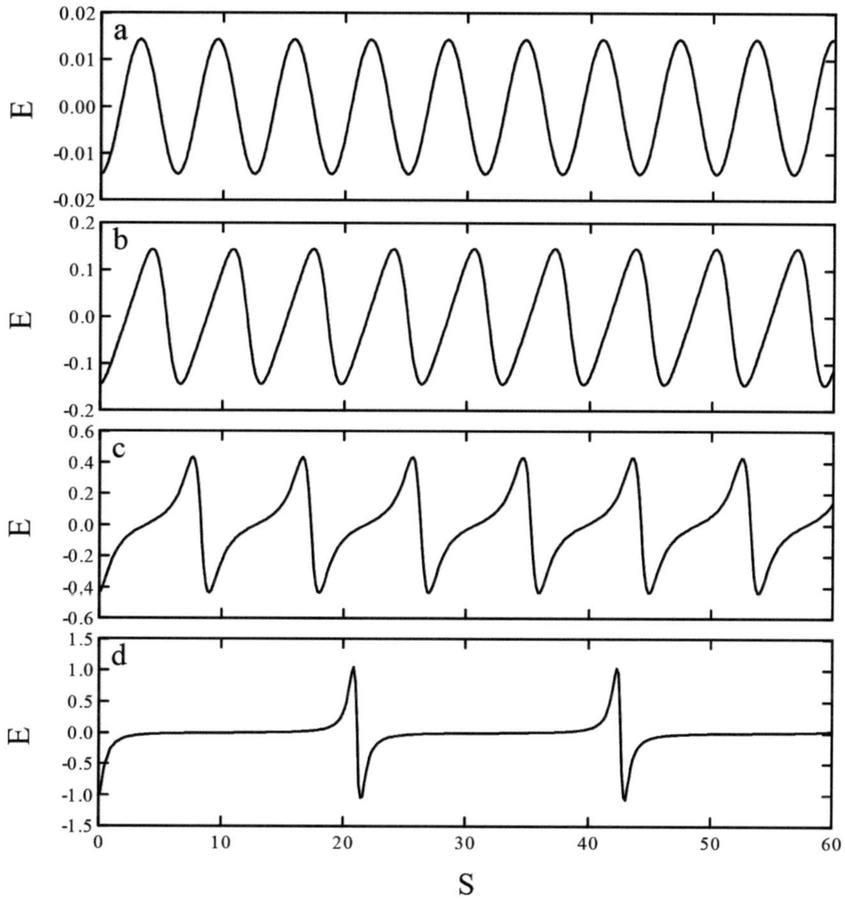

Figure 3.
Numerical solution of the normalized electric field for the parameters $M = 3.5$, $\theta = 2^o$, $R = 10.0$, $\delta_c = \delta_h = 0.0$, $n_{ec0}/n_0 = n_{pc0}/n_0 = 0.5$, $T_c/T_h = 0.0$, *and* E_0 = *(a) 0.05 [linear waveform], (b) 0.5 [sinusoidal waveform], (c) 1.5 [sawtooth waveform] and (d) 3.5 [bipolar waveform].*

higher E_0 value of 3.5, the potential structure has a spiky bipolar form showing that as the period of the wave increases and the frequency of the wave decreases.

5. Conclusion

Linear and nonlinear electrostatic waves in a magnetized four component two-temperature electron-positron plasma have been investigated. In the linear analysis fluid and kinetic theory approaches are employed to describe the wave motion. The fluid theory approach focused on the wave dynamics of both the acoustic and cyclotron branches. Solutions of the dispersion relation from fluid theory yielded electron-acoustic, upper hybrid, electron plasma and electron cyclotron branches. Perpendicular and parallel wave propagation was examined showing its influence on the dispersive properties of the wave. The kinetic theory approach further examined Landau damping effects on the acoustic mode, analyzing the frequency and growth rate of the wave. The analysis shows that a large enough drift velocity (V_{oh}) is required to produce wave growth. Both fluid and kinetic theory show excellent agreement for the real frequencies of the acoustic mode and solutions of the corresponding dispersion relation can be explored as a function of several plasma parameters. In the nonlinear analysis, the two-fluid model is used to derive a set of differential equations for the electrostatic solitary waves in a magnetized two-temperature electron-positron plasma. In particular, electrostatic solitary waves and their electric fields, similar to those found in the Broadband Electrostatic Noise are explored. For the onset of spiky ESWs, it is noted that as the wave speed increases, a larger driving electric field is required.

Acknowledgements

Thank you to Prof. Ramesh Bharuthram (University of the Western Cape, South Africa), Prof. Gurbax Lakhina and Prof. Satyavir Singh (Indian Institute of Geomagnetism, Navi Mumbai, India and Dr. Suleman Moolla (University of KwaZulu-Natal, Durban, South Africa) for your valuable contributions.

Author details

Ian Joseph Lazarus
Department of Physics, Durban University of Technology, Durban, South Africa

*Address all correspondence to: lazarusi@dut.ac.za

References

[1] Weinberg S. Gravitation and Cosmology. New York: Wiley; 1972

[2] Rees MJ. In: Gibbons GW, Hawking SW, Siklas S, editors. The Very Early Universe. Cambridge: Cambridge University Press; 1983

[3] Miller HR, Witta PJ. Active Galactic Nuclei. Berlin: Springer-Verlag; 1987. p. 202

[4] Piran T. The physics of gamma-ray bursts. Reviews of Modern Physics. 2005;**76**:1143

[5] Goldreich P, Julian WH. Pulsar Electrodynamics. The Astrophysical Journal. 1969;**157**:869

[6] Michel FC. Theory of pulsar magnetospheres. Reviews of Modern Physics. 1982;**54**:1

[7] Tandberg EH, Emslie AG. The Physics of Solar Flares. Cambridge: Cambridge University Press; 1988. p. 124

[8] Beskin VS, Gurevich AV, Istomin YN. Electrodynamics of pulsar magnetospheres. Soviet Physics—JETP. 1983;**58**:235

[9] Bharuthram R. Arbitrary Amplitude Double Layers in a Multi-Species Electron-Positron Plasma. Astrophysics and Space Science. 1992;**189**:213

[10] Sturrock PA. A Model of Pulsars. The Astrophysical Journal. 1971;**164**:529

[11] Berezhiani VI, Mahajan SM. Large relativistic density pulses in electron-positron-ion plasmas. Physical Review E. 1995;**bf52**:1968

[12] Shatashvili NL, Javakhishvili JI, Kaya H. Nonlinear Wave Dynamics in Two Temperature Electron-Positron-Ion Plasma. Astrophysics and Space Science. 1997;**250**:109

[13] Bhattacharyya R, Janaki MS, Dasgupta B. Relaxation in electron-positron plasma: a possibility. Physics Letters A. 2003;**315**:120

[14] Stenflo L, Shukla PK, Yu MY. Nonlinear Propagation of Electromagnetic Waves in Magnetized Electron-Positron Plasmas. Astrophysics and Space Science. 1985;**117**:303

[15] Mofiz UA, de Angelis U, Forlani A. Solitons in weakly nonlinear electron-positron plasmas and pulsar microstructure. Physical Review A. 1985;**31**:951

[16] Yu MY, Shukla PK, Stenflo L. Alfven Vortices in a Strongly Magnetized Electron-Positron Plasma. The Astrophysical Journal. 1986;**309**: L63

[17] Pillay R, Bharuthram R. Large Amplitude Solitons in a Multi-Species Electron-Positron Plasma. Astrophysics and Space Science. 1992;**198**:85-93

[18] Verheest F, Lakhina GS. Oblique Solitary Alfv *e*′ n Modes in Relativistic Electron-Positron Plasmas. Astrophysics and Space Science. 1996;**240**:215

[19] Lazarus IJ, Bharuthram R, Hellberg MA. Modified Korteweg-de Vries-Zakharov-Kuznetsov solitons in symmetric two-temperature electron-positron plasmas. Journal of Plasma Physics. 2008;**74**:519

[20] Zank GP, Greaves RG. Linear and nonlinear modes in nonrelativistic electron-positron plasmas. Physics Review. 1995;**51**:6079

[21] Iwamoto N. Collective modes in nonrelativistic electron-positron

plasmas. Physical Review E. 1993;**47**: 604

[22] Mozer FS, Ergun R, Temerin M, Cattell CA, Dombeck J, Wygant JR. New Features of Time Domain Electric-Field Structures in the Auroral Acceleration Region. Physical Review Letters. 1997; **79**:1281

[23] Andre M, Koskinen H, Gustafsson G, Lundin R. Ion waves and upgoing ion beams observed by the Viking satellite. Geophysical Research Letters. 1987;**14**: 463

[24] Matsumoto H, Kojima H, Miyatake T, Omura Y, Okada M, Nagano I, et al. Electrostatic Solitary Waves (ESW) in the magnetotail: BEN wave forms observed by GEOTAIL. Geophysical Research Letters. 1994;**21**:2915

[25] Franz JR, Kintner PM, Pichett JS. POLAR Observations of Coherent Electric Field Structures. Geophysical Research Letters. 1998;**25**:1277

[26] Ergun RE, Carlson CW, McFadden JP, Mozer FS, Delroy GT, Peria W, et al. FAST satellite observations of large-amplitude solitary wave structures. Geophysical Research Letters. 1998;**25**: 2041

[27] Gary SP, Tokar RL. The electron acoustic Mode. Physics of Fluids. 1985; **28**:2439

[28] Lazarus IJ, Bharuthram R, Singh SV, Pillay SR, Lakhina GS. Linear electrostatic waves in two-temperature electron-positron plasmas. Journal of Plasma Physics. 2012;**78**:621

[29] Bharuthram R, Pather T. The kinetic dust-acoustic instability in a magnetized dusty plasma. Planetary and Space Science. 1996;**44**:137

[30] Fried BD, Conte SD. The Plasma Dispersion Function. New York: Academic Press; 1961

[31] Rosenberg M. Ion- and dust-acoustic instabilities in dusty plasmas. Planetary and Space Science. 1993;**41**: 229

[32] Temerin MA, Cerny K, Lotko W, Mozer FS. Observations of double layers and solitary waves in the auroral plasma. Physical Review Letters. 1982;**48**:1175

[33] Omura Y, Kojima H, Matsumoto H. Computer simulation of Electrostatic Solitary Waves: A nonlinear model of broadband electrostatic noise. Geophysical Research Letters. 1994;**21**: 2923

[34] Greaves RG, Surko CM. An Electron-Positron Beam-Plasma Experiment. Physical Review Letters. 1995;**75**:3846

[35] Bharuthram R. Arbitrary Amplitude Double Layers in a Multi-Species Electron-Positron Plasma. Astrophysics and Space Science. 1992;**189**:213

[36] Press WH, Teukolsky SA, Vetterling WT, Flannery BP. Numerical Receipes in Fortran 90-The Art of Parallel Scientific Computing. 1996;**2**:702, 731, 1297-704, 740, 1308

[37] Lazarus IJ, Bharuthram R, Moolla S, Singh SV, Lakhina GS. Nonlinear electrostatic solitary waves in electron-positron plasmas. Journal of Plasma Physics. 2016;**82**:1

Section 4

Application of Charge Particles

Chapter 6

Biological Effects of Negatively Charged Particle-Dominant Indoor Air Conditions

Suni Lee, Yasumitsu Nishimura, Naoko Kumagai-Takei, Hidenori Matsuzaki, Megumi Maeda, Nagisa Sada, Kei Yoshitome and Takemi Otsuki

Abstract

To identify health-promoting indoor air conditions, we developed negatively charged particle-dominant indoor air conditions (NCPDIAC). Experiments assessing the biological effects of NCPDIAC comprised (1) 2.5-h stays in NCPDIAC or control rooms, (2) 2-week nightly stays in control followed by NCPDIAC rooms, (3) 3-month OFF to ON and ON to OFF trials in individual living homes equipped with NPCDIAC in their sleeping or living rooms, and (4) in vitro assays comparing the immune effects between negatively charged particle-dominant and control cell culture incubators. The most significant difference examined between NCPDIAC and control rooms in the 2.5-h stays was an increase in interleukin (IL)-2 with occupancy of the NCPDIAC room. For the 2-week nightly stay experiments, natural killer (NK) cell activity increased with occupancy of the NCPDIAC room. The 3-month OFF to ON trial showed an increase in NK cell activity, while the ON to OFF trial yielded a decrease in NK cell activity. Additionally, the in vitro assays also showed an increase in NK cell activity. The use of NCPDIAC resulted in increased NK cell activity, which has the effect of enhancing immune surveillance for the occurrence of cancer and improving symptoms associated with viral infections.

Keywords: indoor air, negatively charged particle, natural killer cell activity

1. Introduction

Indoor air conditions can sometimes affect human health. For example, sick building syndrome (SBS) is one of the most well-known health impairments caused by indoor air conditions [1–3]. Volatile organic compounds (VOCs) are considered to be the cause of SBS [1–3]. SBS can induce a variety of signs or symptoms such as headache; eye, nose, and throat irritation; fatigue; dizziness; and nausea. The condition of patients with SBS may worsen following exposure to certain VOCs, with individual patients revealing specific hypersensitivity to particular chemicals. It is considered that some pathophysiological alterations in the psycho-neuro-immune-endocrine network at the level of genes, molecules, proteins, cells, and organs may occur in SBS patients, which then defines or determines their sensitivity to such low concentrations of VOCs [1–3]. However, the precise nature of these alterations is yet to be delineated. Consequently, the only advice presently available to SBS patients is to avoid exposure to VOCs for which the patient shows particular sensitivity [1–3].

Furthermore, most homes in Japan possess air-conditioning units in each room. During the winter season, the room adjacent to the bathroom which is utilized for changing clothes is narrow and particularly cold. Moreover, the lavatory is also cold. Consequently, there is a risk of changes in blood pressure and the onset of cardiovascular events caused by drastic changes in room temperature in these areas [4, 5].

Thus, decreasing the amount and use of chemicals and maintaining appropriate room temperatures at home are things that can be considered with respect to the task of establishing health-promoting indoor conditions. Additionally, although there are few reports detailing the use of indoor air under negatively charged conditions, a consideration of air electrical charges may assist with this task [6–9].

2. Development of NCPDIAC

The development of NCPDIAC has previously been reported [10].

As shown in **Figure 1**, NCPDIAC was established using extraporous charcoal paint and loading an electric voltage (approximately 72–100 V) behind the room walls [10]. The charcoal paints were mainly used for deodorization and dehumidification. As a result, the surface of the walls acquired a slightly negative charge, and small positively charged particles 20–30 nm in diameter collected on the surface of walls [10]. Thus, although negatively charged particles were not introduced into the

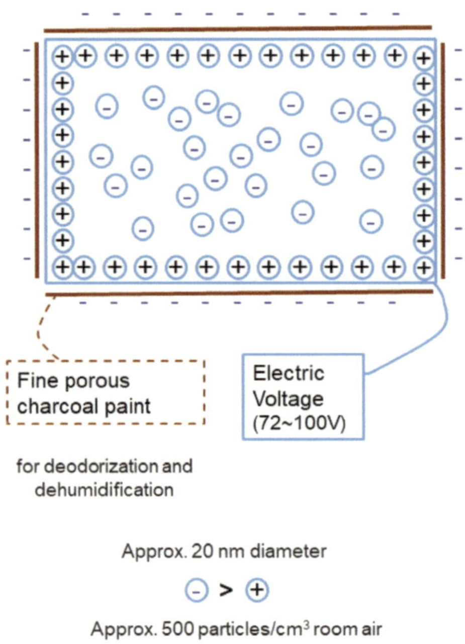

Figure 1.
NCPDIAC was established using extraporous charcoal paint and loading an electric voltage behind the walls.

indoor conditions, the balance between positively and negatively charged particles was such that negatively charged particles were predominant [10].

3. 2.5H stay experiments

Results of the 2.5H (2.5-h) stay experiments have previously been reported [10].

Three control rooms and three NCPDIAC rooms were built in the wide sub-underground laboratory. Both types of rooms were built in a large subunderground laboratory in the Comprehensive Housing R&D Institute, SEKISUI HOUSE, Ltd., at Kizu-town, Kyoto prefecture, Japan. The area and volume of the laboratory were approximately 539 m^2 and 1564 m^3, respectively, and those of the experimental rooms were 9.1 m^2 and 22.8 m^3, respectively. The appearance of the control and NCPDIAC rooms is shown in **Figure 2A**. All of the healthy volunteers (HV) referred to in **Figure 2B** were unaware of the room type (control or NCPDIAC) they were to occupy during the experimental period. The following items were measured immediately prior to (prestay) and following (poststay) entry into the rooms as previously reported [10].

1. General conditions: blood chemistry including liver [alanine aminotransferase (ALT), aspartate aminotransferase (AST), and gamma-glutamyl transferase (γGT)] and kidney functions [creatinine, blood urea nitrogen (BUN) and uric acid], blood sugar and lactic acid levels, and peripheral blood counts (white blood cell, red blood cell, hemoglobin, hematocrit and platelet) were measured using peripheral venous blood. Blood pressure and pulse rate were also measured.

2. Stress markers: Levels of blood cortisol and salivary cortisol, chromogranin A, amylase, and secretory immunoglobulin A were measured as stress markers.

3. Parameters related to the autonomic nervous system: The autonomic nervous system was examined using the Flicker test, a stabilometer, and heart rate monitor for 3 min. The Flicker test and flicking frequencies of red, green, and yellow colors were monitored. A Gravicoder GS-7TM instrument (Anima Inc., Tokyo, Japan) was used as a stabilometer and the Romberg ratio was used as the parameter for body sway. The ratio was calculated from the whole trajectory of the body sway during a 30-s standing period with eyes closed divided by that with eyes open. The heart rate was monitored using a Heart Rate Monitor S810iTM instrument (Polar Electro, Kempele, Finland) for 3 min. During monitoring, HV sat on chairs and were kept at rest. The R wave intervals in the electrocardiogram were estimated and the standard deviation (SD) or R wave interval was considered as an index of heart rate fluctuation.

4. Immunological parameters: Serum levels of immunoglobulin (Ig) E and Ig A, and cytokines related to the Th1/Th2 balance [Interferon (IFN)-γ, tumor necrosis factor (TNF)-α, Interleukin (IL)-2, IL-4, IL-6 and IL-10] were evaluated. Individual samples for cytokine measurement were applied to the Cytometric Bead Array of Human Th1/Th2 cytokine kit II (CBA, BD Bioscience, San Jose, CA, USA) and measurements were made using FACSCalibur flow cytometry (BD Bioscience) according to the manufacturer's instructions. For samples that revealed less than the lower limit of the values from analytical methods for cytokines and immunoglobulins, the 1/10 value of the minimum values among the entire measurable samples was substituted instead of 0 or left as "unmeasurable," as previously reported [10].

5. Blood viscosity: Blood viscosity was measured using a Micro-Channel Flow Analyzer MC-FAN (MC Laboratory Inc., Tokyo, Japan) according to the manufacturer's instructions. Briefly, peripheral heparinized blood sample (100 μl) was placed into the instrument and allowed to flow through the microchannel chips, which are a model for capillary vessels, and the flowing time was recorded. The flowing blood sample was visualized using a CCD camera equipped to the microscope.

These experiments were approved by the institutional ethical committee (#114). Samples were only taken from HV who provided written informed consent.

The electrical charge in these rooms was measured using an Ion Counter EB-1000TM instrument made by Eco Holistic Inc., Suita, Japan.

Differences in the positively and negatively charged particles in control and NCPDIAC rooms are shown in **Figure 2C**. The number of positively charged particles in the rooms with NCPDIAC was reduced. However, the number of negatively charged particles in control and NCPDIAC rooms did not differ. Negatively charged air conditions were therefore formed by reducing the number of positively charged particles in NCPDIAC rooms.

Differences between control and NCPDIAC room values for all items measured were determined by calculating [poststay]-[prestay]. As shown in **Figure 2D**, differences were found in IL-2 [10]. The increase in IL-2 levels (by approximately 1 pg/ml)

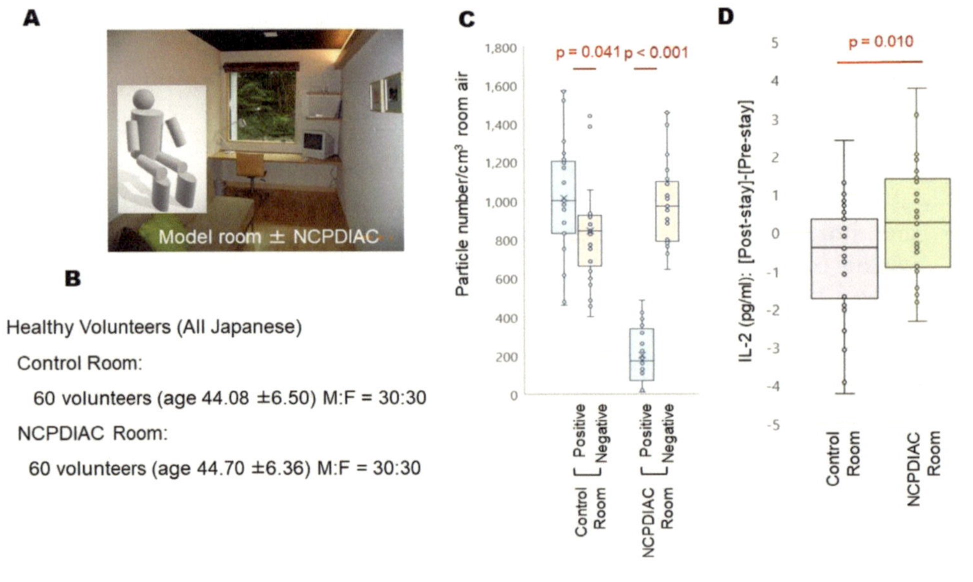

Figure 2.
(A) Control and NCPDIAC rooms were constructed as shown. Healthy volunteers occupied rooms for 2.5 h, being unaware of the room type (control or NCPDIAC). Volunteers remained within the rooms in a stable state, without sleeping or excitement. (B) A total of 60 Japanese volunteers participated in experiments for each room type. The gender ratio and average age of the volunteers were almost identical in both groups (occupants of control and NCPDIAC rooms). (C) The box-and-whisker plots show the number of positively and negatively charged particles per 1 cm³ air in control and NCPDIAC rooms. Although there was no difference in the level of positively and negatively charged particles in control rooms, NCPDIAC rooms possessed significantly lower levels of positively charged particles compared with control rooms. Furthermore, there was no difference in the level of negatively charged particles between control and NCPDIAC rooms. Statistical differences were assayed using the student T test. (D) The box-and-whisker plots show a comparison of IL-2 levels ([Post-Stay]-[Pre-Stay], pg/ml) in volunteers who occupied the control and NCPDIAC rooms during the 2.5-h stay experiments. The most significant difference found in the examined values using [poststay]-[prestay] related to IL-2 levels, which increased significantly following NCPDIAC room stays compared with control room stays. This difference was analyzed using the Mann-Whitney U test.

was considered not to be caused by any pathophysiological conditions. Additionally, it was considered that levels could return to base values in individual HV. As a result, it appears that NCPDIAC affected the immune system without any adverse effects with respect to the signs, symptoms, or measured items in the present experiments [10].

4. 2W nightly stay experiments

In the next step, 2W (2-week) nightly stay experiments were performed as previously reported [11].

Approximately 1 year after obtaining the results of the 2.5H experiments and with subsequent discussions, new participants were recruited for our 2W nightly stay experiments. This study was approved by the institutional ethical committee (#176) and samples were only taken from HV who provided written informed consent.

The dormitory belonging to SEKISUI HOUSE, Ltd., Kizugawa-City, Kyoto Prefecture, Japan, and ordinarily used for the training of employees was utilized for our 2W nightly stay experiments. These employees usually receive training for 3 months as shown in **Figure 3A**. Then, in the case of volunteers for our experiments, employees were moved at the second month from their original room into a control

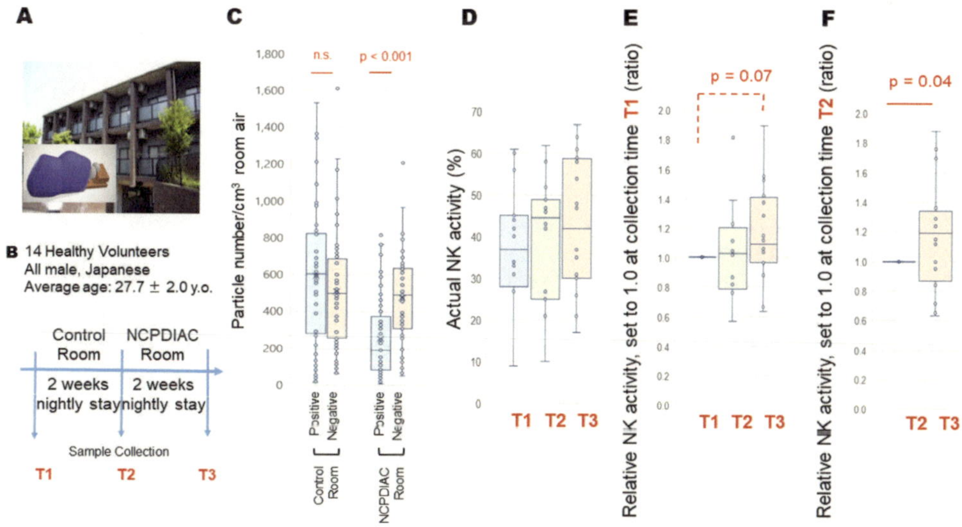

Figure 3.
(A) For the 2W (2-week) nightly stay experiments the dormitory for training employees was used. (B) All volunteers initially occupied the control room every night for 2 weeks. Thereafter, volunteers occupied NCPDIAC rooms without being aware of which room type (control or NCPDIAC) they had initially occupied. Sample collections were performed at T1, T2, and T3 time points. T1 represented the time point prior to volunteers occupying the control room. T2 represented the time point immediately following the 2W nightly stay in control rooms, and just prior to occupancy of the NCPDIAC room, while T3 represented the time point following the 2W nightly stay in the NCPDIAC room. Although HV agreed to be recruited in this study, they were unaware of the room type (control or NCPDIAC) they had initially entered. (C) The box-and-whisker plots show the number of positively and negatively charged particles per 1 cm³ air in control and NCPDIAC rooms. Although there was no difference between the number of positively and negatively charged particles in control rooms, NCPDIAC rooms possessed significantly lower levels of positively charged particles compared with control rooms. Furthermore, there was no difference in the level of negatively charged particles between control and NCPDIAC rooms. Statistical analyses were performed using the student T test. (D) The box-and-whisker plots show the actual NK cell activity (%) at T1, T2, and T3 time points. There were no differences among the three time points. Statistical analyses were performed using the ANOVA test. (E) The box-and-whisker plots show the relative NK activity after setting T1 of individual volunteers to 1.0. Statistical analyses comparing T1 and T2 or T3 were performed using the student T test. (F) Relative NK activity after setting T2 of individual volunteers to 1.0. There was a significant increase in relative NK activity after the 2W nightly stay in the NCPDIAC rooms. Statistical analyses were performed using the student T test.

(both were the same in terms of NCPDIAC). After occupying the control room every night for 2 weeks, they were then moved into NCPDIAC rooms, without being aware of the room type (control or NCPDIAC) they had initially occupied. As shown in **Figure 3B**, sample collections were performed prior to volunteers occupying the control room (T1), after their 2W nightly stay in the control room (T2), and finally after occupying the NCPDIAC room (T3). The measured items remained unchanged during the 2.5H stay experiments. Additionally, several parameters that had not changed during the hour-based time period but were altered during the week-based time period when environmental factors or physiological conditions had changed were included as biological parameters. Those parameters comprised NK cell activity, along with urine 17 hydroxycorticosteroid (OHCS) and 8-oxo-2′-deoxyguanosine (OHdG) levels.

All volunteers initially occupied the control room every night for 2 weeks. Thereafter, volunteers occupied NCPDIAC rooms without being aware of which room type (control or NCPDIAC) they had initially occupied. Sample collection was performed at T1, T2, and T3 time points. T1 represented the time point prior to volunteers occupying the control room. T2 represented the time point immediately following the 2W nightly stay in control rooms, and just prior to occupancy of the NCPDIAC room, while T3 represented the time point following the 2W nightly stay in the NCPDIAC room.

Among all the items measured, significant change was only found in the NK activity. The NK cell activity was determined using a ^{51}Cr-release assay according to a method outlined in previous reports [12, 13]. The effector cell (mononuclear cell)-to-target cell (K562 cell) ratio was 10:1. However, there was a wide range of individual actual NK cell activity in the vicinity of approximately 10% to greater than 50%. In the 2W nightly stay experiments, the NK cell activity among HV ranged from less than 30% to near 60% at T1 (**Figure 3B**). Although actual NK activity measurements did not reveal any statistical significance (**Figure 3D**) due to the wide variation in individual volunteers, when the relative NK activity was set to 1.0 at T1, there was a tendency toward increased NK cell activity at T3 (**Figure 3E**) [11]. Since the room conditions prior to T1 and T2 collections were basically the same, the relative NK cell activity between T2 and T3 was compared, with T2 being set to 1.0 in individual HV [11]. As shown in **Figure 3F**, there was a significant increase in relative NK activity at T3, after the 2W nightly stay with NCPDIAC [11].

5. In vitro experiments

The in vitro experiments were performed to examine cellular alterations under negative particle dominant conditions. Results have previously been reported [14].

Freshly isolated peripheral blood mononuclear cells derived from seven healthy volunteers were cultured in a standard CO_2 incubator at 37°C with 5% CO_2 under humidified conditions (standard cell culture conditions comprised a humidity of 95%). For these experiments, it was impossible to use charcoal paint or to load an electric voltage. Thus, negatively charged particles were forced in and circulated. The difference between positively and negatively charged particles in 1 cm^3 of incubator air was approximately 3000 (**Figure 4A**) [14].

This study was approved by the institutional ethical committee (#883) and samples were only taken from the HV who provided written informed consent. Peripheral blood mononuclear cells were derived from samples obtained from HV.

For the NK cell activity, K562 cells, a human immortalized myelogenous leukemia cell line, were stained with Vybrant™ Dio Cell-Labeling Solution by incubation for 20 min at room temperature [14]. Dio-stained cells were then washed with phosphate buffered slain (PBS), and peripheral blood mononuclear cells (PBMC)

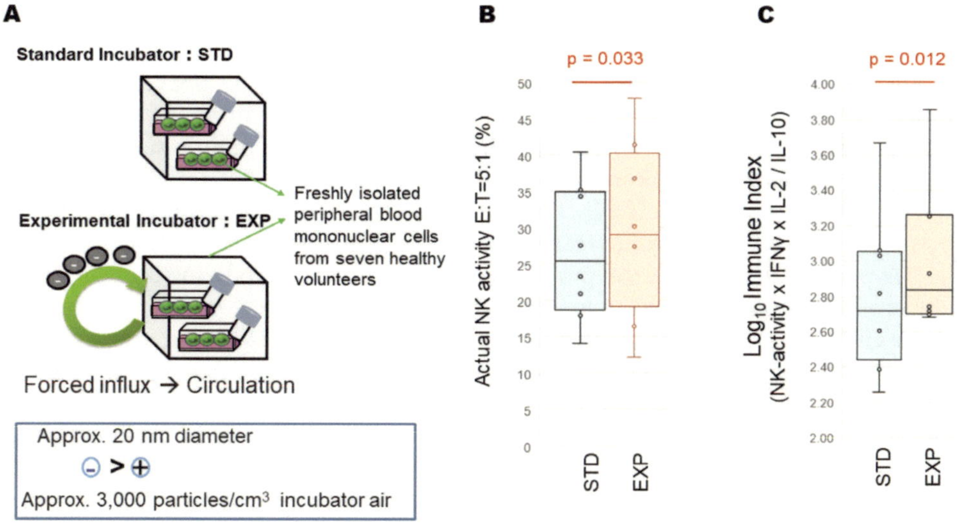

Figure 4.
(A) Peripheral blood mononuclear cells from HV were incubated in standard (STD) or experimental (EXP) incubators. Negatively charged particles (20–30 nm in diameter) were forced into incubators and then circulated. The experimental incubator was set to generate negatively charged particles using a neutralizing instrument that created negatively charged particles (SJ-M200, Keyence Co. Ltd., Osaka, Japan). This instrument yielded negatively charged particles that were set to directly enter the inside of the incubator (by making a hole). Since the interior volume of the incubator was 49 l, the negatively charged particles entered and passed out at a rate of approximately 3000 particles/cm^3. (B) The box-and-whisker plots show the increase in actual NK cell activity, represented by the E/T (effector vs. target cells) ratio, which, being 5:1, showed greater increase with experimental incubators and negatively charged particle-dominant culture conditions compared with standard incubators. (C) The LOG$_{10}$ value of the "immune index" as [NC-activisty X IFN-γ concentration X IL-2 concentration]/[IL-10 concentration] was compared between standard and experimental incubator conditions. Experimental incubator conditions showed a significant increase in this index.

were incubated with 5000 Dio-labeled K562 cells in 96-well round bottom plates at an effector cell-to-target cell (E/T) ratio of 2.5:1, 5:1, or 10:1 for 5 h in experimental or standard incubators. Following incubation, cells were collected and stained with propidium iodide (PI) at 5 μg/ml and the percentage of PI+ Dio-labeled cells among the total Dio-labeled cells, representing the percentage of lysed cells, was examined using FACSCalibur flow cytometry. The percentage of specific lysis induced by effector cells was calculated after analyzing the substrate and spontaneous dead cell numbers expressed in wells without effector [14].

Additionally, peripheral blood mononuclear cells were cultured in RPMI1640 culture medium with antibiotics just as for standard cell cultures for 1 or 2 weeks without any stimulants such as cytokines. The concentrations of IFN-γ, IL-2, and IL-10 (as well as IL-6, TNF-α, and IL-4) in supernatants were measured. Additionally, other items such as surface CD25, CD69, programmed death-1 (PD1), and CD44 expression in CD4+ T helper cells, CD8+ T cells, and NK cells were measured [14].

It is noteworthy that NK cell activity was significantly higher when incubations were performed in the experimental incubator compared with those performed in the standard incubator (**Figure 4B**). Additionally, if we calculate the "immune index" as [NC-activity X IFN-γ concentration X IL-2 concentration]/[IL-10 concentration] and compare this index from standard and experimental incubators using log$_{10}$ titer (**Figure 4C**), there is a significantly greater increase in this "Log$_{10}$ Immune Index" associated with the use of experimental incubators compared with the use of standard incubators. The higher value of this item was assumed to reflect stimulation of immune status [14].

These results indicated that a predominance of negatively charged particles induces immune stimulation at non-pathophysiological levels [14].

6. 3M ON and OFF trials

Finally, long-term (3-month, or 3M) stay experiments were performed and the results have previously been reported [15].

Following the aforementioned experiments (2.5H and 2W nightly stay experiments and evaluations), it seemed that NCPDIAC stimulated NK activity with no adverse effects on HV [15]. The increase in NK activity could be accounted for by the short-term, albeit slight, increase in IL-2, which may activate NK cells during the 2W period. The following experiments were then applied in living homes. The homes of seven volunteers were modified for NCPDIAC, targeting mainly sleeping rooms and living rooms. A switch panel approximately 4 × 22 × 29 cm in size had been fitted. Volunteers would then switch this panel ON and OFF every 3 months. Then, prior to and following every 3-month ON or OFF living period, clinical measurements including NK cell activity and others (as measured in the 2W nightly stay experiments) were performed. A total of 16 OFF to ON (3M ON) and 13 ON to OFF (3M OFF) trials were performed as shown in **Figure 5A** [15].

Blood samples were taken just before switching ON or OFF. Thus, during the OFF to ON (3M ON) period, HV stayed at home with NCPDIAC (sleeping room and living room). During the ON to OFF (3M OFF) period, HV occupied rooms in their homes without NCPDIAC [15].

All seven HV comprised Japanese living in Japan and were asked to join this project by first-class registered architects who are colleagues of the authors. The average age of the volunteers was 54.86 ± 9.15 years and included five males and two females.

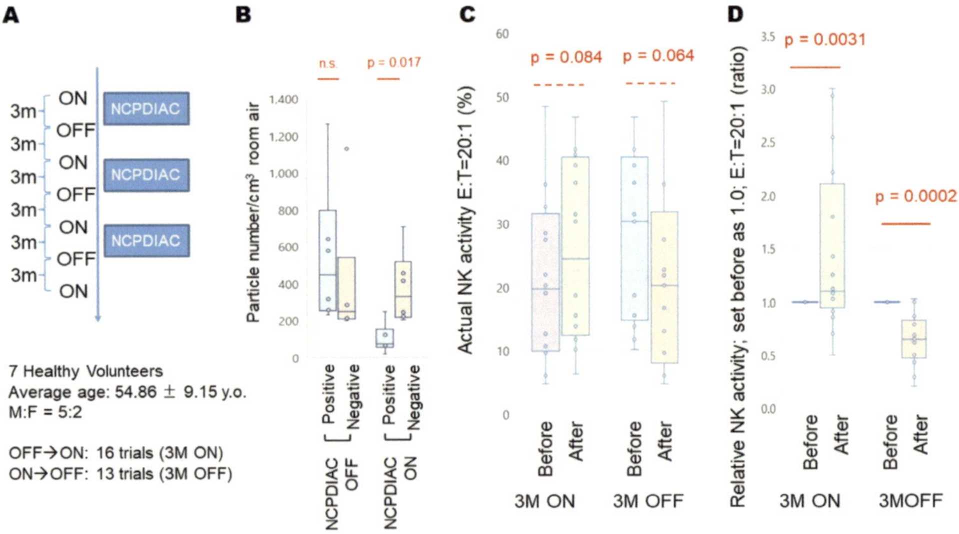

Figure 5.
(A) Seven healthy volunteers utilized sleeping and living rooms with NCPDIAC in their homes. During this period, the occupants themselves would switch NCPDIAC ON and OFF. Sixteen trials of OFF to ON (3M ON) and thirteen trials of ON to OFF (3M OFF) were executed. (B) The levels of positively and negatively charged particles in the representative six rooms including the homes of HV where the NCPDIAC apparatus was set were measured using as ion counter (EM-1000, Eco Holistic Inc., Suita, Japan) during the OFF and ON periods. During the OFF period, there was no difference between positively and negatively charged particles; however, during the ON period, the level of positively charged particles was significantly reduced, thereby establishing a difference between the number of positively and negatively charged particles. Statistical analyses were performed using the student T test. (C) Changes in actual NK cell activity prior to and following the 3M ON and 3M OFF periods. Values tended to increase during the 3M ON and decrease during the 3M OFF periods. (D) Relative NK cell activities with NK cell activity previously set to 1.0. There was a significant increase in relative NK cell activity during the 3M ON period and a significant decrease during the 3M OFF period. Statistical analyses were performed using the student T test.

All volunteers had built or renovated their residential homes prior to being recruited to this project and agreed to set up an NCPDIAC device for the experiments [15].

This study was approved by the institutional ethical committee (#854) and samples were only taken from HV who provided written informed consent.

The clinical parameters measured were similar to those determined for the 2W nightly stay experiments with additional cytokines being measured. Twenty-nine cytokines were measured using the Luminex 26 Cytokine Plex Kit Human Cytokine/Chemokine Panel (MPXHCYTO60KPMX26, Merck Millipore, Billerica, MA) [15]. Additionally, adipokines and cytokines related to oxidative stress in serum from 16 ON and 13 OFF trials were measured using the Human Adipokine Magnetic 14-Plex Panel with Luminex instruments (Bio-rad, Hercules, CA, USA). Fourteen of the cytokines examined comprised IL-1β, IL-10, IL-6, monocyte chemotactic protein (MCP)-1, leptin, SAA (serum amyloid A), hepatocyte growth factor (HGF), insulin, lipocalin-2, TNF-α, B cell activating factor (BAFF: belonging to the tumor necrosis factor family), resistin, plasminogen activator inhibitor (PAI)-1, and IL-8 [16].

After analyzing all of the results, it was determined that all items measured except for NK activity revealed no significant difference between 3M ON and 3M OFF conditions. NK activity was measured as the E/T ratio and was calculated to be 10:1 and 20:1. With an actual NK activity of 20:1, there was a tendency of NK activity to increase in the 3M ON period and decrease in the 3M OFF period [15]. Additionally, the relative changes in NK activity set before as 1.0 for individual values of 3M ON and 3M OFF periods revealed a significant increase during the 3M ON periods and a decrease during the 3M OFF periods in the 20:1 E/T ratio as shown in **Figure 5C** [with an E/T ratio of 10:1, similar significant results were obtained in the 3M ON ($p = 0.017$) and 3M OFF ($p = 0.012$) periods] [15].

Taken together, it was shown that NCPDIAC cause enhancement of NK cell activity even in living homes. The apparatus utilized for establishing NCPDIAC may possess advantages in reducing the occurrence of cancers, as well as reducing signs or symptoms associated with virus-infected diseases such as influenza [15].

For adipokines and cytokines related to oxidative stress, there were no significant changes observed [16]. However, with the exception of one case, serum amyloid A (SAA) levels decreased significantly during the ON trials (data not shown) [16]. Considering that SAA is an acute phase-reactive protein like C-reactive protein (CRP), this observed decrease may indicate a prevention of cardiovascular and atherosclerotic changes, since an increase in high-sensitive CRP is associated with subsequent detection of these events [16].

7. Discussion

Initial assessment of the biological effects of NCPDIAC began with evaluations of 2.5H stay experiments since similar experiments had yet to be reported and investigations should involve collaboration with HV. Thus, our initial 2.5H stay experiments demonstrated a small but significant increase in IL-2. Additionally, it was assumed that there were no adverse effects during the 2.5H period. These experiments were then followed with 2W nightly stay experiments. However, to set up NCPDIAC in the homes of HV was very difficult since sleeping and/or living rooms required alterations. Thus, we used a dormitory that belonged to the collaborating house company where many trainees (company employees) would be staying during the 3M period. With these experiments, we found an enhancement of NK cell activity during the 2W nightly stay. Additionally, there were no adverse effects as determined by the various parameters examined.

During these studies employing HV, in vitro assays were performed. From the in vitro experiments, we confirmed the increased NK cell activity and slight immune-stimulatory effects of negatively charged particles. Again, there were no adverse effects to the human body as determined by the various parameters examined.

These results encouraged us to apply NCPDIAC in the actual homes (sleeping and living rooms) of volunteers. Following approval by the institutional ethical committee and obtaining written informed consent, the homes of 7 HV were set up with NCPDIAC. Results showed that the 3M ON period enhanced and the 3M OFF period reduced NK cell activity.

It was important to proceed with these experiments in a methodical, step-by-step fashion as careful consideration and confirmation of the results were required to preserve the health and well-being of the volunteers.

8. Conclusions

In this chapter, the establishment of NCPDIAC and results of the experiments for short-term (2.5H, 2.5-h), mid-term (2W, 2-week), and relatively long-term (3M, 3-month in actual living homes) stays were shown.

Long-term monitoring in actual living homes comprising 6-month, 12-month, and 3-year duration periods has commenced, and the results will be reported in due course.

It is extremely important that appropriate living environments are investigated and created that mitigate or prevent the onset of many diseases such as cancers and virus-infected diseases. In addition to NCPDIAC, other devices that maintain stable temperature as well as increase air tightness may prevent acute accidents related to cardiovascular events. Our recent long-term trial monitoring experiments have included these devices. We hope that these health-promoting indoor environments enable people to live healthier and happier lives.

Notes and Acknowledgements

All experiments described in this chapter were approved by the Ethics Committee of the Kawasaki Medical School, Kurashiki, Japan.

All authors thank Ms. Shoko Yamamoto and Tamayo Hatayama for their skillful technical assistance.

Conflicts of interest

For the short-term (2.5H) and mid-term (2W) stay experiments and the in vitro experiments, the Department of Hygiene, Kawasaki Medical School, obtained research funding from SEKISUI HOUSE Ltd., Osaka, Japan. For the long-term (3M) stay experiments, the Department of Hygiene obtained research funding from Yamada SXL Hone Co. Ltd., Takasaki, Japan. Additionally, the extraporous charcoal paints were provided by Artech Kohboh, Co. Ltd., Omura, Nagasaki, Japan.

Author details

Suni Lee[1], Yasumitsu Nishimura[1], Naoko Kumagai-Takei[1], Hidenori Matsuzaki[2], Megumi Maeda[3], Nagisa Sada[1,4], Kei Yoshitome[1] and Takemi Otsuki[1]*

1 Department of Hygiene, Kawasaki Medical School, Kurashiki, Japan

2 Department of Life Science, Faculty of Life and Environmental Science, Prefectural University of Hiroshima, Shobara, Japan

3 Department of Biofunctional Chemistry, Graduate School of Environmental and Life Science, Okayama University, Okayama, Japan

4 Department of Biophysical Chemistry, Graduate School of Medicine, Dentistry and Pharmaceutical Sciences, Okayama University, Okayama, Japan

*Address all correspondence to: takemi@med.kawasaki-m.ac.jp

References

[1] Hodgson MJ. Clinical diagnosis and management of building-related illness and the sick-building syndrome. Occupational Medicine. 1989;**4**:93-606

[2] Redlich CA, Sparer J, Cullen MR. Sick-building syndrome. Lancet. 1997;**349**:1013-1016

[3] Menzies D, Bourbeau J. Building-related illnesses. The New England Journal of Medicine. 1997;**337**:1524-1531

[4] Saeki K, Obayashi K, Tone N, Kurumatani N. A warmer indoor environment in the evening and shorter sleep onset latency in winter: The HEIJO-KYO study. Physiology & Behavior. 2015;**149**:29-34. DOI: 10.1016/j.physbeh.2015.05.022

[5] Saeki K, Obayashi K, Tone N, Kurumatani N. Daytime cold exposure and salt intake based on nocturnal urinary sodium excretion: A cross-sectional analysis of the HEIJO-KYO study. Physiology & Behavior. 2015;**152**:300-306. DOI: 10.1016/j.physbeh.2015.10.015

[6] Krueger AP, Reed EJ. Biological impact of small air ions. Science. 1976;**193**:1209-1213

[7] Misiaszek J, Gray F, Yates A. The calming effects of negative air ions on manic patients: A pilot study. Biological Psychiatry. 1987;**22**:107-110

[8] Weber C, Henne B, Loth F, Schoenhofen M, Falkenhagen D. Development of cationically modified cellulose adsorbents for the removal of endotoxins. ASAIO Journal. 1995;**41**:M430-M434

[9] Yamada R, Yanoma S, Akaike M, Tsuburaya A, Sugimasa Y, Takemiya S, Motohashi H, Rino Y, Takanashi Y, Imada T. Water-generated negative air ions activate NK cell and inhibit carcinogenesis in mice. Cancer Letters. 2006;**239**:190-197

[10] Takahashi K, Otsuki T, Mase A, Kawado T, Kotani M, Ami K, et al. Negatively-charged air conditions and responses of the human psycho-neuro-endocrino-immune network. Environment International. 2008; **34**:765-772. DOI: 10.1016/j.envint.2008.01.003

[11] Takahashi K, Otsuki T, Mase A, Kawado T, Kotani M, Nishimura Y, et al. Two weeks of permanence in negatively-charged air conditions causes alteration of natural killer cell function. International Journal of Immunopathology and Pharmacology. 2009;**22**:333-342

[12] Domzig W, Stadler BM, Herberman RB. Interleukin 2 dependence of human natural killer (NK) cell activity. Journal of Immunology. 1983;**130**:1970-1973

[13] Ishikawa H, Saeki T, Otani T, Suzuki T, Shimozuma K, Nishino H, Fukuda S, Morimoto K. Aged garlic extract prevents a decline of NK cell number and activity in patients with advanced cancer. The Journal of Nutrition. 2006;**136**:816S-820S

[14] Nishimura Y, Takahashi K, Mase A, Kotani M, Ami K, Maeda M, Shirahama T, Lee S, Matsuzaki H, Kumagai-Takei N, Yoshitome K, Otsuki T. Exposure to negatively charged-particle dominant air-conditions on human lymphocytes in vitro activates immunological responses. Immunobiology. 2015;**220**:1359-1368. DOI: 10.1016/j.imbio.2015.07.006

[15] Nishimura Y, Takahashi K, Mase A, Kotani M, Ami K, Maeda M, Shirahama T, Lee S, Matsuzaki H, Kumagai-Takei N, Yoshitome K, Otsuki T. Enhancement of NK cell

cytotoxicity induced by long-term living in negatively charged-particle dominant indoor air-conditions. PLoS One. 2015;**10**:e0132373. DOI: 10.1371/journal.pone.0132373

[16] Lee S, Yamamoto S, Nishimura Y, Matsuzaki H, Yoshitome K, Hatayama T, Ikeda M, Yu M, Sada N, Kumagai-Takei N, Otsuki T. Decrease in serum amyloid A (SAA) protein following three-month stays under negatively charged particle-dominant indoor air conditions. Biomedical and Environmental Sciences. 2018;**31**:335-342. DOI: 10.3967/bes2018.044

For EU product safety concerns:
IN TECH d.o.o., Prolaz Marije Krucifikse Kozulić 3, 51000 Rijeka, Croatia,
info@intechopen.com or visit our website at intechopen.com.

Meet the editors

Fernanda Lasakosvitsch has a degree in Biological Sciences and a PhD in Sciences from the Department of Microbiology, Immunology and Parasitology with emphasis in Molecular Biology at the Universidade Federal de São Paulo. She has a postdoctoral degree in Molecular Biology from the Universidade de São Paulo. Nowadays she is a researcher and scientific producer of the Funzionali nutrology group, developing studies in enteral nutrition, parenteral nutrition and metabolism in critically ill hospitalized patients.

Sergio dos Anjos Garnes graduated in Medicine from the Federal University of Mato Grosso do Sul, post graduate in Clinical Nutrition, specialist in Nutrology expertise in Parenteral and Enteral nutrition. He is the coordinator of the multi-professional therapy group at the German hospital Oswaldo Cruz in São Paulo and medical director of Funzionali nutrology group, Brazil.

Contents

Preface

In the evolutionary process, physiological mechanisms are developed as a response to the demands of the environment and are in accordance with the ecological factors in which the organisms exist. Diseases such as cancer, type 2 diabetes, autoimmunity, allergy, atherosclerosis, psychic disorder and obesity are a reflection of the mismatch between the environment and human evolutionary history. These diseases have at least one characteristic in common: the disturbance of homeostasis.

The earliest records of an organism having the ability to adjust to maintain its stability are from Hippocrates. He believed that diseases were healed by natural powers, that is, within organisms, there would be mechanisms that would tend to adjust functions when deviated from their natural state.

Most physiological processes operate under a narrow range of conditions, which are maintained by specialized homeostatic mechanisms in response to variations in the environment and are adjusted for changes in functional demands and biological priorities. All types of adaptation are based on homeostatic mechanisms. An organism functioning as an open system may seem extremely susceptible to environmental fluctuations and become unstable but in living beings, the concepts of stability and change assume a prominent role that is essential for the maintenance of life. If there was no stability, there would be no evolution since there would also be no guarantee of maintaining the biochemical characteristics essential to life.

Organisms stay alive because of their ability to stabilize their internal organization, preventing disintegration into an increasingly disorganized state. Maintaining "stability" in the midst of an ever-changing environment requires efficient mechanisms to overcome difficulties, so the living being is an organization in which each disruptive influence induces, per se, the increment of a compensatory activity to neutralize the disturbance.

This book aims to provide the reader an up-to-date view of the self-regulatory mechanisms that are activated to achieve homeostasis, the pathways that are altered during the disease process, and how medicine can intervene to restore balance in critical patients.

Fernanda Lasakosvitsch, PhD
Funzionali Serviços Médicos Ltda
São Paulo, Brazil

Universidade Federal de Sao Paulo
São Paulo, Brazil

Sergio Dos Anjos Garnes
Universidade Federal de Sao Paulo
São Paulo, Brazil

Section 1

Introduction

Chapter 1

Introductory Chapter: Homeostasis

Fernanda Lasakosvitsch

Additional information is available at the end of the chapter

http://dx.doi.org/10.5772/intechopen.81130

1. Introduction

The human body is composed of several systems and organs, consisting of millions of cells that need relatively stable conditions to function and contribute to the survival of the body as a whole. The maintenance of stable conditions for the cells against the variations of the external environment is an essential function of the body and configures the homeostasis.

In 1870, Claude Bernard described the basic principles of physiological regulation, evidencing the body's need to maintain a stable internal environment. This allows life processes to function optimally during periods of environmental fluctuations and disturbances, and to this end, an array of reflexes and regulatory systems that allow the body to thrive and reproduce despite environmental challenges [1]. Later, Cannon [2] coined the term homeostasis to describe the collective activity of bodily systems that monitor and protect vital life parameters against harmful detours. Many of these processes are unnoticed at the conscious level and depend on behavioral, autonomous, endocrine responses among others to prevent the everyday challenges of the internal environment. Cannon defined homeostasis as "a condition that may vary but remains relatively constant." This balance is guaranteed thanks to the physiological processes that act in a coordinated way in the body and that prevent the changes in the environment interfere in its functioning. Factors such as pH, temperature, plasma osmolality, glucose and calcium are critical for the normal functioning of most organisms and therefore are controlled within narrow limits [3].

The importance of these factors is due to its universal biochemical effects. Temperature and pH may alter the spatial conformation of the structure and function of enzymes and functional RNAs; temperature and plasma osmolality are important in the integrity and function of the membranes; glucose is a key currency for moving and storing energy; and calcium plays key roles in muscle physiology and multiple signaling pathways. The variation of these factors constitutes a noise that affects all physiological systems, including the homeostatic systems themselves. Since the noise is global (affects all other physiological components in other homeostatic systems), cells cannot filter it or develop private channels to avoid it; instead, the options are to send redundant signals, minimize the spatial and temporal proximity between the transmitter and the receiver, or minimize the variation in the noise-inducing factors. Of all these events, minimizing noise is a key factor that drives the evolution of homeostasis [4].

Until the last decades, homeostasis was considered as a reaction to continuous disturbances in vital systems through negative feedback responses. According to Langley [5], homeostasis is a self-regulating negative feedback system that keeps the internal environment constant. Current views on the subject assume that negative feedback systems that are not activated until a disturbance of the regulated variable occurs may be inefficient and fail in its regulatory role. Negative feedback was the first underlying process used to explain how homeostasis works. This was a reactive strategy through which the perturbation of a regulated variable far from its optimal value was detected and, consequently, provoked corrective responses that served to return the variable back to the pre-disturbance levels. For example, a sudden drop in oxygen content in the blood is detected by sensors that are synaptically linked to areas of the brain that control respiratory rate, and respiration and oxygen in the blood are consequently increased. Although negative feedback is certainly an essential component of regulation, based on past experience, organisms learn to prepare early responses to impending disruption, thereby reducing their impact [4].

A fundamental principle of homeostatic regulation is that the answers work in a coordinated way to defend the body parameters critical to the well-being of an animal; that is, that homeostatic regulation is purposeful. Strategies that may appear to be anticipatory use the initial sensory information in an area of the body as the signal of an event that could potentially disrupt a critical regulated variable, activating corrective effectors and preventing the critical variable from being disturbed. For example, when entering a cold environment, skin temperature sensors are activated before there is any change in core temperature (the variable that is presumably homeostatically regulated); and an effect of increased activity of skin thermometers is to trigger a thermopreflection reflex cycle that induces increased metabolic rate or peripheral vasoconstriction that can stabilize central temperature [6].

If the body cannot maintain the homeostasis of the critical variables, normal function is interrupted and a state, or pathological condition, can develop. Diseases are characterized in two general groups according to their origin: those in which the problem arises from an internal insufficiency or failure of some normal physiological process, and those that originate from some external source. The critical patient develops a condition where the loss of organic homeostasis puts him at immediate risk of death. Therapeutic priorities are focused on correcting these imbalances, such as hemodynamic stabilization, restoration and maintenance of blood volume, stabilization of ventilation conditions and correction of other abnormalities such as electrolyte imbalance, metabolic changes and pain control. Thus, medicine can intervene to restore balance through the activation of self-regulating mechanisms, restoration of the pathways that are altered during the disease process and, finally, to achieve homeostasis in critical patients.

http://dx.doi.org/10.5772/intechopen.81130

Author details

Fernanda Lasakosvitsch[1,2*]

*Address all correspondence to: flcastanho@gmail.com

1 Funzionali Serviços Médicos Ltda, São Paulo, SP, Brazil

2 Universidade Federal de Sao Paulo, Brazil

References

[1] Bernard C, Noble D. Experimental Physiology-Paton Lecture the first systems biologist, and the future of physiology. Experimental Physiology [Internet]. 2007;**93**:16-26. Available from: https://pdfs.semanticscholar.org/9049/44f321fef4a0996b00c36a1ce643f7ddce96.pdf [cited 2018 Aug 21]

[2] Cannon WB. Organization for physiological homeostasis. Physiological Reviews. 1929;**9**(3):399-431

[3] Carlson AJ. The Wisdom of the Body. Walter B. Cannon. American Journal of Sociology [Internet]. 1933 Jan;**38**(4):651-651. Available from: https://www.journals.uchicago.edu/doi/10.1086/216202 [cited 2018 Aug 21]

[4] Ramsay DS, Woods SC. Clarifying the roles of homeostasis and allostasis in physiological regulation. Physiological Reviews [Internet]. 2014 Apr;**121**(2):225-247. Available from: http://www.ncbi.nlm.nih.gov/pubmed/24730599 [cited 2018 Aug 21]

[5] Langley LL. Homeostasis: Origins of the Concept. 1st ed. In: Langley LL, editor. Michigan: Dowden, Hutchinson & Ross; 1973. 362 p

[6] Morrison SF, Nakamura K. Central neural pathways for thermoregulation. Frontiers in Bioscience [Internet]. 2011;**16**:74-104. Available from: http://www.pubmedcentral.nih.gov/articlerender.fcgi?artid=3051412&tool=pmcentrez&rendertype=abstract

Section 2

Physiological Mechanisms of Homeostasis

Chapter 2

Circadian Body Temperature Rhythm and the Interaction with Energy State

Kei Nagashima, Ken Tokizawa, Shuri Marui and Yuki Uchida

Additional information is available at the end of the chapter

http://dx.doi.org/10.5772/intechopen.76229

Abstract

We have revealed that circadian body temperature (T_b) rhythm is significantly influenced by fasting/fasting-related hormones. The effect of circadian mechanism and fasting/fasting-related hormones on thermoregulation was examined. Fasting decreases T_b during the light phase in rodents. For the regulation, the suprachiasmatic nucleus (SCN) and clock genes, such as *Cry* and *Clock*, are necessary. In addition, ghrelin and several hypothalamic nuclei, that is, the medial preoptic area, paraventricular nucleus (PVN), and arcuate nucleus (ARC), play a key role in the T_b rhythm. During the light phase, fasting and ghrelin affect the hypothalamic areas. The activity of the SCN increases and that of the ARC decreases. The SCN sends inhibitory signals to the PVN, which may result in a lower heat production in the interscapular brown adipose tissue (iBAT) and T_b. By contrast, during the dark phase, the activity of the SCN decreases and that of the ARC increases. The inhibitory signal from the SCN is less, and the PVN is activated. Heat production of the iBAT increases and T_b is maintained. There are functional and anatomical connections between the circadian and thermoregulation systems. The circadian system modulates thermoregulatory response to hypothermia and/or cold depending on time and feeding condition.

Keywords: fasting, ghrerin, leptin, brown adipose tissue, paraventricular nucleus, arcuate nucleus, suprachiasmatic nucleus

1. Introduction

The temperature of the cell(s), tissues, organs, and body is an important factor that determines the biological functions and survival of organisms, from prokaryotes to vertebrates, although the preferred temperature varies. Homeothermic animals can constantly regulate body temperature (T_b). Thermoregulation is the balance between heat loss and heat production in the body. The temperature regulation is different among species, and the range of temperature regulation is fairly narrow compared to the larger temperature ranges in the living environment [1].

Homeothermic animals show a circadian T_b rhythm, although the goal of thermoregulation is to maintain a constant T_b. The T_b is higher in the active phase and lower in the inactive phase in both diurnal and nocturnal animals. The T_b rhythm is also observed under conditions wherein the influence of physical activity (i.e., heat production) is minimized. For example, circadian T_b rhythm is observed even in individuals forced to be on complete bed rest [2]. Results suggested that the circadian T_b rhythm can also be regulated.

Despite the usual small amplitude of the circadian T_b rhythm (e.g., less than 1°C in human beings [3]), the rhythm is still important for preserving energy during the inactive phase. The maintenance of higher T_b is energy-costly because more energy from the total daily intake is used for heat production [4]. Therefore, the circadian T_b rhythm may be important in saving energy in homeothermic animals when energy is not needed.

Environmental temperature significantly affects the thermoregulation system. We assessed the circadian T_b change in rats that were placed in an ambient temperature of 18, 25, or 32°C (unpublished data). Laboratory rats are usually housed at 25°C. However, such cold or heat stress did not alter the circadian T_b rhythm. Results suggested that the T_b rhythm is not simply a result of circadian change in heat loss or production in the body. Rather, it is a result of the coordinated thermoregulatory processes that maintain a specific T_b at a given time of the day. Moreover, the T_b rhythm is probably generated by the association between the circadian and thermoregulation system. However, the mechanism is not yet known.

As mentioned previously, the T_b rhythm remains unchanged even when environmental temperature is altered. However, our previous studies have revealed that the T_b rhythm is remarkably influenced by fasting/fasting-related hormones. It is hypothesized that a lack of energy affects thermoregulation and the T_b rhythm. However, several previous studies do not support this hypothesis. In this study, a review of the literature on the mechanism and physiological effects of the circadian T_b rhythm was carried out.

2. Association between fasting and circadian core body temperature rhythm

Fasting is a strong stimulus that changes the amplitude of the circadian T_b rhythm in both mammals and birds [5–8]. In a previous study by Tokizawa et al. [9], T_b was obtained with a thermometer placed in the abdominal cavity of mice, and it was continuously and noninvasively monitored with a telemetry. The most significant reduction was during the light (inactive) phase, whereas the reduction in the dark (active) phase was not remarkable (**Figure 1A**). In addition, no difference was observed in the spontaneous activity during both phases (**Figure 1B**). Results suggested that less heat production due to a decreased physical activity was not associated with the mechanism involved in the reduction of T_b during fasting. Moreover, a remarkable change in T_b was observed at a specific time, which may indicate an association with the circadian rhythm, although the phase shift of the rhythm was not observed.

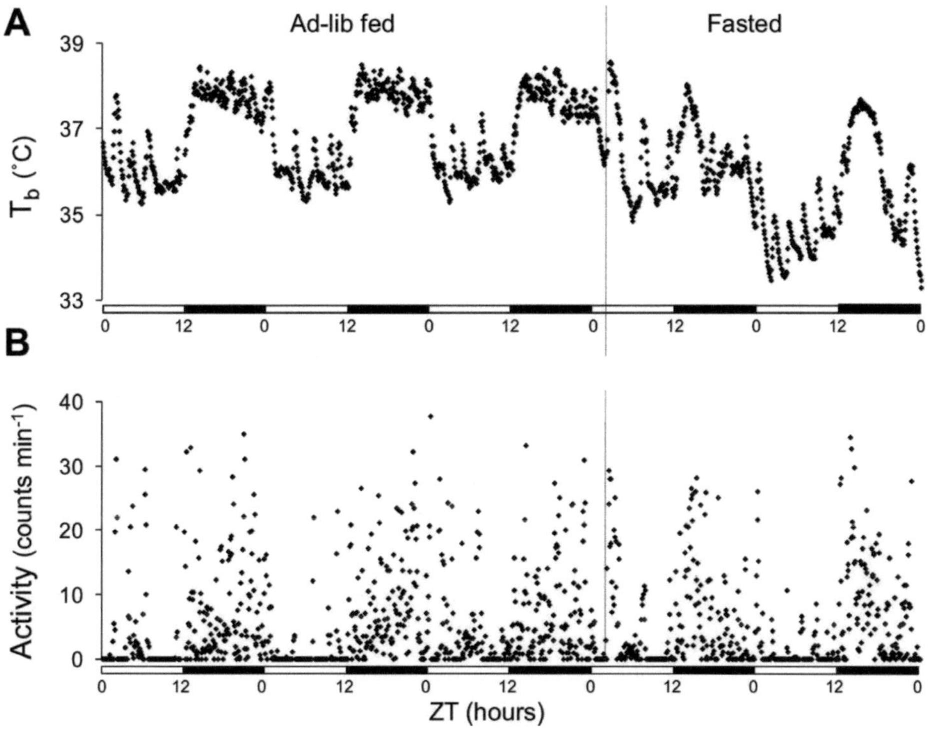

Figure 1. Circadian changes in T_b and spontaneous activity during *ad lib* feeding and fasting in settings with normal temperature (27°C) in mice. The x-axis indicates ZT. The lighting condition is shown at the bottom with the open bar denoting the light phase and the closed bar the dark phase. Data were plotted in 5-min bins. The vertical dashed line shows the time when the 48-h fasting period was started. The modified figure was obtained from the manuscript by Tokizawa et al. [9].

Compared to that during a normal T_b rhythm, the significant decrease in T_b during the light phase (i.e., from 36.0°C during feeding to 34.1°C during fasting) may lessen heat transfer from the body to the environment and preserve energy. In contrast, the maintenance of T_b during the dark phase may be essential in the maintenance of physical activity. On the basis of previous studies, the following questions might be raised: (1) Is the reduction in T_b during the light phase a regulated phenomenon by the thermoregulation system? (2) Is the circadian system involved in the phenomenon? and (3) Which factor that is associated with fasting causes the decrease in T_b, and how?

3. Physiological mechanism involved in the decrease in body temperature during fasting

T_b in homeothermic animals is controlled by heat production and heat loss mechanisms. However, whether such control is maintained and generates the circadian T_b rhythm during fasting is still not verified. Fasting is a strong stimulus that decreases metabolic heat production in rodents and pigeons [5, 6, 10–15]. In addition, fasting decreases thermal conductance (i.e., opposite of thermal resistance) from the body core to the environment (heat loss [16–18]). If these physiological responses occurred in the same manner during the day of fasting, the reduction of T_b, specifically during the light phase, would not be observed. Therefore, the thermoregulatory responses must be different during the two phases.

Nagashima et al. [19] assessed the T_b, counts of locomotor activity, and oxygen consumption rate ($\dot{V}O_2$) in rats that were made to fast for 3 days. **Figure 2** illustrates the data before fasting and on the last day of fasting and recovery day. T_b and counts of activity were obtained as the

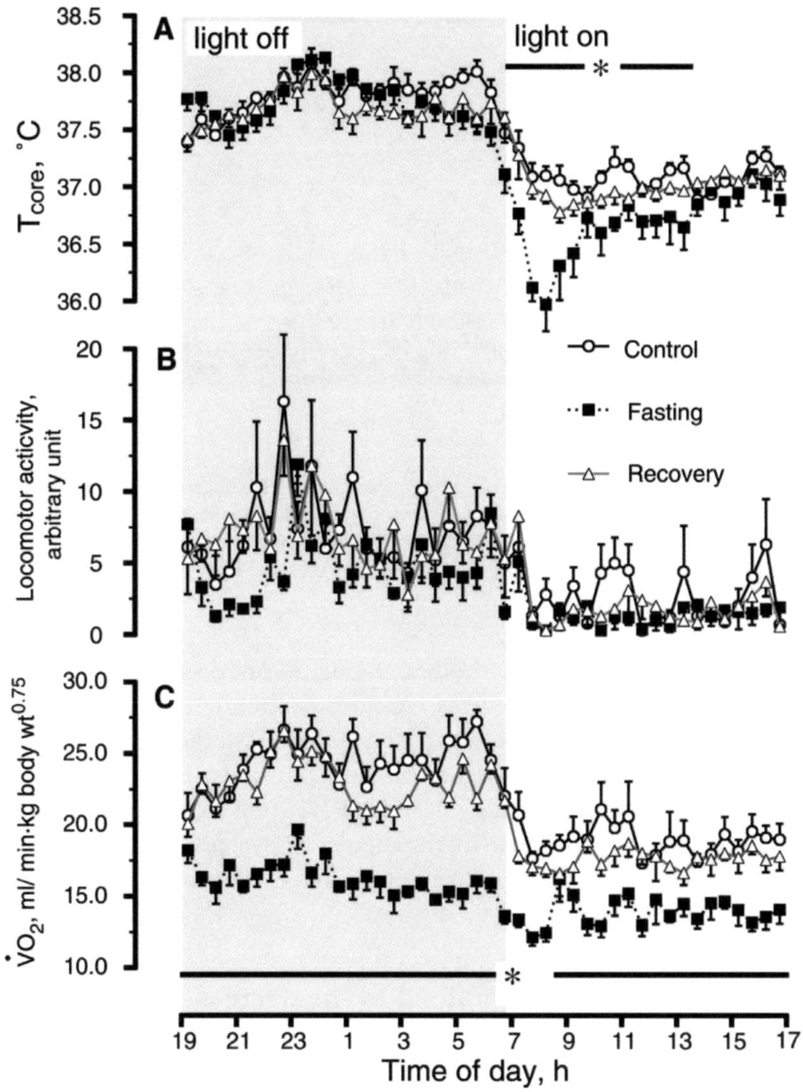

Figure 2. Body core temperature (*A*, T_{core}), counts of locomotor activity (*B*), and oxygen consumption rate (*C*, $\dot{V}o_2$) during feeding, day 3 of fasting, and day 4 of recovery. Stippled area, dark phase (1900–0700). Each point has an average of 30 min. The values are the means ± SE in six rats. *Significant difference from controlled feeding, $P < 0.05$. The figure was obtained from the manuscript by Nagashima et al. [19].

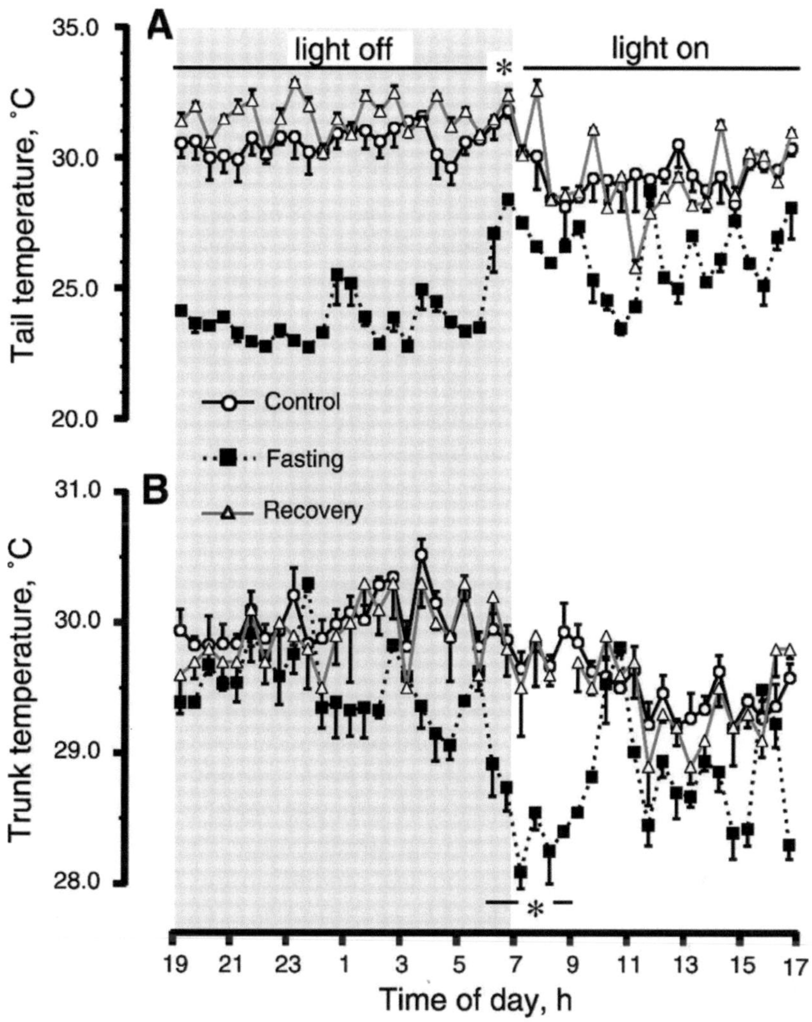

Figure 3. Surface temperatures of the tail (*A*) and trunk (*B*) determined by thermography in a controlled feeding condition, on day 3 of fasting, and day 4 of recovery. T_{tail} is the average of surface temperatures at one-third of the length of the tail from the root and the tip. T_{trunk} is the average of surface temperatures of the head and middle parts of the upper and lower back. Each point has an average of 30 min. The values are the means ± SE in five rats. *Significant differences from controlled feeding conditions, $P < 0.05$. The figure was obtained from the manuscript by Nagashima et al. [19].

study by Tokizawa et al. [9], and $\dot{V}O_2$ was assessed by indirect calorimetry. $\dot{V}O_2$ decreased during fasting in both dark and light phases. Although the rhythms of T_b and $\dot{V}O_2$ were significantly associated during feeding, the amplitude of the $\dot{V}O_2$ rhythm decreased and that of T_b increased during the fasting. The result suggested that T_b was maintained by the suppression of heat loss during fasting and the dark. The estimated thermal conductance during the dark phase of the fasting period decreased from that in the fed condition (1.10 and 1.65 ml·min^{-1}•kg body wt$^{-0.75}$·°C^{-1}, respectively), which supported the hypothesis. On the contrary, such a change was not observed during the light phase.

Animals use several mechanisms to change the efficiency of heat loss. Among the mechanisms, the tail is a crucial site for the regulation of heat loss in rats and mice, with its physiological and anatomic characteristics, that is, a high density of arteriovenous anastomosis [6], the absence of fur, and a remarkable surface-to-volume ratio. Young and Dawson [20] reported that rats could dissipate 25% of basal heat production by changing the blood flow in the tail.

To assess the contribution of the tail in thermoregulation, Nagashima et al. [19] estimated the tail surface temperature by thermography (**Figure 3**). Tail temperature during the fasting day was lower than that during the fed control day. However, body trunk temperature during the fasting was lower than that during the fed control only around the light-onset period. Interestingly, different from the fed control, tail temperature in the fasting condition increased after light onset and remained at a higher level during the light phase. The results strongly suggested an attenuation of tail blood flow in the dark phase during the fasting condition and were blunted at the beginning of the light phase. Thus, the tail may help in determining the process of heat loss from the body in the fasting condition, which was the factor regulating the T_b rhythm.

4. Involvement of suprachiasmatic nucleus in T_b rhythm during fasting

The suprachiasmatic nucleus (SCN) in the hypothalamus is thought to be the master clock of the circadian rhythm. In addition, electrical or chemical lesions of the SCN destroy the T_b rhythm in rodents [21–26]. However, these findings may not indicate that the circadian clock regulates the T_b rhythm. Behavioral and/or physiological responses, such as locomotor activity, eating, and the secretion of several hormones, are also inhibited due to the SCN lesion, which may have direct influences on heat production and T_b [17, 27–29]. Simply put, SCN lesion inhibits such behavioral and/or physiological responses [3, 21, 24, 30], and the T_b rhythm is subsequently destroyed.

Liu et al. [23] assessed the circadian T_b rhythm within 4 days of fasting in rats, of which SCN was electrically lesioned. The rats had arrhythmia of T_b and spontaneous activity, and no difference was observed during the light and dark phases and the non-fasting and fasting periods. Although the experiment did not answer the question of whether the circadian clock directly regulates the T_b rhythm, the result suggested that SCN may be important for the changes in thermoregulation and T_b due to fasting.

5. Importance of molecular circadian mechanism on T_b rhythm during fasting

Several genes in the central and peripheral tissues have expression rhythms with a periodicity of ~24 h. Among these genes, those with autoregulatory transcription-translation loops with a periodicity of ~24 h may be responsible for the core molecular mechanisms of the circadian clock (i.e., clock genes). These genes are observed in both the central and the peripheral tissues [31–33].

Nagashima et al. [34] assessed the circadian T_b rhythm in mice that lack the cryptochrome 1 and 2 genes (*Cry1* and *Cry2*), two of the core clock genes [35–38]. The mice loose the transcription-translation rhythms of the *Cry1* and *Cry2* and other clock genes and periodicity in a wheel-running behavior [38], as well as the electrophysiological activity of the SCN cells under constant dark conditions [39]. However, the behavior is suppressed under the light condition, and the daily rhythm is observed under both light and dark conditions. Therefore, despite the lack of the internal rhythm, lighting rhythm could induce T_b rhythm.

Table 1 summarizes the cosinor rhythm analysis [34] of T_b, spontaneous activity, and $\dot{V}O_2$ in normal wild-type mice that lack *Cry1*/*Cry2* (*Cry1*$^{-/-}$/*Cry2*$^{-/-}$). The mice were placed under three different conditions: (1) constant darkness (DD) with ad lib feeding, (2) 12–12-h light-dark (LD; lights on at 1900) cycle with ad lib feeding, and (3) LD with food restriction. The food restriction protocol aimed to enhance the eating rhythm. Ordinary chow was given at 1900, which is the end of the light period, at 70–80% of the normal daily intake. Both *Cry1*$^{-/-}$/*Cry2*$^{-/-}$ and wild-type mice finished eating 10–12 h after the appearance of chow in the food-restriction regimen.

In the wild-type mice, no significant difference was observed between the mean, amplitude, and peak phase of the T_b rhythm under the LD condition with ad lib feeding and DD condition. Differences were found in the T_b and $\dot{V}O_2$ rhythms between ad lib feeding and food-restriction days (i.e., greater amplitude of the rhythms during the food-restriction day). The activity rhythm did not change throughout the three conditions. In *Cry1*$^{-/-}$/*Cry2*$^{-/-}$ mice, T_b, $\dot{V}O_2$, and spontaneous activity were arrhythmic under the DD condition. However, T_b, $\dot{V}O_2$, and spontaneous activity became rhythmic in the same peak phases during the LD conditions. As for the T_b and $\dot{V}O_2$ rhythms in the LD condition, the daily means were lower, and the amplitudes of the rhythms were higher in the food-restriction condition than those under the ad lib feeding condition. The study showed that the circadian T_b rhythm is observed even in mice that lack the internal circadian mechanism, when an external lighting and feeding stimuli that alter heat production are observed. The result showed that the heat production rhythm may be a key component for the T_b rhythm.

	T_{core}, °C			**$\dot{V}O_2$, ml · min^{-1} · kg body wt$^{-0.75}$**			**Activity, Au**		
	Fed	**Fast**	**Rec**	**Fed**	**Fast**	**Rec**	**Fed**	**Fast**	**Rec**
Mesor	37.5 ± 0.1	37.3 ± 0.1*	37.4 ± 0.1	21.63 ± 0.84	15.02 ± 0.55*	20.11 ± 0.78‡	5.0 ± 1.0	3.2 ± 0.3	5.0 ± 0.9
Amplitude	0.5 ± 0.1	0.7 ± 0.1*	0.4 ± 0.1*,‡	3.74 ± 0.44	1.57 ± 0.27*	3.06 ± 0.38*,‡	3.4 ± 0.6	2.6 ± 0.3	3.5 ± 0.5‡
Acrophase	18.2 ± 0.3	16.4 ± 0.3*	17.5 ± 0.3*,‡	18.2 ± 0.2	16.4 ± 0.3*	17.6 ± 0.4*,‡	17.7 ± 0.4	17.9 ± 0.4	17.3 ± 0.6
r	0.76–0.91†	0.80–0.85†	0.56–0.91†	0.51–0.85†	0.36–0.61†	0.45–0.74†	0.38–0.67†	0.39–0.63†	0.43–0.67†

Values are means ± SE. $\dot{V}O2$, oxygen consumption rate; Au, arbitrary unit; T_{core}, core temperature; Fed, *day* 4 of fed control period; Fast, *day* 3 of fasting period; Rec, *day* 4 of recovery from fasting. Acrophase is shown in zeitgeber time (ZT, ZT 0=0700).
*Significantly different from control, P < 0.05.
†Significant regression coefficient (r) for fitted cosine curve for daily change of variable, P < 0.05.
‡Significantly different from fasting period, P< 0.05.

Table 1. Analysis of cosinor rhythmometry for daily changes in core temperature,Vo_2, and locomotor activity.

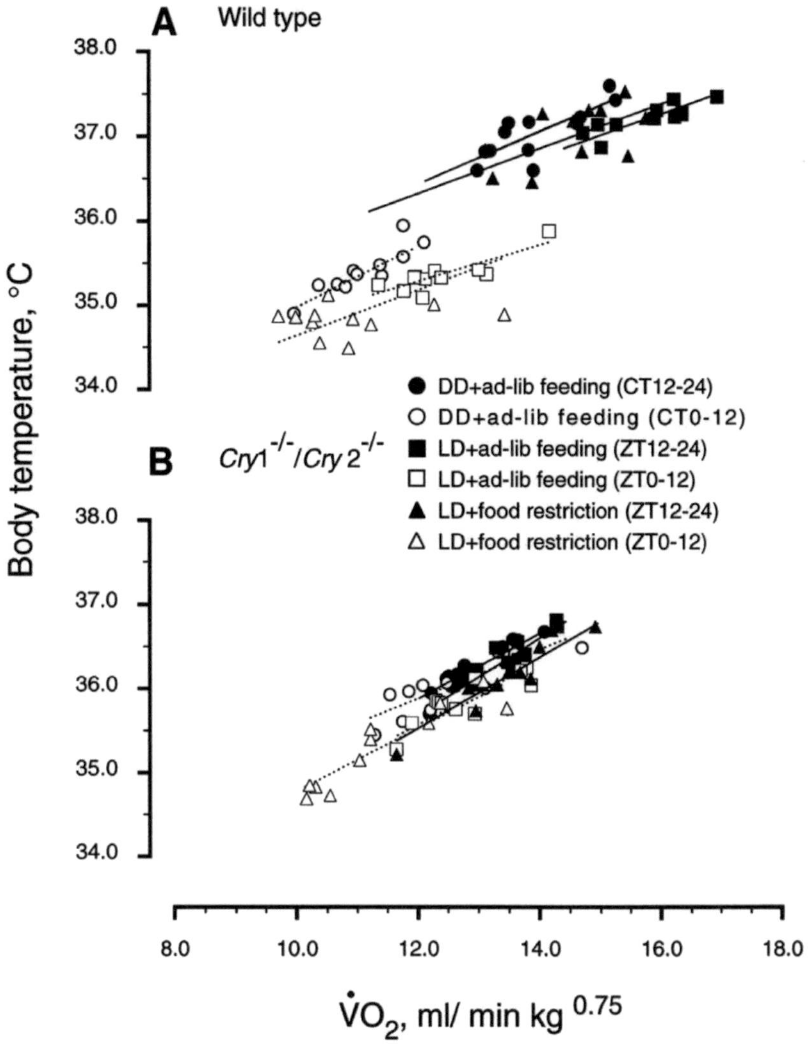

Figure 4. Relationship between metabolic heat production ($\dot{V}O_2$) and T_b in wild-type mice (*A*) and *Cry1*$^{-/-}$/*Cry2*$^{-/-}$ (*B*) based on three trials. The values are the means (±SE) in five mice for 30 min, and data corresponding to each hour are used for this illustration. Regression lines were applied for the averaged values in each group. The figure was obtained from the manuscript by Nagashima et al. [34].

Figure 4 shows the results of the regression analysis of the $\dot{V}O_2$ and T_b of each group. If heat production was the sole determinant of circadian T_b rhythm, these regressions must be identical regardless of trial, circadian phase, and group. In addition, model equations fitted to all the data in the three trials within the same phase and the group was established. Model equations were also created for all possible associations among the three trials. The analysis showed that the data could be classified into four groups: those at circadian time (CT; a standard of time based on the free-running period of a rhythm). The onset of activity of diurnal organisms defines circadian time 0, CT 0. The onset of activity of nocturnal organisms defines circadian time 12, CT 12, or zeitgeber time (ZT; under standard light–dark cycles, the time of lights on defines zeitgeber time 0 and the time of lights off defines zeitgeber time 12), 12–24 and CT or ZT 0–12 in *Cry1*$^{-/-}$/*Cry2*$^{-/-}$ mice and wild-type mice, respectively. Moreover, the model indicated that the two regressions in the wild-type mice were different (i.e., for the active and inactive phases), and those in the *Cry1*$^{-/-}$/*Cry2*$^{-/-}$ mice were similar. The results suggested that the circadian phase can be a significant factor in determining the T_b rhythm in wild-type mice but not in the *Cry1*$^{-/-}$/*Cry2*$^{-/-}$ mice. Thus, T_b was kept higher in the active phase than in the inactive phase regardless of the $\dot{V}O$ level in wild-type mice. Moreover, the regression slopes of $\dot{V}O$ and T_b in both phases were smaller than those of the *Cry1*$^{-/-}$/*Cry2*$^{-/-}$ mice (wild-type: 0.24 and 0.25 in the active and inactive phases, respectively; *Cry1*$^{-/-}$/*Cry2*$^{-/-}$: 0.44). Results showed that the wild-type mice can maintain their T_b within a narrower range than the *Cry1*$^{-/-}$/*Cry2*$^{-/-}$ mice over the same variation of $\dot{V}O$ in each phase. Thus, the circadian T_b rhythm in wild-type animals is not a simple byproduct of the heat production rhythm but a phenomenon regulated by the circadian system.

The mammalian sirtuins (SIRT) modulate the circadian epigenome and provide specificity in transcriptional control [40–43]. It was reported that the circadian clock regulates mitochondrial oxygen consumption rate in the oxidoreductase factor nicotinamide adenine dinucleotide (NAD^+)/SIRT3-dependent manner. In addition, NAD^+-dependent enzymes are important in fasting and oxidative metabolism. Peek et al. [40] evaluated NAD^+ biosynthesis, lipid and glucose oxidation, and acetylation of mitochondrial proteins in normal and circadian (Bmal1)-mutant mice. They reported that lipid oxidation and mitochondrial protein acetylation exhibited circadian oscillations that corresponded with the clock-driven NAD^+ cycle in the liver; however, rhythmic NAD^+ and oxidative cycles were self-sustained in fasted mice. These results suggest a strong interaction between circadian and metabolic rhythms and its destruction during fasting but do not explain for the change in circadian T_b rhythm during fasting.

6. Role of the circadian system and associated brain areas in controlling the circadian T_b rhythm during fasting

Tokizawa et al. [9] tested the hypothesis that thermoregulation is modulated by the circadian system (including the SCN and clock genes), depending on the time of day and feeding condition. Moreover, the physiological and neural responses of the mice during exposure to cold (20°C) under ad lib feeding and during 48-h fasting conditions and the dark and light phases were compared. The differences in the responses between wild-type and *Clock*-mutant mice were also examined. *Clock* is also a gene that organizes the core loop of molecular circadian oscillation. The mutation of *Clock* causes the disappearance of oscillation [44]. However, mutant mice show a T_b rhythm under light-dark conditions because of the masking effect of light.

During ad lib feeding at a low temperature setting (20°C), the T_b of the wild-type mice was similar to that of the wild-type mice during both the dark and light phases at normal temperature settings (27°C) (**Figure 5A** and **B**). However, $\dot{V}O_2$ was higher during the light phase at settings with a temperature of 27°C (ZT2.5–3.5; **Figure 5E**). During fasting at low temperature settings, the reduction in T_b increased during both the light and dark phases (ZT1–4 and 14.5–16, respectively; **Figure 5A** and **B**). However, T_b increased during the light phase. $\dot{V}O_2$ increased in settings with a temperature of 27°C at ZT13.5–14.5 during the dark phase (**Figure 5F**) but remained unchanged during the light phase (**Figure 5E**).

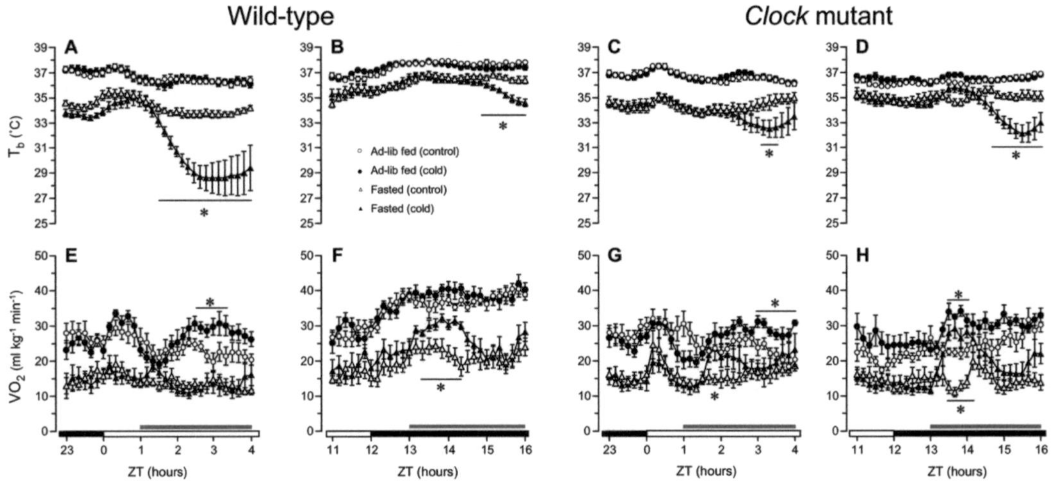

Figure 5. T_b and $\dot{V}O_2$ at low-temperature settings (27 or 20°C) with *ad lib* feeding or fasting in the wild-type and *Clock*-mutant mice (*A–H*). The values given are the means ± SEM (n = 8). $^*P < 0.05$, 20°C (low-temperature setting) versus 27°C (control setting) in each feeding condition. $^{\#}P < 0.05$, versus the dark phase in wild-type mice and both phases in *Clock*-mutant mice; $^{\S}P < 0.05$, versus dark phase in wild-type mice and light phase in *Clock*-mutant mice. The modified figure was obtained from the manuscript by Tokizawa et al. [9].

During ad lib feeding at low temperature settings, T_b in *Clock*-mutant mice was also maintained at the 27°C level (**Figure 5C** and **D**). $\dot{V}O_2$ increased above the 27°C level during both the dark and light phases (ZT3–4 and 13.5–14; **Figure 5G** and **H**), and no difference was observed between the two phases. During fasting at low temperature conditions, T_b decreased below the 27°C level at ZT3–3.5 during the light phase and ZT14.5–16 during the dark phase (**Figure 5C** and **D**), and no significant difference was observed between the phases. $\dot{V}O_2$ was higher than the 27°C level at ZT2 and 13.5–14 (**Figure 5G** and **H**). This increase in $\dot{V}O_2$ was higher during the dark phase than during the light phase.

The study indicated that the physiological response of mice to low temperature conditions is different during ad lib feeding and fasting in wild-type animals and during the dark and light phases. However, such differences were not observed or significantly decreased in *Clock*-mutant mice. The thermoregulatory mechanism of heat production is attenuated or inhibited during fasting and light phases. For such response, the circadian system is important. To support this result, the expression of uncoupling protein 1 (UCP1) mRNA [45–47] in the interscapular brown adipose tissue (iBAT, one of the effector organs for thermoregulatory heat production) was suppressed during the light phase and fasting in wild-type animals. However, cold exposure increased the expression during both dark and light phases in *Clock*-mutant mice.

7. Histological evidence on functional and anatomical connection between the circadian and thermoregulation systems

On the basis of the physiological findings, the association between the circadian and thermoregulation systems may be significant during fasting. The center of thermoregulation is thought to be in the hypothalamus [48, 49]. However, whether functional and anatomical neural connections between the SCN exist (the center of the circadian system) and the hypothalamic subregion is involved in the thermoregulation remains unclear.

Liu et al. [23] assessed the cFos (i.e., early gene expression protein) expression in the SCN of rats. cFos can be a marker of neuronal activation in the brain [50]. The number of cFos immunoreactive (cFos-IR) cells changes daily, which is higher during the light phase and smaller during the dark phase. A 4-day fasting did not change the phase difference in the number of cFos-IR cells. However, the number increased during the dark phase and decreased during the light phase. Whether such change in the SCN causes altered T_b rhythm during fasting remains unknown. However, results showed that the SCN receives some information that is associated with fasting and alters T_b rhythm and/or the thermoregulatory responses.

Tokizawa et al. [9] evaluated cFos expressions in the SCN and other hypothalamic areas in wild-type and *Clock*-mutant mice. The mice were exposed to settings with a temperature of 20°C and/or 48-h fasting. Neural associations were also observed between the SCN and the hypothalamic areas involved in thermoregulation. Data are summarized in **Figure 6**. Fasting increased the number of cFos-IR cells in the SCN in both wild-type and *Clock*-mutant mice. The number was smaller in *Clock*-mutant mice than in wild-type mice. cFos-IR cells also increased during fasting in other hypothalamic areas, such as the medial preoptic nucleus (MPO), dorsomedial hypothalamus (DMH), paraventricular nucleus (PVN), and dorsal subparaventricular zone (dSPZ), in wild-type mice. Differences were observed in the number of cFos-IR cells during the dark and light phases (**Figure 6A**, **B**, **D**, **E**). Small increases were also observed in the DMH, ARC, and PVN in *Clock*-mutant mice. However, no phase differences were observed. Neural outputs from the SCN, including the SPZ, reached these hypothalamic areas [51, 52]. Thus, the activation of the SCN during fasting may be linked with the activation of the hypothalamic areas.

The preoptic area in the hypothalamus is thought to be important in thermoregulation because it has several thermosensitive neurons in the core body and skin temperatures [53]. In addition, stimulatory and inhibitory signals are sent from the area to other brain areas, which regulate the effector organs of thermoregulation, such as vasodilation, shivering, and non-shivering thermogenesis [48, 49]. The sympathetic outflow may originate from the PVN [54–56]. The DMH receives thermal input from the skin and is associated with the control of BAT thermogenesis [57]. Whether all cFos-IR cells are associated with circadian change of T_b and/or thermoregulation is not verified. However, the changes in cFos expression between the phases may, in part, be responsible for the decrease in T_b that was observed during fasting and the light phase.

Fasting increased the cFos expression in the ARC in mice [58, 59]. The ARC is involved in the regulation of food intake and energy expenditure, which responds to peripheral nutritional signals, such as the levels of leptin and insulin [60]. However, no study on the direct association between ARC and thermoregulation was conducted. The ARC had phase differences in cFos expression in wild-type mice (**Figure 5C**), which increased during fasting. The ARC has neural input from the SCN [61–63]. Therefore, the fasting signals received by the ARC increased during the dark phase and were attenuated during the light phase in wild-type mice, and this may be related to the signals from the SCN.

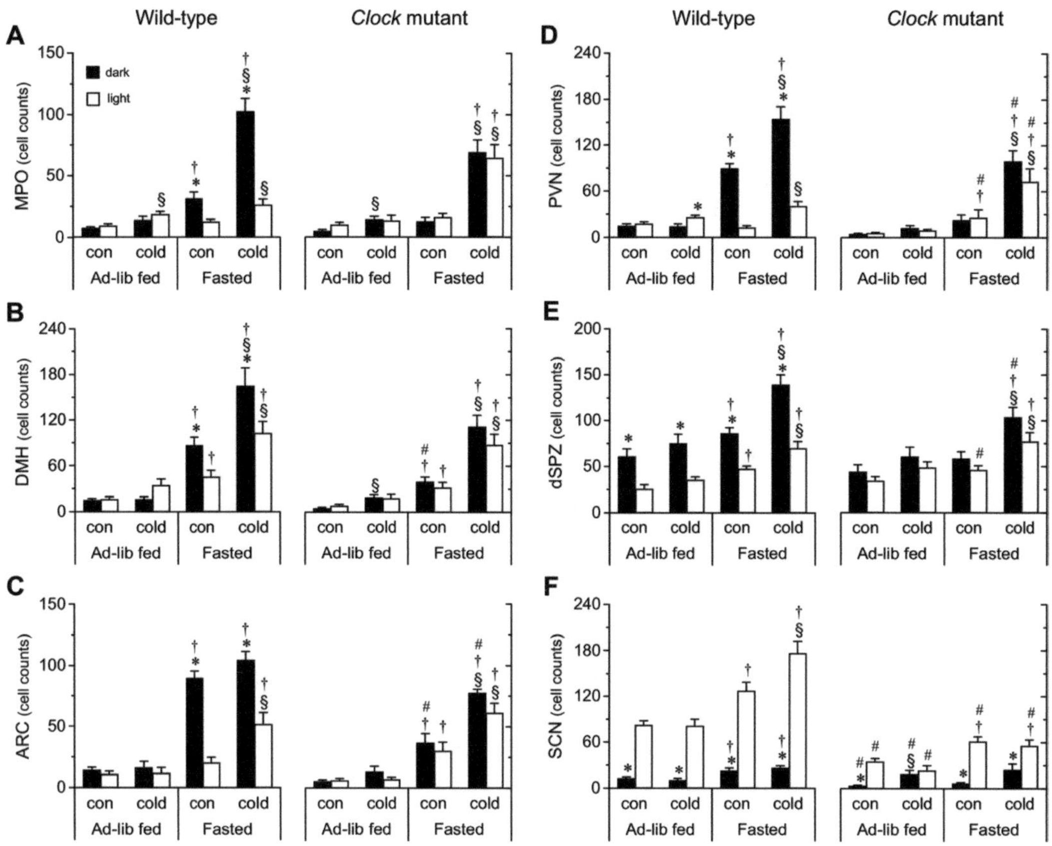

Figure 6. Counts of cFos-IR cells in the MPO (*A*), dorsomedial hypothalamus (*B*, DMH), ARC (*C*), PVN (*D*), dorsal subparaventricular zone (*E*, dSPZ), and SCN (*F*) under various conditions. Values are the means ± SEM ($n = 8$). $^{*}P < 0.05$, dark versus light phase; $^{§}P < 0.05$, 20°C (low-temperature setting) versus 27°C (control setting); $^{†}P < 0.05$, fasting versus *ad lib* feeding; $^{\#}P < 0.05$, wild-type versus *Clock*-mutant mice. The figure was obtained from the manuscript by Tokizawa et al. [9].

The increase in cFos expression in *Clock*-mutant mice that was attributed to low temperature during fasting was also observed in all the hypothalamic areas besides the SCN. However, no differences were observed between the two phases. In wild-type mice, the number of cFos expression in the SCN increased during the light phase. However, the increases in the MPO and PVN in wild-type mice were lower than those in Clock-mutant mice. Therefore, exposure to cold while fasting increases neural activity in the SCN during the light phase. In addition, normal molecular circadian mechanisms may be necessary for the response. On the basis of the results of cFos expression in the hypothalamic areas, the SCN may send inhibitory signals to the MPO and PVN, which may result in attenuated thermoregulatory responses. Moreover, since the inhibitory signals are stronger in some conditions, such as in fasting, the attenuation of thermoregulation becomes stronger.

On the basis of the speculated neural connection between the SCN and MPO and/or PVN, Tokizawa et al. [9] conducted an experiment, which directly evaluated the associations. A cholera toxin b-subunit was injected (CTb; monosynaptic retrograde neural tracer) to the MPO or PVN during the light phase, and the presence in the SCN was assessed 3 days later. Cold exposure

http://dx.doi.org/10.5772/intechopen.76229

and fasting were also conducted in the same manner as the previous cFos study. The injection of CTb in the PVN resulted in a widely spread labeling in both the dorsomedial and the ventrolateral parts of the SCN of both wild-type and *Clock*-mutant mice. In both mice, 5–10% of CTb-labeled neurons in the SCN were also cFos-positive at 27°C during the light phase and ad lib feeding. In the cold exposure during fasting, the ratio of the double-labeled neurons of CTb and cFos increased to 25–30% only in the wild-type mice. Moreover, the double-labeled neurons are GABAergic. When CTb was injected to the MPO, the dorsomedial part of the SCN was labeled. However, the ratio of the double-labeled neurons remained at the same level (15–20%) during ad lib feeding and during fasting and cold exposure. These results suggested that the SCN may send inhibitory signals to the PVN and MPO during the light phase. Fasting and/or cold exposure increased the inhibitory signals only in the PVN. Moreover, in the process, a normal molecular circadian mechanism would be necessary. The sympathetic outflow may originate from the PVN [54–56]. Therefore, such inhibitory signals may attenuate metabolic heat production and/or skin vasoconstriction (i.e., cold defense mechanisms), thus decreasing T_b during fasting.

8. Signals attenuating thermoregulatory responses during fasting

In Section 6 of the cFos study, the ARC also seems to play a key role in the circadian change of thermoregulation during fasting, which decreases plasma leptin and increases plasma ghrelin [64]. A reduction of leptin results in hypothermia [65]. Gluck et al. [66] showed that ghrelin induces hypothermia. Decreases in both hormones are signals activating neuropeptide Y (NPY) neurons in the ARC [67], which strongly reduces heat production. The receptors for leptin and ghrelin are found in NPY neurons [68, 69]. These experimental results suggest that leptins and/or ghrelins are the factors that modulate thermoregulatory heat production and decrease T_b during fasting. Moreover, the neurons in the SCN have leptin and ghrelin receptors [68, 70, 71]. In addition, the neural activity of the SCN is modulated in the presence of leptin and ghrelin [72, 73].

Tokizawa et al. [74] reported that the thermoregulatory response to cold was attenuated in *ob/ob* mice (genetically deficient of leptin). However, the response was not different between the light and dark phases. On the contrary, ghrelin injection to normal mice inhibited thermoregulatory heat production during cold exposure and reduced T_b. Such a response was observed only during the light phase. Therefore, ghrelin plays a key role in the phase-specific (light phase) modulation of thermoregulation and T_b during fasting.

The administration of ghrelin suppresses sympathetic nerve activity and iBAT temperature in rats [75, 76]. In the study by Tokizawa et al., ghrelin levels after the injection and 48-h fasting did not differ between the two phases. Therefore, a central mechanism that modulates the sensitivity to plasma ghrelin levels must exist, which may affect thermoregulatory responses to cold exposure. Ghrelin induced phase-specific changes in cFos expression in the hypothalamic areas: increased cFos-IR cells in the SCN during the light phase, the ARC during the dark phase, and the PVN during the cold exposure in the dark phase. Ghrelin injection activates NPY neurons in the ARC in both phases, and cFos-IR cell counts are higher in the dark than in the light phase. In the SCN, the ghrelin effect was limited to the light phase, and 25% of the cFos-IR cells were NPY neurons. NPY acts as nonphotic stimuli in the SCN [77]. The activation of the hypothalamic nuclei, that is, the SCN, ARC, and PVN, seems to be involved in the changes in the thermoregulatory metabolic heat production. **Figure 7** shows the summary of the findings.

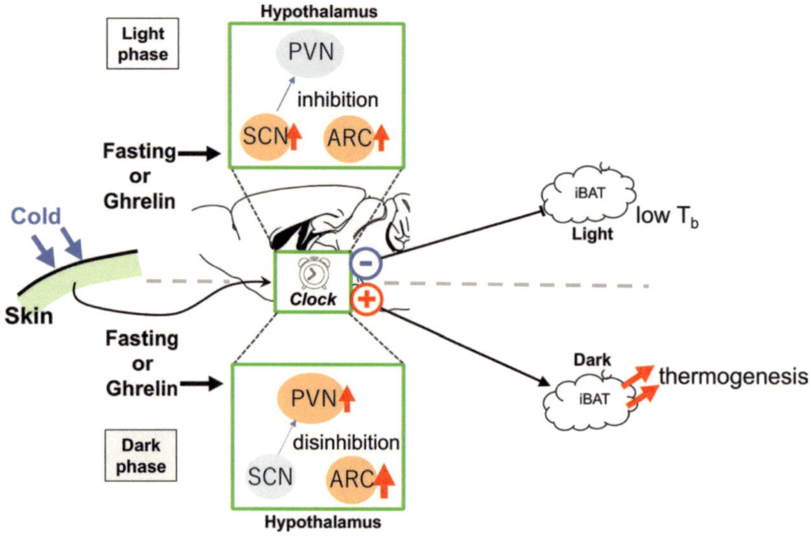

Figure 7. Summary of the histological findings together with physiological studies. During the light phase, fasting and an increase in plasma ghrelin level affect the hypothalamic areas. The activity of the suprachiasmatic nucleus (SCN) increases and that of the arcuate nucleus (ARC) relatively decreases. The SCN sends inhibitory signals to the paraventricular nucleus (PVN), which may result in a lower metabolic heat production of the interscapular brown adipose tissue (iBAT) and a lower body temperature. On the contrary, during the dark phase, the activity of the SCN decreases and that of the ARC relatively increases. The inhibitory signal from the SCN is less, and the PVN is activated. Metabolic heat production of the iBAT increases and body temperature is maintained.

9. Conclusion

This is a review article showing our previous series of studies involved in fasting-induced change in T_b rhythm. Interestingly, the change in T_b rhythm is a regulated phenomenon by molecular mechanisms such as *Cry and Clock* and neural mechanisms such as the SCN, MPO, and ARC in the hypothalamus. These studies are important in considering physiological importance and mechanism of the circadian body temperature and metabolic rhythms.

Acknowledgements

This research was partially supported by the Ministry of Education, Science, Sports, and Culture, Japan; Grant-in-Aid for challenging Exploratory Research, No. 16 K13055; and Ibuka Foundation, Waseda University.

Conflict of interest

There is no conflict of interest in this review.

Author details

Kei Nagashima[1*], Ken Tokizawa[2], Shuri Marui[1] and Yuki Uchida[3]

*Address all correspondence to: k-nagashima@waseda.jp

1 Body Temperature and Fluid Laboratory, Faculty of Human Sciences, Waseda University, Tokorozawa, Saitama, Japan

2 National Institute of Occupational Safety and Health, Kiyose, Tokyo, Japan

3 Women's Environmental Science Laboratory, Faculty of Human Life and Environment, Department of Health Sciences, Nara Women's University, Nara-City, Nara, Japan

References

[1] Tokizawa K, Yoda T, Uchida Y, et al. Estimation of the core temperature control during ambient temperature changes and the influence of circadian rhythm and metabolic conditions in mice. Journal of Thermal Biology. 2015;**51**:47-54. DOI: 10.1016/j.jtherbio.2015.03.005

[2] Yu X, Chang S, Song D, et al. Body temperature changes during 7 d −6 degrees head-down bed rest. Space Medicine & Medical Engineering (Beijing). 2002;**15**:167-169

[3] Aschoff J. Circadian control of body temperature. Journal of Thermal Biology. 1983;**8**: 143-147

[4] Ocobock C. Human energy expenditure, allocation, and interactions in natural temperate, hot, and cold environments. American Journal of Physical Anthropology. 2016; **161**:667-675. DOI: 10.1002/ajpa.23071

[5] Graf R, Krishna S, Heller H. Regulated nocturnal hypothermia induced in pigeons by food deprivation. The American Journal of Physiology. 1989;**256**:R733-738. DOI: 10.1152/ajpregu.1989.256.3.R733

[6] Rashotte M, Basco P, Henderson R. Daily cycles in body temperature, metabolic rate, and substrate utilization in pigeons: Influence of amount and timing of food consumption. Physiology & Behavior. 1995;**57**:731-746

[7] Sakurada S, Shido O, Sugimoto N, et al. Autonomic and behavioural thermoregulation in starved rats. The Journal of Physiology. 2000;**526**(Pt 2):417-424

[8] Yoda T, Crawshaw L, Yoshida K, Su L. Effects of food deprivation on daily changes in body temperature and behavioral thermoregulation in rats. The American Journal of Physiology. 2000;**278**:R134-139. DOI: 110.1152/ajpregu.2000.278.1.R134

[9] Tokizawa K, Uchida Y, Nagashima K. Thermoregulation in the cold changes depending on the time of day and feeding condition: Physiological and anatomical analyses

of involved circadian mechanisms. Neuroscience. 2009;**164**:1377-1386. DOI: 10.1016/j.neuroscience.2009.08.040

[10] Döring H, Schwarzer K, Nuesslein-Hildesheim B, Schmidt I. Leptin selectively increases energy expenditure of food-restricted lean mice. International Journal of Obesity and Related Metabolic Disorders. 1998;**22**:83-88

[11] Munch IC, Markussen NH, Óritsland NA. Resting oxygen consumption in rats during food restriction, starvation and refeeding. Acta Physiologica Scandinavica. 1993;**148**:335-340. DOI: 10.1111/j.1748-1716.1993.tb09564.x

[12] Overton JM, Williams TD, Chambers JB, Rashotte ME. Cardiovascular and metabolic responses to fasting and thermoneutrality are conserved in obese Zucker rats. American Journal of Physiology. Regulatory, Integrative and Comparative Physiology. 2001;**280**:R1007-R1015

[13] Rothwell NJ, Stock MJ. Effect of chronic food restriction on energy balance, thermogenic capacity, and brown-adipose-tissue activity in the rat. Bioscience Reports. 1982;**2**:543-549. DOI: 10.1007/BF01314214

[14] Sakurada S, Shido O, Sugimoto N. Autonomic and behavioural thermoregulation in starved rats. Journal of Physiology. 2000;**526**:417-424

[15] Williams TD, Chambers JB, May OL, et al. Concurrent reductions in blood pressure and metabolic rate during fasting in the unrestrained SHR. American Journal of Physiology. Regulatory, Integrative and Comparative Physiology. 2000;**278**:R255-R262

[16] Markussen NH, Oritsland NA. Metabolic depression and heat balance in starving Wistar rats. Comparative Biochemistry and Physiology. A, Comparative Physiology. 1986;**84**:771-776

[17] Phillips DL, Rashotte ME, Henderson RP. Energetic responses of pigeons during food deprivation and restricted feeding. Physiology & Behavior. 1991;**50**:195-203

[18] Severinsen T, Munch IC. Body core temperature during food restriction in rats. Acta Physiologica Scandinavica. 1999;**165**:299-305. DOI: 10.1046/j.1365-201x.1999.00488.x

[19] Nagashima K, Nakai S, Matsue K, et al. Effects of fasting on thermoregulatory processes and the daily oscillations in rats. The American Journal of Physiology-Regulatory, Integrative and Comparative Physiology. 2003;**284**

[20] Young AA, Dawson NJ. Evidence for on-off control of heat dissipation from the tail of the rat. Canadian Journal of Physiology and Pharmacology. 1982;**60**:392-398

[21] Abe K, Kroning J, Greer MA, Critchlow V. Effects of destruction of the suprachiasmatic nuclei on the circadian rhythms in plasma corticosterone, body temperature, feeding and plasma thyrotropin. Neuroendocrinology. 1979;**29**:119-131

[22] Eastman CI, Mistlberger RE, Rechtschaffen A. Suprachiasmatic nuclei lesions eliminate circadian temperature and sleep rhythms in the rat. Physiology & Behavior. 1984;**32**:357-368

[23] Liu S, Chen XM, Yoda T, et al. Involvement of the suprachiasmatic nucleus in body temperature modulation by food deprivation in rats. Brain Research. 2002;**929**:26-36

[24] Moore RY, Eichler VB. Loss of a circadian adrenal corticosterone rhythm following suprachiasmatic lesions in the rat. Brain Research. 1972;**42**:201-206

[25] Refinetti R, Kaufman CM, Menaker M. Complete suprachiasmatic lesions eliminate circadian rhythmicity of body temperature and locomotor activity in golden hamsters. Journal of Comparative Physiology. A. 1994;**175**:223-232

[26] Ruby NF, Dark J, Burns DE, et al. The suprachiasmatic nucleus is essential for circadian body temperature rhythms in hibernating ground squirrels. The Journal of Neuroscience. 2002;**22**:357-364

[27] Martin RD. Body temperature, activity and energy costs. Nature. 1980;**283**:335-336

[28] Refinetti R. Contribution of locomotor activity to the generation of the daily rhythm of body temperature in golden hamsters. Physiology & Behavior. 1994;**56**:829-831. DOI: 10.1016/0031-9384(94)90251-8

[29] Stirling J, Stock M. Metabolic origins of thermogenesis induced by diet. Nature. 1968;**220**:801-802. DOI: 10.1038/220801a0

[30] Ibuka N, Kawamura H. Loss of circadian rhythm in sleep-wakefulness cycle in the rat by suprachiasmatic nucleus lesions. Brain Research. 1975;**96**:76-81

[31] Balsalobre A, Brown SA, Marcacci L, et al. Resetting of circadian time in peripheral tissues by glucocorticoid signaling. Science. 2000;**289**:2344-2347. DOI: 10.1126/SCIENCE.289.5488.2344

[32] Balsalobre A, Damiola F, Schibler U. A serum shock induces circadian gene expression in mammalian tissue culture cells. Cell. 1998;**93**:929-937

[33] Damiola F, Le Minh N, Preitner N, et al. Restricted feeding uncouples circadian oscillators in peripheral tissues from the central pacemaker in the suprachiasmatic nucleus. Genes & Development. 2000;**14**:2950-2961. DOI: 10.1101/GAD.183500

[34] Nagashima K, Matsue K, Konishi M, et al. The involvement of Cry1 and Cry2 genes in the regulation of the circadian body temperature rhythm in mice. American Journal of Physiology. Regulatory, Integrative and Comparative Physiology. 2005;**288**:R329-R335. DOI: 10.1152/ajpregu.00395.2004

[35] Griffin EA, Staknis D, Weitz CJ. Light-independent role of CRY1 and CRY2 in the mammalian circadian clock. Science. 1999;**286**:768-771. DOI: 10.1126/SCIENCE.286.5440.768

[36] Lucas RJ, Foster RG. Circadian clocks: A cry in the dark? Current Biology. 1999;**9**:R825-R828

[37] Stanewsky R, Kaneko M, Emery P, et al. The cryb mutation identifies cryptochrome as a circadian photoreceptor in drosophila. Cell. 1998;**95**:681-692

[38] Hoeijmakers JHJ, Horst GTJ van der, Muijtjens M, et al. Mammalian Cry1 and Cry2 are essential for maintenance of circadian rhythms. Nature. (1999);**398**:627-630. DOI: 10.1038/19323

[39] Bonnefont X, Albus H, Meijer JH, van der Horst GTJ. Light signalling in cryptochrome-deficient mice. Novartis Foundation Symposium. 2003;**253**:56-66-72, 102-9

[40] Peek CB, Affinati AH, Ramsey KM, et al. Circadian clock NAD+ cycle drives mitochondrial oxidative metabolism in mice. Science. 2013;**342**:123417. DOI: 10.1126/science.1243417

[41] Rutter J. Regulation of clock and NPAS2 DNA binding by the redox state of NAD cofactors suplemental online material. Science. 2001;**293**:510-515

[42] Nakahata Y, Sahar S, Astarita G, et al. Circadian control of the NAD+ salvage pathway by CLOCK-SIRT1. Science. 2009;**324**:654-657. DOI: 10.1126/science.1170803

[43] Ramsey KM, Yoshino J, Brace CS, et al. Circadian clock feedback cycle through NAMPT-mediated NAD+ biosynthesis. Science. 2009;**324**:651-654. DOI: 10.1126/science.1171641

[44] Vitaterna MH, King DP, Chang AM, et al. Mutagenesis and mapping of a mouse gene, clock, essential for circadian behavior. Science. 1994;**264**:719-725

[45] Jakus PB, Sipos K, Kispal G, Sandor A. Opposite regulation of uncoupling protein 1 and uncoupling protein 3 in vivo in brown adipose tissue of cold-exposed rats. FEBS Letters. 2002;**519**:210-214

[46] Denjean F, Lachuer J, Geloen A, et al. Differential regulation of uncoupling protein-1, −2 and −3 gene expression by sympathetic innervation in brown adipose tissue of thermoneutral or cold-exposed rats. FEBS Letters. 1999;**444**:181-185

[47] Talan MI, Kirov SA, Clow LA, Kosheleva NA. Cold acclimation-associated changes in brown adipose tissue do not necessarily indicate an increase of nonshivering thermogenesis in C57BL/6J mice. Physiology & Behavior. 1996;**60**:1285-1289

[48] Nagashima K, Nakai S, Tanaka M, Kanosue K. Neuronal circuitries involved in thermoregulation. Autonomic Neuroscience. 2000;**85**:18-25. DOI: 10.1016/S1566-0702(00)00216-2

[49] Kanosue K, Crawshaw LI, Nagashima K, Yoda T. Concepts to utilize in describing thermoregulation and neurophysiological evidence for how the system works. European Journal of Applied Physiology. 2010;**109**:5-11. DOI: 10.1007/s00421-009-1256-6

[50] Sagar SM, Sharp FR, Curran T. Expression of c-fos protein in brain: Metabolic mapping at the cellular level. Science. 1988;**240**:1328-1331

[51] Leak RK, Moore RY. Topographic organization of suprachiasmatic nucleus projection neurons. The Journal of Comparative Neurology. 2001;**433**:312-334

[52] Kalsbeek A, Palm IF, La FSE, et al. SCN outputs and the hypothalamic balance of life. Journal of Biological Rhythms. 2006;**21**:458-469

[53] Yoshida K, Konishi M, Nagashima K, et al. Fos activation in hypothalamic neurons during cold or warm exposure: Projections to periaqueductal gray matter. Neuroscience. 2005;**133**:1039-1046. DOI: 10.1016/j.neuroscience.2005.03.044

[54] Freeman PH, Wellman PJ. Brown adipose tissue thermogenesis induced by low level electrical stimulation of hypothalamus in rats. Brain Research Bulletin. 1987;**18**:7-11. DOI: 10.1016/0361-9230(87)90026-8

[55] Amir S. Stimulation of the paraventricular nucleus with glutamate activates interscapular brown adipose tissue thermogenesis in rats. Brain Research. 1990;**508**:152-155. DOI: 10.1016/0006-8993(90)91129-5

[56] Yoshimatsu H, Egawa M, Bray GA. Sympathetic nerve activity after discrete hypothalamic injections of l-glutamate. Brain Research. 1993;**601**:121-128. DOI: 10.1016/0006-8993(93)91702-T

[57] Nakamura K, Morrison SF. A thermosensory pathway that controls body temperature. Nature Neuroscience. 2008;**11**:62-71. DOI: 10.1038/nn2027

[58] Wang L, Martinez V, Barrachina MDTY. Fos expression in the brain induced by peripheral injection of CCK or leptin plus CCK in fasted lean mice. Brain Research. 1998;**791**:157-166. DOI: 10.1016/S0006-8993(98)00091-2

[59] Morikawa Y, Ueyama E, Senba E. Fasting-induced activation of mitogen-activated protein kinases (ERK/p38) in the mouse hypothalamus. Journal of Neuroendocrinology. 2004;**16**:105-112

[60] Morton GJ, Cummings DE, Baskin DG, et al. Central nervous system control of food intake and body weight. Nature. 2006;**443**:289-295. DOI: 10.1038/nature05026

[61] Saeb-Parsy K, Lombardelli S, Khan FZ, et al. Neural connections of hypothalamic neuroendocrine nuclei in the rat. Journal of Neuroendocrinology. 2000;**12**:635-648

[62] Van der Beek EM, Horvath TL, Wiegant VM, et al. Evidence for a direct neuronal pathway from the suprachiasmatic nucleus to the gonadotropin-releasing hormone system: Combined tracing and light and electron microscopic immunocytochemical studies. The Journal of Comparative Neurology. 1997;**384**:569-579

[63] Yi CX, van der Vliet J, Dai J, et al. Ventromedial arcuate nucleus communicates peripheral metabolic information to the suprachiasmatic nucleus. Endocrinology. 2006;**147**:283-294

[64] Luque RM, Park S, Kineman RD. Severity of the catabolic condition differentially modulates hypothalamic expression of growth hormone-releasing hormone in the fasted mouse: Potential role of neuropeptide Y and corticotropin-releasing hormone. Endocrinology. 2007;**148**:300-309. DOI: 10.1210/en.2006-0592

[65] Freeman DA, Lewis DA, Kauffman AS, et al. Reduced leptin concentrations are permissive for display of torpor in Siberian hamsters. American Journal of Physiology. Regulatory, Integrative and Comparative Physiology. 2004;**287**:R97-R103. DOI: 10.1152/ajpregu.00716.2003

[66] Gluck EF, Stephens N, Swoap SJ. Peripheral ghrelin deepens torpor bouts in mice through the arcuate nucleus neuropeptide Y signaling pathway. American Journal of Physiology.

Regulatory, Integrative and Comparative Physiology. 2006;**291**:R1303-R1309. DOI: 10.1152/ajpregu.00232.2006

[67] Morton GJ, Schwartz MW. The NPY = AgRP neuron and energy homeostasis. International Journal of Obesity. 2001;**25**:56-62

[68] Mercer JG, Hoggard N, Williams LM, et al. Localization of leptin receptor mRNA and the long form splice variant (Ob-Rb) in mouse hypothalamus and adjacent brain regions by in situ hybridization. FEBS Letters. 1996;**387**:113-116

[69] Willesen MG, Kristensen P, Rømer J. Co-localization of growth hormone secretagogue receptor and NPY mRNA in the arcuate nucleus of the rat. Neuroendocrinology. 1999;**70**:306-316

[70] Guan XM, Hess JF, Yu H, et al. Differential expression of mRNA for leptin receptor isoforms in the rat brain. Molecular and Cellular Endocrinology. 1997;**133**:1-7

[71] Zigman JM, Jones JE, Lee CE, et al. Expression of ghrelin receptor mRNA in the rat and the mouse brain. The Journal of Comparative Neurology. 2006;**494**:528-548. DOI: 10.1002/cne.20823

[72] Inyushkin AN, Bhumbra GS, Dyball REJ. Leptin modulates spike coding in the rat suprachiasmatic nucleus. Journal of Neuroendocrinology. 2009;**21**:705-714. DOI: 10. 1111/j.1365-2826.2009.01889.x

[73] Yannielli PC, Molyneux PC, Harrington ME, Golombek DA. Ghrelin effects on the circadian system of mice. The Journal of Neuroscience. 2007;**27**:2890-2895. DOI: 10.1523/ JNEUROSCI.3913-06.2007

[74] Tokizawa K, Onoue Y, Uchida Y, Nagashima K. Ghrelin induces time-dependent modulation of thermoregulation in the cold. Chronobiology International. 2012;**29**:736-746. DOI: 10.3109/07420528.2012.678452

[75] Mano-Otagiri A, Ohata H, Iwasaki-Sekino A, et al. Ghrelin suppresses noradrenaline release in the brown adipose tissue of rats. The Journal of Endocrinology. 2009;**201**:341-349. DOI: 10.1677/JOE-08-0374

[76] Yasuda T, Masaki T, Kakuma T, Yoshimatsu H. Centrally administered ghrelin suppresses sympathetic nerve activity in brown adipose tissue of rats. Neuroscience Letters. 2003;**349**:75-78. DOI: 10.1016/S0304-3940(03)00789-4

[77] Gamble KL, Ehlen JC, Albers HE. Circadian control during the day and night: Role of neuropeptide Y Y5 receptors in the suprachiasmatic nucleus. Brain Research Bulletin. 2005;**65**:513-519. DOI: 10.1016/j.brainresbull.2005.03.004

Chapter 3

Sex and Sex Hormones in Tissue Homeostasis

Judith Lechner and Gerhard Gstraunthaler

Additional information is available at the end of the chapter

http://dx.doi.org/10.5772/intechopen.76177

Abstract

Women are not small men. Sex-specific differences do not only affect the classical target organs of sexual differentiation and reproduction, but have been found to involve most, if not all the organs and tissues in the body. One of the consequences of this dimorphism is that diseases manifest in a sex- and gender-specific way. Key to maintenance of a healthy state is functioning tissue able to cope with insults. Regulated death of damaged cells and replacement with new cells by proliferation is a prerequisite for maintaining tissue function taking place at different pace in the different organs. The intent of this chapter is to review current evidence for sex-specific differences in tissue homeostasis focusing on the variability of hormone exposure characteristic for the female reproductive life stages.

Keywords: tissue maintenance, sex differences, proliferation, cell death, kidney, menstrual cycle

1. Introduction

Living systems are continuously challenged by potentially toxic internal and external processes. The normal metabolic function of the cells produces a plethora of potentially damaging oxidative metabolites inducing damage in DNA, proteins, and lipids. In addition, living cells are exposed to a variety of external factors, which may be internalized as building blocks and/or energy sources. These vital processes put the organism at risk to be harmed. Coping strategies are necessary to avoid damage. There are several lines of cellular defenses induced via cell stress pathways, including compartmentalization processes, enzymatic modification, externalization, degradation, and repair [1, 2].

Ultimately, these processes may not be sufficient to prevent major cellular damage. Therefore, every cell is in addition equipped with internal cell death programs, which can be activated in order to prevent a damaged cell to cause harm to the organism [3]. Cell losses are inevitable and take place continuously in our bodies even without a specific trigger. The rate of cell death may be significantly enhanced at times of increased challenges. Cellular losses are necessary in order to prevent detrimental effects like neoplastic transformation [4]. Thus, programmed cell death needs to be carefully balanced, and this ability is a key determinant for the health and survival of the organism.

Cells lost by cell death need to be replaced in order to maintain cell numbers and ultimately tissue function. Controlled regeneration is, thus, required to cope for cell losses due to toxic challenges derived from internal and external sources, be they derived from normal metabolic processes or damaging environmental stimuli. Cell losses and proliferation need to be carefully balanced in order to guarantee proper function. Thus, the process of tissue homeostasis, that is, the capability to send damaged cells into cell death programs, to replace the cells by proliferation, and to regulate the exact balance of these events are crucial processes for preserving a healthy state. Any distortion of the balance between cell death and cell proliferation—be it by overwhelming damaging events beyond the host's range of tolerance and/or primarily ineffective or maladaptive homeostatic mechanisms by the host—is prone to induce malfunctioning of the organs in the body ultimately causing disease and potentially death.

2. Sex and sex hormones in tissue homeostasis

2.1. Origin of sex differences

During development, sex differences originate from genetic and hormonal influences. Master regulators for male sex differentiation, like SRY, are encoded by DNA of the Y chromosome governing the embryonic development of the male phenotype in mammals. Female or male gonadal development gives rise to a sex-specific hormonal environment [5]. Sex hormones induce organizational effects during the life span causing persistent sex-specific changes within the tissues, for example, by epigenetic modifications [6, 7]. Activational effects further introduce sex differences in tissue structure and function depending on the pattern of exposure to gonadal hormones. All the organs in the body are affected throughout life [8]. In this respect, sex differences are based on the different chromosomal equipment that qualifies every cell in the body as male or female. These basic differences are further shaped by sex hormones depending on previous or current, transient or persistent exposure [9]. This hypothesis was phrased by Arnold [10] as follows: "XX and XY cells are different prior to the secretion of gonadal hormones, and gonadal hormones affect XX and XY cells unequally."

Sex hormones act through receptors widely expressed throughout the cells of the body. The classical estrogen (ERα, ERβ), androgen (AR) and progesterone (PR) receptors belong to the nuclear receptor protein family acting through the nucleus as transcription factor or co-factor. In addition to their nuclear actions, they were found to be localized to the cell membrane and mitochondria inducing fast, non-genomic intracellular signaling pathways, for example, by the interaction with growth factor or cytokine receptors. Estrogen-binding cell membrane-localized receptors of the seven transmembrane receptor family were also characterized, for example, G-protein coupled estrogen receptor, GPR30/GPER-1 [11–13].

Ultimately, differences between the sexes derive from chromosomal and hormonal sex differences, which are further influenced by environmental factors. Thus, differences originating from the biological sex are further shaped by gender, which refers to the perceptions of male or female identity and depends on sex-based social structures [8].

2.2. Influence of sex and sex hormones on cellular proliferation

Tissue homeostasis is guaranteed, when cells lost in physiological tissue turnover or under stress conditions are replaced by proliferation. Organs with high demanding functions have increased regeneration potential and continually renew their cell populations. This is the case for intestine, skin, and blood, for example. Liver, bone, and blood even have the capacity to fully recover to the original size after loss of tissue [14]. Other organs have lower regeneration potential, like the heart, brain, and kidney [15–17]. Many organs contain stem cell niches hosting adult tissue stem cells that are precursor cells maintained in a relatively undifferentiated state ready to replace lost cells by proliferation followed by differentiation [18, 19].

Sex hormones have classically been implicated in regulation of proliferation of cells of reproductive organs and cancer of reproductive tissue [20–24]. Besides these effects, sex hormones were also found to have pronounced effects on the proliferation of different stem cell populations. Cell proliferation of embryonic stem (ES) cells was found to be enhanced by female gonadal hormones [25]. ES cells are derived from the inner cell mass of the embryoid body. They can self-renew in vitro and are pluripotent, that is, they can differentiate into all the cell types of the body [26]. Estrogen appears to act via nuclear and cell surface signaling pathways involving Erk1/2 activation, cyclin-dependent kinases and proto-oncogenes like c-myc, c-fos, c-jun, and pRB in ES cells. In addition, store-operated calcium channels were found to play a role in estrogen-mediated cell proliferation through the transcription factor NF-AT [27].

Differentiation of ES cells into dopaminergic neurons was also shown to be affected by estrogen. ERβ promoted differentiation by crosstalk signaling with insulin like growth factor-1 [28]. Motor neuron differentiation from ES cells was found to be enhanced by 17-β estradiol and progesterone through nuclear ERα and progesterone receptor [29]. Dopaminergic precursors derived from ES cells were found to increase proliferation upon treatment with progesterone in vitro [30].

Induced pluripotent stem (iPS) cells are similar to ES cells with regard to their ability to differentiate into all cell types, providing a promising tool for in vitro research and regenerative medicine. They are derived from adult mature cells by reprogramming through the introduction of specific transcription factors [31]. Similar to ES cells, sex hormones were shown to affect iPS cells. Neuronal cells derived by differentiation of iPS cells showed increased dendritic branching by treatment with 17-β estradiol [32]. Functional integration of dopaminergic neuronal cells from iPS cells into neuronal circuits was found to be enhanced by estradiol [33]. Testosterone was described to enhance differentiation of iPS cells into insulin-producing cells [34].

Sex differences were also described for tissue stem cells in vivo. Adult stem cells are believed to provide a local pool of self-renewing, multipotent cells pivotal in tissue homeostasis and recovery upon damage [35]. Stem cells in many stem cell niches appear to have a higher ability to self-renew, have an increased regeneration potential, and in some cases, show higher proliferative activity in women [36, 37]. Intrinsic sexual dimorphism was described for neural stem cells that hold much promise for potential brain damage repair therapy in the future. Proliferation of neural stem cell was, for example, shown to depend on hormone changes in the adult mouse due to the estrous cycle, pregnancy, reproductive status, and age. Phases of high estrogen exposure like pro-estrus were found to be associated with increased hippocampal adult neurogenesis indicating a role of estrogens [38]. Differential expression of sex steroid receptors and androgen metabolizing enzymes may result in differential outcomes in neural stem cell transplantation [39]. Neural stem cell proliferation was found to be dependent on nuclear ERs, while oligodendroglial differentiation was stimulated by cell membrane-associated ERs [40]. Other researcher also proposed that actions of sex steroids on the brain might be correlated with reduced brain damage. Intact females were found to be less susceptible upon injury than ovariectomized females and males [41]. Similarly, muscle-derived stem cells derived from female mice and transplanted into dystrophic mutant mice showed a better potential to regenerate skeletal muscle than stem cells from males [42].

Hematopoietic stem cells were found to be more abundant and proliferative in female mice in comparison to males dependent on estrogen exposure [43]. 17-β estradiol was found to improve hematopoietic differentiation from human iPS cells and from human umbilical cord blood through ERα signaling suggesting a universal function for estrogen in hematopoietic stem cell differentiation [44, 45].

Estrogens have beneficial effects on bone regeneration [46]. Osteoblasts are stimulated by estrogen to proliferate with distinct roles for ERα and ERβ [47]. In vitro, proliferation of bone marrow mesenchymal stromal cells was found to be enhanced by estrogen [48]. Estrogens enhanced the proliferation and migration of bone marrow-derived endothelial progenitor cells to ischemic regions of the heart facilitating repair and regeneration [49]. Androgens were also described to stimulate the proliferation and angiogenesis/vascular repair capability of circulating endothelial progenitor cells in males, not females [50].

2.3. Sex differences in cell survival

When progenitor cells involved in tissue regeneration enter a cell senescence state, tissue homeostasis may be compromised. The cells are able to permanently halt the cell cycle and persist in a quiescent, but still functional state [51]. This is a possible fate of cells damaged beyond repair. The three major types of senescence are replicative senescence, oncogene-induced senescence, and DNA damage-induced DNA damage. The DNA damage response pathway appears to be eventually involved in the execution of the program independent of the primary stimulus [52]. Furthermore, senescent cells are able to influence their neighboring cells by secretion of a range of activating signals referred to as senescence-associated secretory phenotype. The signals may favor a pro-inflammatory or—alternatively—an immunosuppressive/pro-fibrotic state. Both phases appear to be important for successful tissue repair and the timing of the shift in the secretome might be crucial [53]. The etiology and progression of many cancerous or age-related diseases have been shown to be influenced by the secretome of senescence cells [54–56].

Alternatively, cells may activate a cell death program as a means to ensure physiological tissue renewal or in response to overwhelming damage. The most common are type I cell death programs or apoptosis, type II or autophagy, type III or necrosis, and mitotic catastrophe [57, 58]. The cell death modalities are characterized by different morphological criteria and are executed by specific intracellular signaling cascades. Specific catabolic enzymes are typically associated with specific forms of cell death, for example, caspases with apoptosis. The pathways are interdependent. The intensity of the damage signal is often decisive for the type of cell death program that is executed or the switch from one modality to the other. In addition, autophagy is not primarily regarded as a cell death mechanism. Autophagy describes a process involving the break-down and recycling of specific subcellular organelles. This process may provide a cell survival strategy by reducing damaged organelles and/or shifting internal resources in order to optimize cell survival. Only if the damaging process exceeds the cellular defenses, cells die in the process [59, 60].

Regarding the role of sex in cell fate decisions, several reports have highlighted distinct sex-dependent differences. Sex hormones have been shown to influence the propensity of cells to undergo apoptosis. In general, lower concentrations of estrogen were found to be protective, while higher concentrations were found to promote apoptosis. Androgens were found to enhance, but also to suppress apoptosis depending on the cellular context [61, 62]. For example, estrogen and testosterone were described to reduce apoptosis in skeletal muscle cells [63]. Both hormones also appear to prevent apoptosis in neuronal cells adding to their neuroprotective function [64, 65]. An anti-apoptotic action of testosterone was also described in pancreatic β cells from male rats, but not from female rats [66]. Estrogen and estrogenic compounds, however, appeared to enhance apoptosis in pancreatic β cells in elderly mice, while it reduced apoptosis in young animals [67]. Regarding vascular endothelial cells, several studies have shown that estrogens protect from apoptotic cell death [68, 69], while apoptosis increased in coronary artery endothelia from postmenopausal women [70]. Testosterone was found to induce apoptosis in endothelial cells [71–73]. Treatment with testosterone also induced apoptosis or senescence in human dermal papilla cells, a process implied in inherited male alopecia [74, 75]. In addition, androgens were found to promote apoptosis in renal and intestinal cell lines and bone marrow-derived macrophages [76–78].

Overall, sex hormones appear to influence cell fate decisions depending on the cell context. Hormone independent sex differences are also apparent shaping the cellular response [79]. Thus, female and male cells appear to rely on different coping strategies in response to stressors. For example, vascular smooth muscle cells isolated from aorta of male rats appear to be more inclined to undergo apoptosis in response to UV irradiation, while female cells are more prone to execute the cell senescence program [62, 80]. Female cells showed characteristics of autophagy, which is presumed to help female cells to repair the UV-induced intracellular damages ultimately providing a survival strategy [81]. In addition, female cells were found to better adhere to the growth support, thus avoiding apoptotic cell death initiation by cell detachment, a process called anoikis-resistance. Differences in the intracellular organization of the actin cytoskeleton and increased phosphorylation of focal adhesion kinase were attributed to this higher propensity of female cells to adhere [81]. Apparently, female cells are better equipped to prevent cell death. While autophagic processes were found to protect neuronal cells from cell death due to starving in female rats, male cells were not able to benefit and died more often from autophagic cell death [82]. Organ-specific sex differences were found in constitutive autophagy, a process implicated in physiological tissue turnover. While autophagic marker proteins were increased in the male versus female heart and liver, no such differences were observed in the kidneys [83]. Osteoblasts showed reduced autophagy in aging female mice, while the rate remained constant in males over the life span. This was correlated with higher oxidative stress in female cells, thus potentially enhancing bone loss and playing a role in the pathophysiology of osteoporosis in women [84]. Estrogens alter the redox balance and counteract bone loss [46]. Stem cells involved in generation of osteoblast, namely bone marrow derived mesenchymal stem cells, were found to be influenced by estrogens not only inducing increased proliferation, but also reducing senescence and apoptosis [85].

In general, increased antioxidative cellular defenses were implicated to provide females with better strategies to cope with oxidative stress and prevent cellular losses [62]. Differences in basal redox state and responses to oxidative imbalance were demonstrated between female and male cells [86]. For example, female cells were shown to produce less hydrogen peroxide and superoxide anion. Anti-oxidative enzymes, such as superoxide dismutase (SOD) and catalase, showed higher basic activity in female versus male cells [87]. Thioredoxin reductases and manganese SOD were increased by estrogen in cardiomyocytes [88, 89]. In vascular smooth muscle cells and circulating monocytes, estrogen was found to stimulate manganese and extracellular SOD expression [90]. Estrogen was, furthermore, shown to modulate the expression of other key molecular defense enzymes differently in XX and XY cells, for example, poly-ADP ribose polymerase (PARP), a DNA damage repair enzyme, or RLIP76, a cell-protective transporter protein [86].

PARP was also found to play a major role in sex differences in stroke. Experiments in mice have shown that ischemic neuronal cell death is dependent on intact neuronal nitric oxide synthase (nNOS)/PARP signaling, while in females a protection is provided by estrogen paradoxically also requiring an intact nNOS/PAPR axis [91]. While male neuronal cells appear to die via a PARP-mediated caspase independent pathway, ischemic cell death pathways appear to be dependent on activation of caspase-dependent cell death pathways in females [9, 92]. Such sex-specific differences may be relevant for the sex-specific difference in stroke prevalence [93, 94].

PARP signaling was also implied in sex differences in cell fate decisions in kidney cells. In a mouse model of immune-mediated nephritis, PARP signaling induced necrosis in male cells and inhibition of PARP shifted the pro-inflammatory necrotic cell death to an anti-inflammatory apoptotic pathway. In female cells, by contrast, cell death was independent of PARP and female cells preferentially underwent apoptosis. Estrogen acted in a pro-survival manner in female cells only. In addition to the kidney cells, bone marrow-derived hematopoietic cells showed similar sex differences [95].

Mitochondria play a crucial role in apoptotic cell death programs. Estrogens were described to modulate the propensity for mitochondrial initiation of apoptosis [61, 96]. Estrogen-mediated modulation of mitochondrial function is achieved by hormone effects on the expression of mitochondrial and nuclear genome-encoded mitochondrial proteins [97–101]. Since mitochondria are central in the cellular defense against oxidative stress, mitochondria are especially sensitive to accumulate damage over time. Malfunctioning mitochondria accumulate during aging, a process regarded as a major contributor to the onset of many age-related diseases [102, 103]. Sex differences were observed in this process. Delayed malfunctioning of mitochondria during the aging process might provide females with better strategies to cope with cellular stressors. Maternal transmission of mitochondria appears to provide a more favorable environment in female offspring [104]. Xist, an RNA-coding gene involved in X chromosome inactivation in female cells, appears to be pivotal for mitochondrial maintenance [105]. Mitochondrial biogenesis and degradation by mitophagy are dependent on the transcription factors p53 and FOXO [106, 107]. Sex-specific differences in the activity of these nuclear factors were reported. Males were shown to exhibit relatively greater FOXO activity. Females, on the other hand, had higher p53 activity resulting in sex-specific differences in the ability to maintain healthy mitochondrial functionality during aging [105].

2.4. Potential consequences of sex-specific differences in tissue homeostasis

The abovementioned paragraphs have described examples of sex-specific differences regarding processes involved in tissue maintenance, like the control of cell proliferation and cell death. Such effects may ultimately result in differences in the ability of female and male tissues to cope with stressors affecting the ability to repair and restore function or develop disease. Many diseases show different incidence and prevalence rates in men and women derived from sex and gender specific pathophysiological mechanisms. Sex and gender differences have been studied intensively in the neural system, the cardiovascular system, and the development of cancer, among others [108–110].

Mechanisms underlying differences in kidney diseases between men and women are less well known, despite renal diseases with a high morbidity and mortality risk being a challenging problem for patients, clinicians and society [111, 112]. International registries show that fewer women than men develop kidney failure [113–116]. The underlying causes, however, are widely unknown. The presumed female protective effects appear to be most pronounced in women of reproductive age [117–120]. This finding suggests that female sex hormones might play a key role. Estrogen was proposed to be renoprotective via modulating renal perfusion and effects on the vasculature. Furthermore, a role of estrogen was proposed in the control of the local renal renin-angiotensin system [121–123]. On the other hand, estrogen was implicated in the control of mesangial and tubular cell proliferation and linked to neoplastic transformation of the kidney in hamster kidneys. Low estrogen concentrations were shown to induce proliferation in glomerular mesangial cell, while high concentrations suppressed it [124]. Primary proximal tubular cell explants and subcultured dissociated proximal tubular cells were shown to proliferate, when treated with estrogen at physiologic concentrations [125]. This finding was confirmed in primary rabbit proximal tubular cells, which showed increased proliferation upon estrogen treatment [126].

We have previously shown that renal tubular cell-specific proteins appear at higher rates in the urine of healthy women at specific hormonal transition phases of the natural ovulating menstrual cycle. Urinary samples from healthy probands showed increased rates of urinary excretion of the marker proteins Fructose-1,6-bisphosphatase and Glutathione-S-transferase α, when estrogen levels decreased after a preceding height associated with ovulation and luteal phase [127]. Both enzymes are specifically found in proximal tubular cells, the most populous cell type in the kidney. When proximal tubular cells are damaged, intracellular enzymes are released into the urine, making them clinical markers for kidney injury. In contrast to ovulating women, male probands and postmenopausal women showed consistently low levels of these renal marker proteins over time. Other urinary proteins, for example, albumin, α1-microglobulin, and immunglobulin G, which are markers for functional changes of the glomerular filter and/or tubular protein resorption, showed constant urinary excretion suggesting that the observed increases of proximal tubular marker protein release in ovulating women are not accompanied by major functional distress of the kidneys [127]. This pattern of urinary marker proteins excretion suggests that cyclical changes of female hormones might affect kidney cell health. Tubular enzymes are released into the urine, if proximal tubular cells are sloughed off and/or their plasma membranes become leaky. This could be due to tubular cells being transiently more prone to damage in situ resulting in plasma membrane leakage or to the cells being removed from the tubular epithelium and released into the urinary space, for example, by apoptosis. Both processes lead to increased cell losses. Tissue homeostasis would be maintained, if increased cell removal was accompanied by increased cell proliferation. This could be the case during the high estrogen exposure phases preceding the observed tubular enzyme releases into the urine. The finding that tubular cells are able to proliferate upon estrogen treatment [126] is in line with this hypothesis. Such a periodic interplay between cell proliferation and cell loss brought about by the specific changes in the pattern of sex hormone exposure might result in an increased rate of tissue renewal. If this was the case, then women in their reproductive years would possess an efficient means to easily get rid of potentially injured, dysfunctional or simply older proximal tubular cells by replacing them with fresh new cells. Such a transiently increased repair capacity might provide an efficient means to cyclically renew renal tubular tissue leading to a higher resistance to damage. It is, however, also possible that during the short phases of increased tubular cell death, the kidneys might be especially sensitive to damage. With regard to the potential beneficial action of treatment of renal proximal tubular cells with a proliferation-inducing growth factor, we have previously demonstrated that epidermal growth factor (EGF) treatment was able to accelerate tissue repair after treatment with interferon α (IFNα) in vitro. However, if EGF was present before or during IFNα treatment, epithelial barrier destabilization was intensified [128, 129]. Therefore, the overall effect might be different in other cycle contexts or under hormone therapy, if the vulnerable phases might not be restricted to short periods.

3. Conclusion

In conclusion, it appears that males and females are equipped with stress coping strategies that may differ between the sexes. Sex differences have been demonstrated in the cellular

http://dx.doi.org/10.5772/intechopen.76177

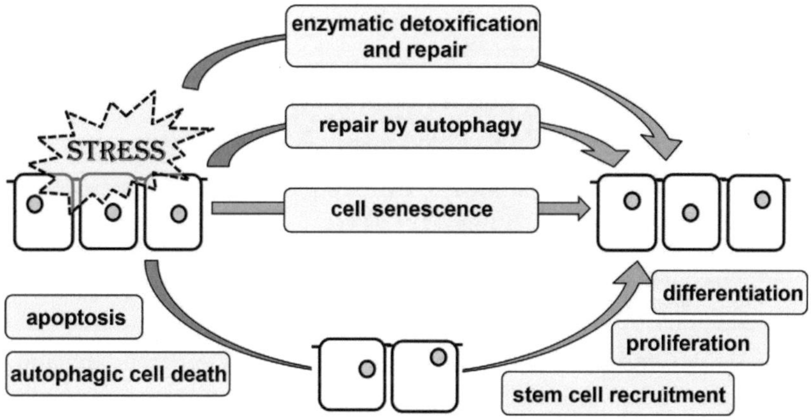

Figure 1. Sex and sex hormone-induced differences in tissue homeostasis. The figure shows strategies involved in tissue maintenance following cellular stress. Sex differences and sex hormone-dependent effects have been shown in these processes in different organs and tissues.

expression levels and activity of detoxifying and repair enzymes, in the propensity to use autophagic processes for repair, in senescence or cell death programs and in the ability to replace cells by proliferation (**Figure 1**). These effects are apparent in isolated cells and are further shaped by exposure to sex hormones. Sex hormone levels cyclically changing in dependence of the female reproductive hormone cycle might enhance physiological tissue regeneration and provide greater damage repair potential. Overall, female tissues appear to be more resistant to cellular stress than their male counterparts.

Acknowledgements

This work was supported by the Austrian Central Bank Fund project No 13671 to J Lechner.

Conflict of interest

The authors confirm that there are no conflicts of interest.

Author details

Judith Lechner* and Gerhard Gstraunthaler

*Address all correspondence to: judith.lechner@i-med.ac.at

Division of Physiology, Medical University of Innsbruck, Innsbruck, Austria

References

[1] Fulda S, Gorman AM, Hori O, Samali A. Cellular stress responses: Cell survival and cell death. International Journal of Cell Biology. 2010;**2010**:23

[2] Jennings P. Stress response pathways, toxicity pathways and adverse outcome pathways. Archives of Toxicology. 2013;**87**(1):13-14

[3] Lockshin RA, Zakeri Z. Cell death in health and disease. Journal of Cellular and Molecular Medicine. 2007;**11**(6):1214-1224

[4] Zhivotovsky B, Orrenius S. Carcinogenesis and apoptosis: Paradigms and paradoxes. Carcinogenesis. 2006;**27**(10):1939-1945

[5] Quinn A, Koopman P. The molecular genetics of sex determination and sex reversal in mammals. Seminars in Reproductive Medicine. 2012;**30**(5):351-363

[6] Wang Y, Liu H, Sun Z. Lamarck rises from his grave: Parental environment-induced epigenetic inheritance in model organisms and humans. Biological Reviews. 2017;**92**(4): 2084-2111

[7] Milagre I, Stubbs TM, King MR, Spindel J, Santos F, Krueger F, Bachman M, Segonds-Pichon A, Balasubramanian S, Andrews SR, Dean W, Reik W. Gender differences in global but not targeted demethylation in iPSC reprogramming. Cell Reports. 2017;**18**(5): 1079-1089

[8] Regitz-Zagrosek V, Kararigas G. Mechanistic pathways of sex differences in cardiovascular disease. Physiological Reviews. 2017;**97**(1):1-37

[9] Siegel C, Turtzo C, McCullough LD. Sex differences in cerebral ischemia: Possible molecular mechanisms. Journal of Neuroscience Research. 2010;**88**(13):2765-2774

[10] Arnold AP. The organizational–activational hypothesis as the foundation for a unified theory of sexual differentiation of all mammalian tissues. Hormones and Behavior. 2009; **55**(5):570-578

[11] Arnal JF, Lenfant F, Metivier R, Flouriot G, Henrion D, Adlanmerini M, Fontaine C, Gourdy P, Chambon P, Katzenellenbogen B, Katzenellenbogen J. Membrane and nuclear estrogen receptor alpha actions: From tissue specificity to medical implications. Physiological Reviews. 2017;**97**(3):1045-1087

[12] Banerjee S, Chambliss KL, Mineo C, Shaul PW. Recent insights into non-nuclear actions of estrogen receptor alpha. Steroids. 2014;**81**:64-69

[13] Boese AC, Kim SC, Yin KJ, Lee JP, Hamblin MH. Sex differences in vascular physiology and pathophysiology: Estrogen and androgen signaling in health and disease. American Journal of Physiology—Heart and Circulatory Physiology. 2017;**313**(3):H524-Hh45

[14] Lindahl A. Chapter 3—Tissue Homeostasis. Tissue Engineering. Burlington: Academic Press; 2008. pp. 73-87

[15] Ellison GM, Nadal-Ginard B, Torella D. Optimizing cardiac repair and regeneration through activation of the endogenous cardiac stem cell compartment. Journal of Cardiovascular Translational Research. 2012;**5**(5):667-677

[16] Lin R, Iacovitti L. Classic and novel stem cell niches in brain homeostasis and repair. Brain Research. 2015;**1628**(Pt B):327-342

[17] Park HC, Yasuda K, Kuo MC, Ni J, Ratliff B, Chander P, Goligorsky MS. Renal capsule as a stem cell niche. American Journal of Physiology—Renal Physiology. 2010;**298**(5):F1254-F1262

[18] Rojas-Rios P, Gonzalez-Reyes A. Concise review: The plasticity of stem cell niches: A general property behind tissue homeostasis and repair. Stem Cells (Dayton, Ohio). 2014; **32**(4):852-859

[19] Wabik A, Jones PH. Switching roles: The functional plasticity of adult tissue stem cells. The EMBO Journal. 2015;**34**(9):1164-1179

[20] Fox EM, Andrade J, Shupnik MA. Novel actions of estrogen to promote proliferation: Integration of cytoplasmic and nuclear pathways. Steroids. 2009;**74**(7):622-627

[21] Bluemn EG, Nelson PS. The androgen/androgen receptor axis in prostate cancer. Current Opinion in Oncology. 2012;**24**(3):251-257

[22] Maximov PY, Abderrahman B, Curpan RF, Hawsawi YM, Fan P, Jordan VC. A unifying biology of sex steroid-induced apoptosis in prostate and breast cancers. Endocrine-related Cancer. 2018;**25**(2):R83-r113

[23] Russo J, Russo IH. The role of estrogen in the initiation of breast cancer. The Journal of Steroid Biochemistry and Molecular Biology. 2006;**102**(1–5):89-96

[24] Finlay-Schultz J, Sartorius CA. Steroid hormones, steroid receptors, and breast cancer stem cells. Journal of Mammary Gland Biology and Neoplasia. 2015;**20**(1–2):39-50

[25] Han HJ, Heo JS, Lee YJ. Estradiol-17beta stimulates proliferation of mouse embryonic stem cells: Involvement of MAPKs and CDKs as well as protooncogenes. American Journal of Physiology—Cell Physiology. 2006;**290**(4):C1067-C1075

[26] Thomson JA, Itskovitz-Eldor J, Shapiro SS, Waknitz MA, Swiergiel JJ, Marshall VS, Jones JM. Embryonic stem cell lines derived from human blastocysts. Science (New York, NY). 1998;**282**(5391):1145

[27] Wong CK, So WY, Law SK, Leung FP, Yau KL, Yao X, Huang Y, Li X, Tsang SY. Estrogen controls embryonic stem cell proliferation via store-operated calcium entry and the nuclear factor of activated T-cells (NFAT). Journal of Cellular Physiology. 2012;**227**(6): 2519-2530

[28] Li H, Ding C, Ding Z, Ling M, Wang T, Wang W, Huang B. 17β-Oestradiol promotes differentiation of human embryonic stem cells into dopamine neurons via cross-talk between insulin-like growth factors-1 and oestrogen receptor β. Journal of Cellular and Molecular Medicine. 2017;**21**(8):1605-1618

[29] Lopez-Gonzalez R, Camacho-Arroyo I, Velasco I. Progesterone and 17beta-estradiol increase differentiation of mouse embryonic stem cells to motor neurons. IUBMB Life. 2011;**63**(10):930-939

[30] Díaz NF, Díaz-Martínez NE, Velasco I, Camacho-Arroyo I. Progesterone increases dopamine neuron number in differentiating mouse embryonic stem cells. Journal of Neuroendocrinology. 2009;**21**(8):730-736

[31] Yamanaka S. Induced pluripotent stem cells: Past, present, and future. Cell Stem Cell. 2012;**10**(6):678-684

[32] Shum C, Macedo SC, Warre-Cornish K, Cocks G, Price J, Srivastava DP. Utilizing induced pluripotent stem cells (iPSCs) to understand the actions of estrogens in human neurons. Hormones and Behavior. 2015;**74**:228-242

[33] Nishimura K, Doi D, Samata B, Murayama S, Tahara T, Onoe H, Takahashi J. Estradiol facilitates functional integration of iPSC-derived dopaminergic neurons into striatal neuronal circuits via activation of integrin $\alpha 5\beta 1$. Stem Cell Reports. 2016;**6**(4):511-524

[34] Liu H, Guo D, Ruzi A, Chen Y, Pan T, Yang F, Li J, Xu K, Zhou T, Qin D, Li YX. Testosterone improves the differentiation efficiency of insulin-producing cells from human induced pluripotent stem cells. PLoS One. 2017;**12**(6):e0179353

[35] Tower J. Stress and stem cells. Wiley Interdisciplinary Reviews—Developmental Biology. 2012;**1**(6):789-802

[36] Dulken B, Brunet A. Stem cell aging and sex: Are we missing something? Cell Stem Cell. 2015;**16**(6):588-590

[37] Ray R, Novotny NM, Crisostomo PR, Lahm T, Abarbanell A, Meldrum DR. Sex steroids and stem cell function, Molecular Medicine (Cambridge, Mass). 2008;**14**(7–8):493-501

[38] Pawluski JL, Brummelte S, Barha CK, Crozier TM, Galea LAM. Effects of steroid hormones on neurogenesis in the hippocampus of the adult female rodent during the estrous cycle, pregnancy, lactation and aging. Frontiers in Neuroendocrinology. 2009; **30**(3):343-357

[39] Lecanu L. Sex, the underestimated potential determining factor in brain tissue repair strategy. Stem Cells and Development. 2011;**20**(12):2031-2035

[40] Okada M, Makino A, Nakajima M, Okuyama S, Furukawa S, Furukawa Y. Estrogen stimulates proliferation and differentiation of neural stem/progenitor cells through different signal transduction pathways. International Journal of Molecular Sciences. 2010; **11**(10):4114-4123

[41] Helen M, Bramlett W, Dalton Dietrich. Neuropathological protection after traumatic brain injury in intact female rats versus males or ovariectomized females. Journal of Neurotrauma 2001;**18**(9):891-900

[42] Deasy BM, Schugar RC, Huard J. Sex differences in muscle-derived stem cells and skeletal muscle. Critical Reviews in Eukaryotic Gene Expression. 2008;**18**(2):173-188

[43] Nakada D, Oguro H, Levi BP, Ryan N, Kitano A, Saitoh Y, Takeichi M, Wendt GR, Morrison SJ. Oestrogen increases haematopoietic stem-cell self-renewal in females and during pregnancy. Nature. 2014;**505**:555

[44] Kim HR, Lee JH, Heo HR, Yang SR, Ha KS, Park WS, Han ET, Song H, Hong SH. Improved hematopoietic differentiation of human pluripotent stem cells via estrogen receptor signaling pathway. Cell & Bioscience. 2016;**6**(1):50

[45] Heo HR, Chen L, An B, Kim KS, Ji J, Hong SH. Hormonal regulation of hematopoietic stem cells and their niche: A focus on estrogen. International Journal of Stem Cells. 2015; **8**(1):18-23

[46] Manolagas SC. Steroids and osteoporosis: The quest for mechanisms. The Journal of Clinical Investigation. 2013;**123**(5):1919-1921

[47] Galea GL, Meakin LB, Sugiyama T, Zebda N, Sunters A, Taipaleenmaki H, Stein GS, van Wijnen AJ, Lanyon LE, Price JS. Estrogen receptor alpha mediates proliferation of osteoblastic cells stimulated by estrogen and mechanical strain, but their acute down-regulation of the Wnt antagonist Sost is mediated by estrogen receptor beta. The Journal of Biological Chemistry. 2013;**288**(13):9035-9048

[48] Hong L, Zhang G, Sultana H, Yu Y, Wei Z. The effects of 17-beta estradiol on enhancing proliferation of human bone marrow mesenchymal stromal cells in vitro. Stem Cells and Development. 2011;**20**(5):925-931

[49] Hamada H, Kim MK, Iwakura A, Ii M, Thorne T, Qin G, Asai J, Tsutsumi Y, Sekiguchi H, Silver M, Wecker A, Bord E, Zhu Y, Kishore R, Losordo DW. Estrogen receptors alpha and beta mediate contribution of bone marrow-derived endothelial progenitor cells to functional recovery after myocardial infarction. Circulation. 2006;**114**(21):2261-2270

[50] Torres-Estay V, Carreño DV, San Francisco IF, Sotomayor P, Godoy AS, Smith GJ. Androgen receptor in human endothelial cells. Journal of Endocrinology. 2015;**224**(3):R131-R1R7

[51] Childs BG, Durik M, Baker DJ, van Deursen JM. Cellular senescence in aging and age-related disease: From mechanisms to therapy. Nature Medicine. 2015;**21**(12):1424-1435

[52] Bielak-Zmijewska A, Mosieniak G, Sikora E. Is DNA damage indispensable for stress-induced senescence? Mechanisms of Ageing and Development. March 2018;**170**:13-21

[53] Ito Y, Hoare M, Narita M. Spatial and temporal control of senescence. Trends in Cell Biology. 2017;**27**(11):820-832

[54] Rao SG, Jackson JG. SASP: Tumor suppressor or promoter? Yes! Trends in Cancer. 2016; **2**(11):676-687

[55] Byun HO, Lee YK, Kim JM, Yoon G. From cell senescence to age-related diseases: Differential mechanisms of action of senescence-associated secretory phenotypes. BMB Reports. 2015;**48**(10):549-558

[56] Davalli P, Mitic T, Caporali A, Lauriola A, D'Arca D. ROS, cell senescence, and novel molecular mechanisms in aging and age-related diseases. Oxidative Medicine and Cellular Longevity. 2016;**2016**:3565127

[57] Galluzzi L, Maiuri MC, Vitale I, Zischka H, Castedo M, Zitvogel L, Kroemer G. Cell death modalities: Classification and pathophysiological implications. Cell Death and Differentiation. 2007;**14**(7):1237-1243

[58] Kroemer G, Galluzzi L, Vandenabeele P, Abrams J, Alnemri ES, Baehrecke EH, Blagosklonny MV, El-Deiry WS, Golstein P, Green DR, Hengartner M, Knight RA, Kumar S, Lipton SA, Malorni W, Nunez G, Peter ME, Tschopp J, Yuan J, Piacentini M, Zhivotovsky B, Melino G. Classification of cell death: Recommendations of the nomenclature committee on cell death 2009. Cell Death and Differentiation. 2009;**16**(1):3-11

[59] Portal-Nunez S, Esbrit P, Alcaraz MJ, Largo R. Oxidative stress, autophagy, epigenetic changes and regulation by miRNAs as potential therapeutic targets in osteoarthritis. Biochemical Pharmacology. 2016;**108**:1-10

[60] Kroemer G, Levine B. Autophagic cell death: The story of a misnomer. Nature Reviews Molecular Cell Biology. 2008;**9**(12):1004-1010

[61] Vasconsuelo A, Pronsato L, Ronda AC, Boland R, Milanesi L. Role of 17beta-estradiol and testosterone in apoptosis. Steroids. 2011;**76**(12):1223-1231

[62] Maselli A, Matarrese P, Straface E, Canu S, Franconi F, Malorni W. Cell sex: A new look at cell fate studies. The FASEB Journal. 2009;**23**(4):978-984

[63] Pronsato L, Ronda A, Milanesi L, Vasconsuelo A, Boland R. Protective role of 17β-estradiol and testosterone in apoptosis of skeletal muscle. Actualizaciones en Osteologia. 2010;**6**(2):65-80

[64] Nilsen J, Chen S, Irwin RW, Iwamoto S, Brinton RD. Estrogen protects neuronal cells from amyloid beta-induced apoptosis via regulation of mitochondrial proteins and function. BMC Neuroscience. 2006;**7**:74

[65] Pike CJ, Nguyen TV, Ramsden M, Yao M, Murphy MP, Rosario ER. Androgen cell signaling pathways involved in neuroprotective actions. Hormones and Behaviour. 2008;**53**(5):693-705

[66] Palomar-Morales M, Morimoto S, Mendoza-Rodriguez CA, Cerbon MA. The protective effect of testosterone on streptozotocin-induced apoptosis in beta cells is sex specific. Pancreas. 2010;**39**(2):193-200

[67] Ackermann S, Hiller S, Osswald H, Losle M, Grenz A, Hambrock A. 17beta-estradiol modulates apoptosis in pancreatic beta-cells by specific involvement of the sulfonylurea receptor (SUR) isoform SUR1. The Journal of Biological Chemistry. 2009;**284**(8):4905-4913

[68] Ling S, Zhou L, Li H, Dai A, Liu JP, Komesaroff PA, Sudhir K. Effects of 17beta-estradiol on growth and apoptosis in human vascular endothelial cells: Influence of mechanical strain and tumor necrosis factor-alpha. Steroids. 2006;**71**(9):799-808

[69] Lu A, Frink M, Choudhry MA, Hubbard WJ, Loring W, Rue I, Bland KI, Chaudry IH. Mitochondria play an important role in 17β-estradiol attenuation of H_2O_2-induced rat endothelial cell apoptosis. American Journal of Physiology—Endocrinology and Metabolism. 2007;**292**(2):E585-EE93

[70] Seli E, Guzeloglu-Kayisli O, Cakmak H, Kayisli UA, Selam B, Arici A. Estradiol increases apoptosis in human coronary artery endothelial cells by up-regulating Fas and Fas ligand expression. The Journal of Clinical Endocrinology & Metabolism. 2006;**91**(12):4995-5001

[71] Ling S, Dai A, Williams MR, Myles K, Dilley RJ, Komesaroff PA, Sudhir K. Testosterone (T) enhances apoptosis-related damage in human vascular endothelial cells. Endocrinology. 2002;**143**(3):1119-1125

[72] Alvarez RJ, Gips SJ, Moldovan N, Wilhide CC, Milliken EE, Hoang AT, Hruban RH, Silverman HS, Dang CV, Goldschmidt-Clermont PJ. 17beta-estradiol inhibits apoptosis of endothelial cells. Biochemical and Biophysical Research Communications. 1997;**237**(2): 372-381

[73] Florian M, Magder S. Estrogen decreases TNF-alpha and oxidized LDL induced apoptosis in endothelial cells. Steroids. 2008;**73**(1):47-58

[74] Yang YC, Fu HC, Wu CY, Wei KT, Huang KE, Kang HY. Androgen receptor accelerates premature senescence of human dermal papilla cells in association with DNA damage. PLoS One. 2013;**8**(11):e79434

[75] Winiarska A, Mandt N, Kamp H, Hossini A, Seltmann H, Zouboulis CC, Blume-Peytavi U. Effect of 5alpha-dihydrotestosterone and testosterone on apoptosis in human dermal papilla cells. Skin Pharmacology and Physiology. 2006;**19**(6):311-321

[76] Verzola D, Villaggio B, Procopio V, Gandolfo MT, Gianiorio F, Fama A, Tosetti F, Traverso P, Deferrari G, Garibotto G. Androgen-mediated apoptosis of kidney tubule cells: Role of c-Jun amino terminal kinase. Biochemical and Biophysical Research Communications. 2009;**387**(3):531-536

[77] Gu S, Papadopoulou N, Gehring E-M, Nasir O, Dimas K, Bhavsar SK, Föller M, Alevizopoulos K, Lang F, Stournaras C. Functional membrane androgen receptors in colon tumors trigger pro-apoptotic responses in vitro and reduce drastically tumor incidence in vivo. Molecular Cancer. 2009;**8**:114

[78] Jin L, Ai X, Liu L, Wang Z, Cheng Y, Qiao Z. Testosterone induces apoptosis via Fas/FasL-dependent pathway in bone marrow-derived macrophages. Methods and Findings in Experimental and Clinical Pharmacology. 2006;**28**(5):283-293

[79] Penaloza C, Estevez B, Orlanski S, Sikorska M, Walker R, Smith C, Smith B, Lockshin RA, Zakeri Z. Sex of the cell dictates its response: Differential gene expression and sensitivity to cell death inducing stress in male and female cells. FASEB Journal. 2009;**23**(6):1869-1879

[80] Straface E, Gambardella L, Brandani M, Malorni W. Sex differences at cellular level: "Cells have a sex". In: Regitz-Zagrosek V, editor. Sex and Gender Differences in Pharmacology. Berlin, Heidelberg: Springer Berlin Heidelberg; 2012. pp. 49-65

[81] Straface E, Vona R, Gambardella L, Ascione B, Marino M, Bulzomi P, Canu S, Coinu R, Rosano G, Malorni W, Franconi F. Cell sex determines anoikis resistance in vascular smooth muscle cells. FEBS Letters. 2009;**583**(21):3448-3454

[82] Du L, Hickey RW, Bayir H, Watkins SC, Tyurin VA, Guo F, Kochanek PM, Jenkins LW, Ren J, Gibson G, Chu CT, Kagan VE, Clark RSB. Starving neurons show sex difference in autophagy. Journal of Biological Chemistry. 2009;**284**(4):2383-2396

[83] Campesi I, Straface E, Occhioni S, Montella A, Franconi F. Protein oxidation seems to be linked to constitutive autophagy: A sex study. Life Sciences. 2013;**93**(4):145-152

[84] Camuzard O, Santucci-Darmanin S, Breuil V, Cros C, Gritsaenko T, Pagnotta S, Cailleteau L, Battaglia S, Panaia-Ferrari P, Heymann D, Carle GF, Pierrefite-Carle V. Sex-specific autophagy modulation in osteoblastic lineage: A critical function to counteract bone loss in female. Oncotarget. 2016;**7**(41):66416-66428

[85] Lee WJ, Lee SC, Lee JH, Rho GJ, Lee SL. Differential regulation of senescence and in vitro differentiation by 17beta-estradiol between mesenchymal stem cells derived from male and female mini-pigs. Journal of Veterinary Science. 2016;**17**(2):159-170

[86] Ortona E, Matarrese P, Malorni W. Taking into account the gender issue in cell death studies. Cell Death & Disease. 2014;**5**:e1121

[87] Malorni W, Straface E, Matarrese P, Ascione B, Coinu R, Canu S, Galluzzo P, Marino M, Franconi F. Redox state and gender differences in vascular smooth muscle cells. FEBS Letters. 2008;**582**(5):635-642

[88] Satoh M, Matter CM, Ogita H, Takeshita K, Wang C-Y, Dorn GW, Liao JK. Inhibition of apoptosis-regulated signaling Kinase-1 and prevention of congestive heart failure by estrogen. Circulation. 2007;**115**(25):3197-3204

[89] Liu H, Yanamandala M, Lee TC, Kim JK. Mitochondrial p38beta and manganese superoxide dismutase interaction mediated by estrogen in cardiomyocytes. PLoS One. 2014; **9**(1):e85272

[90] Strehlow K, Rotter S, Wassmann S, Adam O, Grohé C, Laufs K, Böhm M, Nickenig G. Modulation of antioxidant enzyme expression and function by estrogen. Circulation Research. 2003;**93**(2):170

[91] McCullough LD, Zeng Z, Blizzard KK, Debchoudhury I, Hurn PD. Ischemic nitric oxide and poly (ADP-ribose) polymerase-1 in cerebral ischemia: Male toxicity, female protection. Journal of Cerebral Blood Flow and Metabolism. 2005;**25**(4):502-512

[92] Liu F, Li Z, Li J, Siegel C, Yuan R, McCullough LD. Sex differences in caspase activation after stroke. Stroke. 2009;**40**(5):1842-1848

[93] Benjamin EJ, Blaha MJ, Chiuve SE, Cushman M, Das SR, Deo R, de Ferranti SD, Floyd J, Fornage M, Gillespie C, Isasi CR, Jimenez MC, Jordan LC, Judd SE, Lackland D, Lichtman JH, Lisabeth L, Liu S, Longenecker CT, Mackey RH, Matsushita K, Mozaffarian D, Mussolino ME, Nasir K, Neumar RW, Palaniappan L, Pandey DK, Thiagarajan RR,

Reeves MJ, Ritchey M, Rodriguez CJ, Roth GA, Rosamond WD, Sasson C, Towfighi A, Tsao CW, Turner MB, Virani SS, Voeks JH, Willey JZ, Wilkins JT, Wu JH, Alger HM, Wong SS, Muntner P. Heart disease and stroke Statistics-2017 update: A report from the American Heart Association. Circulation. 2017;**135**(10):e146-e603

[94] Koellhoffer EC, McCullough LD. The effects of estrogen in ischemic stroke. Translational Stroke Research. 2013;**4**(4):390-401

[95] Jog NR, Caricchio R. Differential regulation of cell death programs in males and females by poly (ADP-ribose) polymerase-1 and 17beta estradiol. Cell Death & Disease. 2013;**4**:e758

[96] Klinge CM. Estrogens regulate life and death in mitochondria. Journal of Bioenergetics and Biomembranes. 2017;**49**(4):307-324

[97] Sanchez MI, Shearwood AM, Chia T, Davies SM, Rackham O, Filipovska A. Estrogen-mediated regulation of mitochondrial gene expression. Molecular Endocrinology. 2015; **29**(1):14-27

[98] Nilsen J, Irwin RW, Gallaher TK, Brinton RD. Estradiol in vivo regulation of brain mitochondrial proteome. The Journal of Neuroscience. 2007;**27**(51):14069

[99] Razmara A, Duckles SP, Krause DN, Procaccio V. Estrogen suppresses brain mitochondrial oxidative stress in female and male rats. Brain Research. 2007;**1176**:71-81

[100] Stirone C, Duckles SP, Krause DN, Procaccio V. Estrogen increases mitochondrial efficiency and reduces oxidative stress in cerebral blood vessels. Molecular Pharmacology. 2005;**68**(4):959

[101] Velarde MC. Mitochondrial and sex steroid hormone crosstalk during aging. Longevity & Healthspan. 2014;**3**(1):2

[102] Salmon AB, Richardson A, Perez VI. Update on the oxidative stress theory of aging: Does oxidative stress play a role in aging or healthy aging? Free Radical Biology & Medicine. 2010;**48**(5):642-655

[103] Haigis MC, Yankner BA. The aging stress response. Molecular Cell. 2010;**40**(2):333-344

[104] Tower J. Sex-specific gene expression and life span regulation. Trends in Endocrinology and Metabolism. 2017;**28**(10):735-747

[105] Tower J. Mitochondrial maintenance failure in aging and role of sexual dimorphism. Archives of Biochemistry and Biophysics. 2015;**576**:17-31

[106] Park JH, Zhuang J, Li J, Hwang PM. p53 as guardian of the mitochondrial genome. FEBS Letters. 2016;**590**(7):924-934

[107] Kim S, Koh H. Role of FOXO transcription factors in crosstalk between mitochondria and the nucleus. Journal of Bioenergetics and Biomembranes. 2017;**49**(4):335-341

[108] Oertelt-Prigione S, Regitz-Zagrosek V. Sex and Gender Aspects in Clinical Medicine. Springer-Verlag London Ltd: London, UK; 2012

[109] Schenck-Gustafsson K, DeCola PR, Pfaff DW, Pisetsky DS. Handbook of Clinical Gender Medicine. Ergolding, Germany. Basel (CH): Karger Medical and Scientific Publishers; 2012

[110] Legato MJ. Principles of Gender-Specific Medicine: Gender in the Genomic Era. Chennai, India: Elsevier Inc, Academic Press; 2017

[111] Kramer A, Pippias M, Stel VS, Bonthuis M, Abad Diez JM, Afentakis N, Alonso de la Torre R, Ambuhl P, Bikbov B, Bouzas Caamaño E, Bubic I, Buturovic-Ponikvar J, Caskey FJ, Castro de la Nuez P, Cernevskis H, Collart F, Comas Farnés J, Garcia Bazaga MdlÁ, De Meester J, Ferrer Alamar M, Finne P, Garneata L, Golan E, Heaf JG, Hemmelder M, Ioannou K, Kantaria N, Kolesnyk M, Kramar R, Lassalle M, Lezaic V, Lopot F, Macário F, Magaz A, Martín-Escobar E, Metcalfe W, Ots-Rosenberg M, Palsson R, Piñera Celestino C, Resić H, Rutkowski B, Santiuste de Pablos C, Spustová V, Stendahl M, Strakosha A, Süleymanlar G, Torres Guinea M, Varberg Reisæter A, Vazelov E, Ziginskiene E, Massy ZA, Wanner C, Jager KJ, Noordzij M. Renal replacement therapy in Europe: A summary of the 2013 ERA-EDTA Registry Annual Report with a focus on diabetes mellitus. Clinical Kidney Journal. 2016;**9**(3):457-469

[112] Saran R, Robinson B, Abbott KC, Agodoa LY, Ayanian J, Bragg-Gresham J, Balkrishnan R, Chen JL, Cope E, Eggers PW, Gillen D, Gipson D, Hailpern SM, Hall YN, Han Y, He K, Herman W, Heung M, Hutton D, Jacobsen SJ, Kalantar-Zadeh K, Kovesdy CP, Li Y, Lu Y, Molnar MZ, Morgenstern H, Nallamothu B, Nguyen DV, O'Hare AM, Obi Y, Plattner B, Pisoni R, Port FK, Rao P, Ravel V, Rhee CM, Sakhuja A, Schaubel DE, Selewski DT, Sim JJ, Song P, Streja E, Kurella Tamura M, Tentori F, White S, Woodside K, Hirth RA, Shahinian VUS. Renal data system 2016 annual data report: Epidemiology of kidney disease in the United States. American Journal of Kidney Disease. 2017;**69**(3S1):A7-A8

[113] USRDS. United States Renal Data System 2016. USRDS annual data report: Epidemiology of kidney disease in the United States. National Institutes of Health, National Institute of Diabetes and Digestive and Kidney Diseases, Bethesda, MD, 2016 http://www.usrds.org. 2016

[114] Kramer A, Stel V, Zoccali C, Heaf J, Ansell D, Grönhagen-Riska C, Leivestad T, Simpson K, Pálsson R, Postorino M, Jager K. An update on renal replacement therapy in Europe: ERA–EDTA registry data from 1997 to 2006. Nephrology Dialysis Transplantation. 2009; **24**(12):3557-3566

[115] Turin TC, Tonelli M, Manns BJ, Ahmed SB, Ravani P, James M, Hemmelgarn BR. Lifetime risk of ESRD. Journal of the American Society of Nephrology. 2012;**23**(9):1569-1578

[116] Kiberd BA, Clase CM. Cumulative risk for developing end-stage renal disease in the US population. Journal of the American Society of Nephrology. 2002;**13**(6):1635-1644

[117] Neugarten J, Golestaneh L. Gender and the prevalence and progression of renal disease. Advances in Chronic Kidney Disease. 2013;**20**(5):390-395

[118] Carrero JJ. Gender differences in chronic kidney disease: Underpinnings and therapeutic implications. Kidney and Blood Pressure Research. 2010;**33**(5):383-392

[119] Kummer S, von Gersdorff G, Kemper M, Oh J. The influence of gender and sexual hormones on incidence and outcome of chronic kidney disease. Pediatric Nephrology. 2012;**27**(8):1213-1219

[120] Seliger SL, Davis C, Stehman-Breen C. Gender and the progression of renal disease. Current Opinion in Nephrology and Hypertension. 2001;**10**(2):219-225

[121] Kang D-H, Yu ES, Yoon K-I, Johnson R. The impact of gender on progression of renal disease: Potential role of estrogen-mediated vascular endothelial growth factor regulation and vascular protection. American Journal of Patholology. 2004;**164**(2):679-688

[122] Iliescu R, Reckelhoff JF. Sex and the kidney. Hypertension. 2008;**51**(4):1000-1001

[123] Maric C. Sex differences in cardiovascular disease and hypertension: Involvement of the renin-angiotensin system. Hypertension. 2005;**46**(3):475-476

[124] Kwan G, Neugarten J, Sherman M, Ding Q, Fotadar U, Lei J, Silbiger S. Effects of sex hormones on mesangial cell proliferation and collagen synthesis. Kidney International. 1996;**50**(4):1173-1179

[125] Oberley TD, Lauchner LJ, Pugh TD, Gonzalez A, Goldfarb S, Li SA, Li JJ. Specific estrogen-induced cell proliferation of cultured Syrian hamster renal proximal tubular cells in serum-free chemically defined media. Proceedings of the National Academy of Sciences. 1989;**86**(6):2107-2111

[126] Han H, Lim M, Lee Y, Kim E, Jeon Y, Lee J. Effects of TCDD and estradiol-17beta on the proliferation and Na+/glucose cotransporter in renal proximal tubule cells. Toxicology In Vitro. 2005;**19**(1):21-30

[127] Seppi T, Prajczer S, Dorler MM, Eiter O, Hekl D, Nevinny-Stickel M, Skvortsova I, Gstraunthaler G, Lukas P, Lechner J. Sex differences in renal proximal tubular cell homeostasis. Journal of the American Society of Nephrology. 2016;**27**(10):3051-3062

[128] Lechner J, Malloth NA, Jennings P, Hekl D, Pfaller W, Seppi T. Opposing roles of EGF in IFN-{alpha}-induced epithelial barrier destabilization and tissue repair. American Journal of Physiology—Cell Physiology. 2007;**293**(6):C1843-C1850

[129] Lechner J, Malloth N, Seppi T, Beer B, Jennings P, Pfaller W. IFN-{alpha} induces barrier destabilization and apoptosis in renal proximal tubular epithelium. American Journal of Physiology—Cell Physiology. 2008;**294**(1):C153-CC60

Section 3

Integrated Systems

Chapter 4

Gateway Reflex: A Neuro-Immune Crosstalk for Organ-Specific Disease Development

Daisuke Kamimura, Yuki Tanaka, Takuto Ohki and
Masaaki Murakami

Additional information is available at the end of the chapter

http://dx.doi.org/10.5772/intechopen.77112

Abstract

Homeostasis of the central nervous system (CNS) is strictly regulated by a unique structure of blood vessels, the blood-brain barrier (BBB). Experimental and clinical evidence has revealed that abnormalities in the BBB in chronic inflammatory diseases such as multiple sclerosis (MS). By using an animal model of MS, we identified novel neuro-immune crosstalk to explain how pathogenic immune cells enter the CNS to disrupt its homeostasis, a phenomenon we named the gateway reflex. Regional neural inputs such as gravity, electricity, pain or chronic stress cause specific neural activation to create a gateway of immune cells, particularly pathogenic ones, at specific blood vessels. Moreover, the recently discovered stress-induced gateway reflex uncovered a stress-induced neural link between the brain, gastrointestine, and heart. Thus, the gateway reflex is critical for the homeostasis of various organs, and aberrant activation of neural pathways by the gateway reflex disrupts normal organ homeostasis. The inflammatory reflex is another mechanism for local neuro-immune interactions. It potently exerts a cholinergic anti-inflammatory effect on various disease conditions. In this section, we discuss emerging roles for local neuro-immune interactions, with a special focus on the gateway reflex.

Keywords: gateway reflex, experimental autoimmune encephalomyelitis, central nervous system, chemokines, pathogenic CD4+ T cells

1. Introduction

A variety of environmental stimulations such as light, temperature, sound, and so on activates specific neurons to trigger biological responses. Gravity is another stimulus on land animals. Without sufficient gravity stimulation, physical functions including bone mass and muscle strength change, as observed in astronauts who stayed in the International Space Station [1, 2]. In addition to these physical stimulations, various events in daily social interactions associated with psychological alterations including anxiety, mental stresses or positive emotions are also stimulatory factors that trigger specific neural pathways to affect body functions. A well-studied mechanism to cope with these stimulations includes the release of glucocorticoids via the hypothalamus-pituitary-adrenal grand (HPA) axis, which systemically changes various physiological functions including the immune system [3, 4]. Besides this systemic regulation, we and other groups have identified local regulations of the inflammatory status by various environmental cues that cause specific neural activations. Here we summarize examples of specific neuro-immune interactions in organ homeostasis.

2. Main text

2.1. Gravity gateway reflex

The central nervous system (CNS) is considered an immune privileged site due to structural protections by the blood-brain barrier (BBB) [5]. The BBB is formed by tight cell-cell interactions between blood endothelial cells and tight liner sheets, a structure that lies outside the basement membrane and consists of pericytes, neurons, microglias, and astrocytes. Tight junctions by the interactions of tight-junction molecules including claudins and occludins are critical for the cell-cell interactions in the BBB that sequester cerebrospinal fluid from circulating blood components [6]. However, despite the BBB, there exist a certain number of immune cells in the CNS to prevent brain tumors and viral infections. In pathological conditions of the CNS, excessive immune cells from the blood accumulate from the breached BBB to cause chronic inflammatory diseases including multiple sclerosis (MS). It is also known that brain micro-inflammation is associated with neurodegenerative disorders such as Parkinson's disease and Alzheimer's disease [7]. Inflammatory cytokines including IL-1β, IL-17A, IFNγ and TNFα increase the BBB permeability, while chemokines recruit immune cells from the blood to cause inflammation [8–10]. However, where and how local CNS inflammation is induced in the BBB is poorly understood.

MS is a chronic inflammatory disease of the CNS. It is characterized by impairments in sensory, motor, autonomic, and neurocognitive functions due to autoimmune-mediated demyelination in CNS nerves [11]. Genetic factors strongly contribute to the MS pathogenesis, and genome-wide association studies (GWAS) have revealed that certain alleles of major histocompatibility complex (MHC) class II genes including HLA-DRB1*15:01 and HLA-DRB1*13:03 and genes involved in CD4+ T cell activation and homeostasis such as IL-2Rα and IL-7R are genetically associated with MS development [12–15]. These genetic results strongly suggest that autoreactive CD4+ T cells play a vital role in the MS pathogenesis and are supported by animal models of MS including experimental autoimmune encephalomyelitis (EAE) [16–19]. Using an adaptive transfer model of EAE [20], we investigated how and where myelin-autoreactive pathogenic CD4+ T cells initially invade the CNS. Whole-mount sections of adult mice with pathogenic CD4+ T cell transfer sacrificed just before the onset of

EAE symptoms revealed that pathogenic CD4+ T cells mainly accumulated at the L5 spinal cord, around the dorsal vessels in particular [18]. On the other hand, no accumulation was observed in the brain or upper spinal levels at the preclinical time point of EAE. Various chemokines including CCL20 attract type-17 CD4+ T (Th17) cells, which are known to have a key pathogenic role in EAE [21, 22], to accumulate more in the dorsal vessels of L5 than L1 cord. Indeed, CCL20 neutralization or deficiency of its receptor, CCR6, on pathogenic CD4+ T cells abrogated the cell accumulation in the L5 spinal cord [18].

Interestingly, even without EAE induction, chemokine levels were higher in the L5 dorsal vessels than in the L1 cord. These results suggest a distinct property of L5 dorsal vessels both in the presence and absence of pathogenic CD4+ T cells. The L5 spinal level has the largest dorsal root ganglion (DRG) in both human and mice, and it is known that sensory neurons in the L5 DRG distribute to the soleus muscles, the main anti-gravity muscles of the body [23, 24]. These facts led us to hypothesize a link between gravity, L5 vessels and local inflammation. We examined this possibility using a ground experiment employed by the National Aeronautics and Space Administration (NASA) and Japan Aerospace Exploration Agency (JAXA) [25, 26]. The tail suspension method puts mice in a handstand position, such that the hind limbs are released from gravity stimulation. As hypothesized, after the tail suspension, chemokine expressions of the L5 dorsal vessels were reduced and pathogenic CD4+ T cells hardly invaded the L5 cord. Chemokines were instead upregulated at the cervical cords as though another gateway for immune cells was formed by the greater gravity stimulation on the arm muscles imposed by the tail suspension [18]. Consistently, the tail suspension significantly reduced the expression of a neural activation marker, c-Fos, in the L5 DRG, suggesting a correlation between regional neural activation and local inflammation in the CNS. Consistently, stimulation of the soleus muscles by weak electric pulses, which mimic gravity-mediated sensory activation, during tail suspension restored chemokine expression, pathogenic CD4+ T cell accumulation, and c-Fos levels at the L5 dorsal vessels [18]. That study identified L5 dorsal vessels as a blood vessel gateway for immune cells including pathogenic CD4+ T cells to the CNS, and revealed that regional sensory neural activation by gravity affects CNS homeostasis and causes local inflammation at L5 dorsal blood vessels when CNS-autoreactive pathogenic CD4+ T cells are present. This phenomenon, which represents a novel neuro-immune interaction, was termed the "gateway reflex" [27–34]. Since then, we have found various types of gateway reflexes depending on the neural stimulation. Thus, the abovementioned example is known as the gravity gateway reflex (**Figure 1**).

How the regional neural pathway by gravity stimulation regulates chemokine expressions in L5 blood vessels is an important question. Contribution of the autonomic nervous system was suggested. In fact, neural activation examined by c-Fos expression was higher in sympathetic ganglions at the L5 cord than at the L1 cord. In addition, the blood flow speed of the L5 dorsal vessels but not of other blood vessels including the L1 dorsal vessels, femoral artery or portal vein, slowed after tail suspension. Functionally, the treatment of mice with β-adrenergic receptor antagonists or chemical sympathectomy inhibited chemokine expressions, the accumulation of pathogenic CD4+ T cells at the L5 dorsal vessels, and the clinical scores of EAE [18]. Norepinephrine is one of the major neurotransmitters of sympathetic nerves. Indeed, in vitro experiments suggest that norepinephrine enhances chemokine expressions from

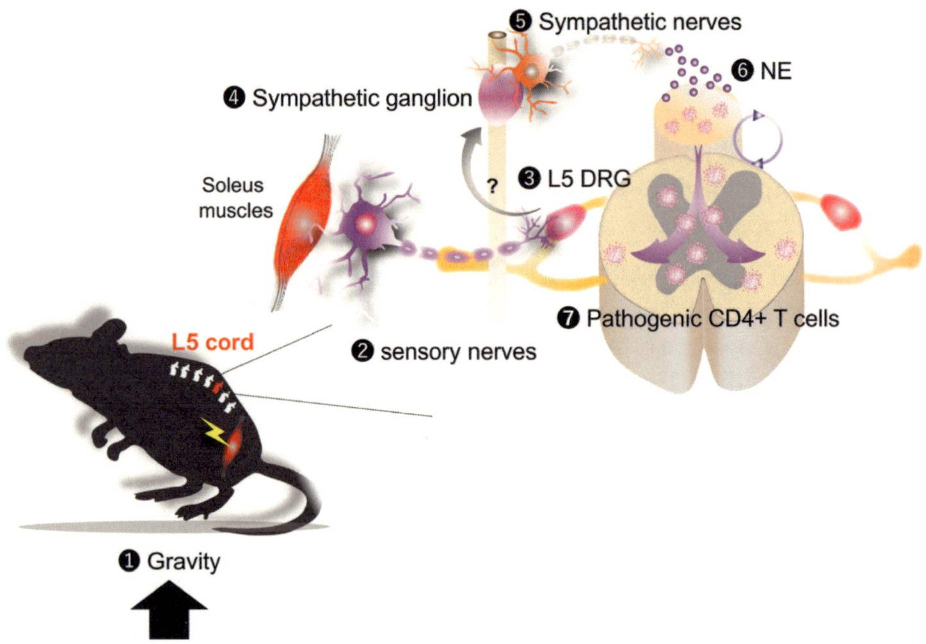

Figure 1. Gravity gateway reflex. Stimulation of the soleus muscles by gravity (1) induces the activation of specific sensory nerves (2). The cell bodies of these sensory neurons are located at the dorsal root ganglion (DRG) of the fifth lumbar (L5) spinal cord (3). Neural activation via the L5 DRG neurons travels to the L5 sympathetic ganglion (4) and induces the activation of sympathetic nerves (5), which results in norepinephrine (NE) secretion (6) at the L5 dorsal vessels. NE enhances the inflammation amplifier in the L5 dorsal vessels, causing an upregulation of chemokines and recruiting pathogenic CD4+ T cells from the vessels (7).

endothelial cell lines with activation of the inflammation amplifier [18]. The inflammation amplifier is a molecular mechanism that operates in nonimmune cells including endothelial cells to produce a large amount of pro-inflammatory mediators including chemokines, cytokines and growth factors upon the concomitant activation of two transcription factors, NF-κB and STAT3 [35–37]. The co-activation of NF-κB and STAT3 by cytokines such as IL-17, IL-6, and TNFα augments NF-κB activity and upregulates NF-κB-target genes including chemokines, which can be further enhanced by neurotransmitters including norepinephrine and ATP (**Figure 2**) [18, 38]. The contribution of nonimmune cells to chronic inflammation via the inflammation amplifier has been found significant in the development of various disease models and evidence of inflammation amplifier activation has been observed in patients with MS, rheumatoid arthritis, atherosclerosis, and chronic rejection after lung transplantation [39–45]. Thus, the gateway reflex represents novel local neuro-immune communication involving the activation of specific sensory and sympathetic neurons for the formation of blood vessel gateways (**Figure 1**). Because chemokine levels are constantly higher at the L5 dorsal vessels under normal conditions without EAE induction, the gateway reflex could have a physiological role as well. Indeed, it is reported in mice that after a learning task, CD4+ T cells, but not CD8+ T cells, accumulated in the CNS, and mice devoid of CD4+ T cells showed impaired learning performance and neurogenesis in the hippocampus [46–48]. The gravity gateway reflex may be utilized by these cells to enter the CNS to maintain and/or control the homeostasis and function of the CNS.

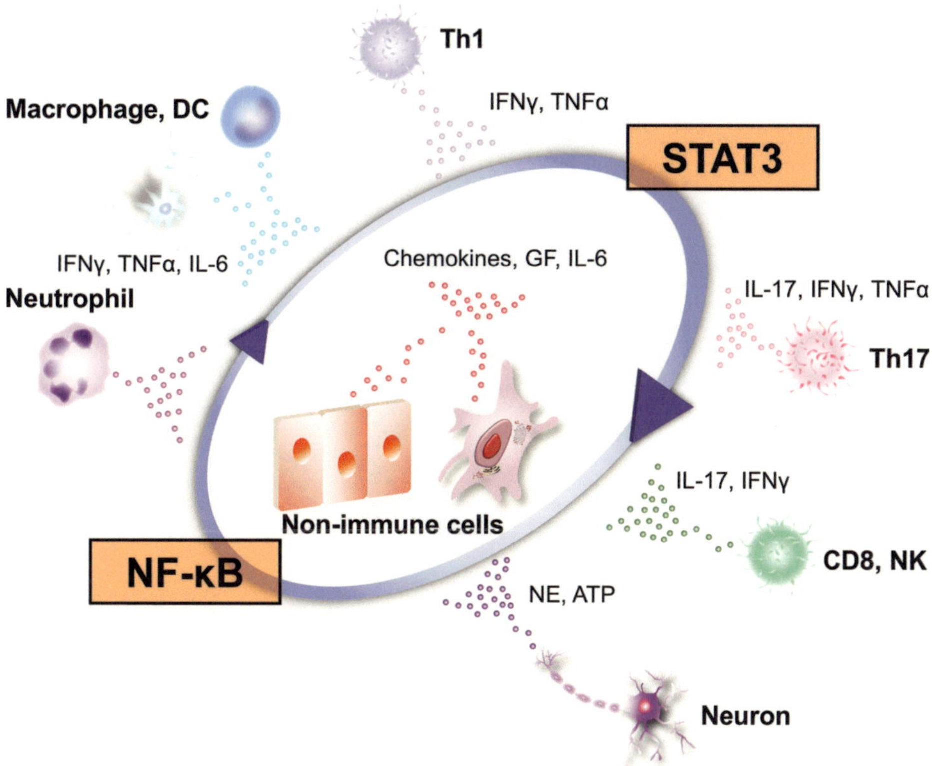

Figure 2. Inflammation amplifier. Co-activation of the transcription factors NF-κB and STAT3 in nonimmune cells such as endothelial cells and fibroblasts induces a synergistic effect on the production of inflammation factors such as chemokines, growth factors, and IL-6. Various factors activating NF-κB and STAT3 including IL-17, TNFα, and IL-6 can drive the amplifier. Secreted IL-6 is thought to act on nonimmune cells to form a positive feedback loop for this synergistic effect. Massive production of chemokines and growth factors by the inflammation amplifier play an essential role in the pathogenesis of many inflammatory diseases. DC, dendritic cells; GF, growth factors; NK, natural killer cells; NE, norepinephrine; Th, helper T cells.

2.2. Electric gateway reflex

The gateway reflex is not specific to the soleus-L5 axis because electrical stimulation of regional sensory neurons in different muscles can induce the formation of gateways in the dorsal vessels of different spinal cord levels. For instance, electric stimulation of the quadriceps, which are controlled by the L3 DRG neurons, induces chemokine expressions at the L3 dorsal vessels. Similarly, electric stimulations of the triceps upregulate chemokines at the dorsal vessels of the cervical to thoracic spinal cords (**Figure 3**) [18]. These results suggest that the electric gateway reflex can be artificially controlled, raising a possibility for a therapeutic application of the gateway reflex to various CNS diseases like MS and brain tumors.

2.3. Pain gateway reflex

We have examined whether other sensory stimulations can also generate the gateway reflex. We focused on pain because it is a tonic sensory stimulation [49, 50] and associated with various diseases that significantly compromise quality of life [51]. Some studies suggest that

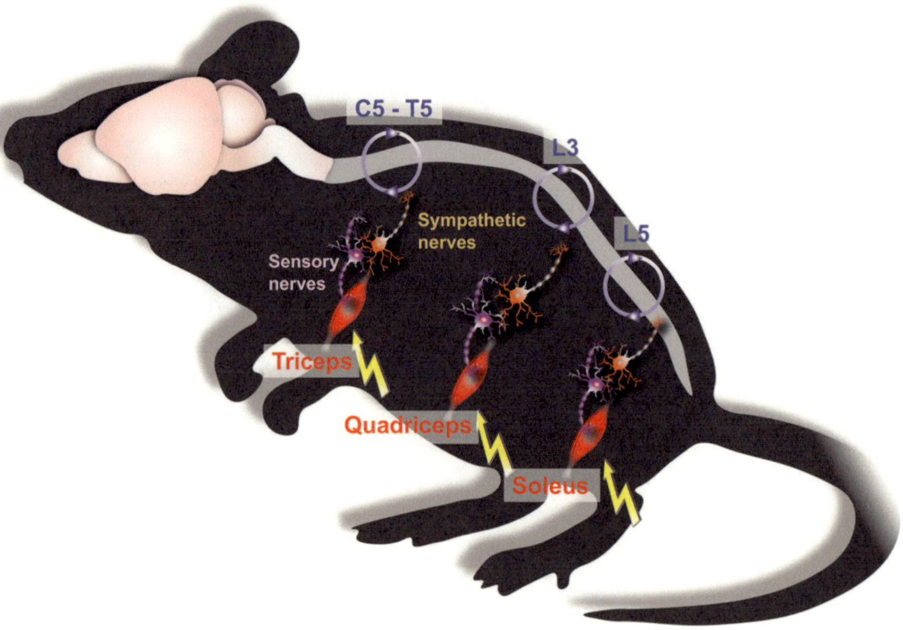

Figure 3. Electric gateway reflex. Electric stimulation to the triceps induces chemokine upregulation at the dorsal vessels of the fifth cervical (C5) to fifth thoracic (T5) spinal cord through sensory-sympathetic neural activation, which leads to activation of the inflammation amplifier at the C5-T5 dorsal vessels. In a similar fashion, electric stimulation of the quadriceps and soleus muscles induces a gateway at the L3 and L5 dorsal vessels, respectively.

pain occurrence is positively correlated with disease severity in MS [52–54], and mechanical allodynia and thermal hyperalgesia are induced in mice during EAE [55, 56]. In the adoptive transfer EAE model we used, recipient mice that received myelin-specific pathogenic CD4+ T cells developed transient paralysis, but then recovered and hardly relapsed even 100 days later. To examine how pain affects EAE symptoms, we induced pain sensation in mice by partial ligation of the middle branch of the trigeminal nerves, which are composed of sensory nerves alone [57]. Pain induction with pathogenic CD4+ T cell transfer significantly worsened the paralysis. By contrast, pain medicines improved EAE development [19]. Thus, pain is not simply an alert of the disease or injury status, but it also has a pathogenic role, triggering EAE relapse. Since a large part of MS patients show relapse and remission and pain is more frequently claimed by MS patients with higher disease scores [52, 53], we examined the effects of pain induction in mice that had recovered from EAE (EAE-recovered mice). As expected, EAE-recovered mice showed a clear sign of relapse after the partial ligation of the trigeminal nerve or injection of pain-inducing chemicals such as substance P and capsaicin [19]. As described above, under normal conditions, the blood vessel gateway for immune cells is the L5 dorsal vessels due to the effect of gravity [18]. To identify specific blood vessels that act as a gateway for immune cells that cause the pain-induced relapse, immunohistological examination of the CNS from EAE-recovered mice was performed. Although the motility of EAE-recovered mice was not significantly different from normal healthy mice, the meningeal region of the L5 cord of EAE-recovered mice contained a high number of periphery-derived monocytes that expressed high levels of MHC class II [18]. These MHC class II high monocytes distributed around the meningeal region of the L5 cord in EAE-recovered mice. After pain induction, however, these cells accumulated at the ventral vessels of the L5 cord bilaterally. Experimental evidence that (1) norepinephrine signaling could be detected around the L5 ventral vessels, (2) MHC class II high monocytes expressed chemokine receptor CX3CR1, and (3) MHC class II high monocytes secreted chemokine CX3CL1, a ligand for CX3CR1, after norepinephrine stimulation suggested an auto/paracrine loop is responsible for the accumulation of MHC class II high monocytes around the L5 ventral vessels via norepinephrine regulation of the CX3CL1-CX3CR1 axis. Because freshly isolated MHC class II high monocytes from EAE-recovered mice have autoantigen (MOG)-presenting capacity without additional peptide loading, it can be suggested that MHC class II high monocytes accumulated around the L5 ventral vessels would activate pathogenic CD4+ T cells through autoantigen presentation, followed by regional inflammation and disease relapse. Suppression of norepinephrine signaling by a β1 blocker or sympathetic nerve ablation by chemical sympathectomy inhibited the pain-mediated accumulation of MHC class II high monocytes around the L5 ventral vessels [19]. These results identified the L5 ventral vessels as the gateway for pain-induced relapse in the EAE model (**Figure 4**). This pain gateway reflex is the third example of the gateway reflex following the aforementioned gravity and electric examples.

2.4. Stress gateway reflex

Chronic stresses deteriorate illness, an effect attributed to the proverbs "Illness starts in the mind" and "Care killed the cat." Chronic stress conditions often cause gastric and intestinal diseases via the brain-gut axis. Although these diseases are well known and often experienced, the underlying molecular mechanisms remain to be elucidated. Stresses generate neural activations involving multiple brain areas such as the paraventricular nucleus (PVN), dorsomedial nucleus of hypothalamus (DMH), dorsal motor nucleus of the vagal nerve (DMX), and vagal nerve pathway [58]. We therefore hypothesized that chronic stresses might induce a specific gateway reflex. Indeed, experiments confirmed the stress gateway reflex. We serendipitously found that micro-inflammation induced by the stress gateway reflex activates an otherwise resting neural circuit to enhance a stress response that causes fatal gastrointestinal and heart failure in mice [38]. Therefore, the stress gateway reflex may explain the mechanisms for the abovementioned proverbs, as described below.

The gravity gateway reflex directs donor pathogenic CD4+ T cells to the dorsal vessels of the L5 spinal cord under normal conditions (**Figure 1**). However, under chronic stress conditions, the cells instead invade specific blood vessels of the boundary area of the third ventricle (3V), thalamus, and part of hippocampus [dentate gyrus, (DG)] to establish micro-inflammation. Therefore, chronic stresses can alter the location of the immune cell gateway from the L5 dorsal vessels to specific vessels in the brain. Immunohistochemistry of c-Fos revealed that the resulting micro-inflammation led to enhanced neural activations in the PVN and DMH, which are stress sensing areas of the hypothalamus. Neural tracing experiments identified direct connections via noradrenergic neurons between the PVN and the specific vessels of the boundary area of 3V, thalamus and DG. Moreover, these experiments showed direct

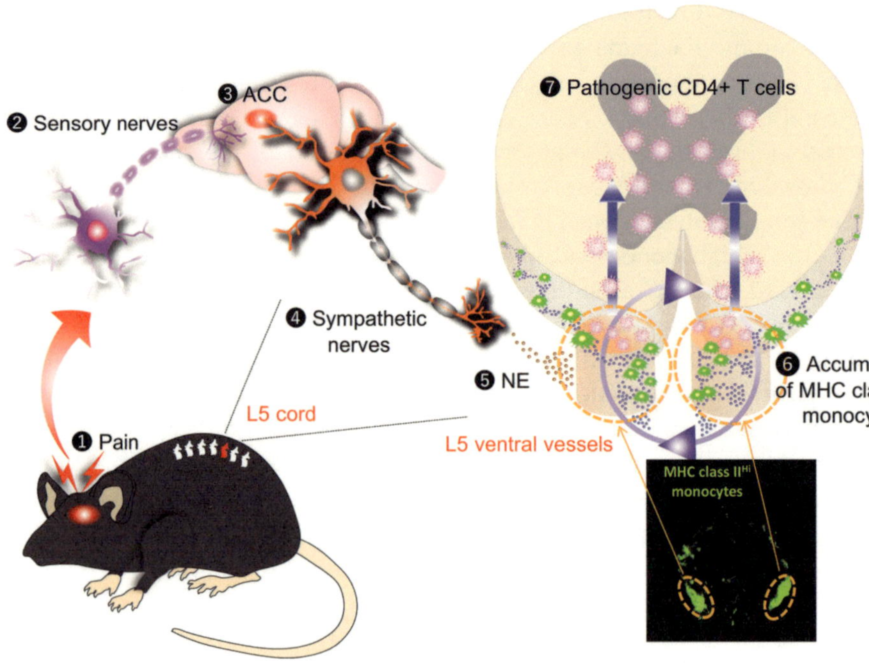

Figure 4. Pain gateway reflex. Pain induces nociceptive sensory nerve stimulation (1, 2), leading to activation of the anterior cingulate cortex (ACC), a pain-processing area of the brain (3). Specific sympathetic nerves are then activated (4) to induce norepinephrine (NE) release around the ventral vessels of the spinal cords (5). NE around the L5 ventral vessels induces the production of chemokine CX3CL1 from MHC class II high (MHC class II^{HI}) monocytes, further recruiting these cells in an auto/paracrine manner (6). MHC class II high monocytes are able to present myelin autoantigens to activate pathogenic CD4+ T cells, leading to disease relapse (7).

connections via nonnoradrenergic neurons between the specific vessels and the DMH. Since the PVN is a key orchestrator of stress signals, its activation is expected to affect specific blood vessels via a new noradrenergic neural pathway. Indeed, the expression of chemokines such as CCL5 was upregulated at specific vessels in mice with chronic stress alone. CNS-reactive pathogenic CD4+ T cells in the blood circulation of the stressed mice sensed the upregulation of CCL5, caused micro-inflammation with peripheral derived MHC class II monocytes at specific vessels of the boundary area of the 3V, thalamus, and DG [38]. ATP is released during an inflammatory response, but it also acts as a neurotransmitter [59, 60]. The injection of an ATP receptor antagonist at specific blood vessels of the boundary area of the 3V, thalamus and DG suppressed both neural activation in the DMH and the mortality rate of EAE mice with chronic stress. Importantly, the direct injection of ATP or cytokines at these specific vessels, which mimicked the micro-inflammation, induced fatal gastrointestinal dysfunction in stressed mice without EAE induction. These results suggest that brain micro-inflammation at specific vessels of the boundary area of the 3V, thalamus, and DG activates a resting neural pathway through ATP production, thus strongly enhancing the stress response to cause fatal gastrointestinal damage via the DMX and vagal nerve activation. These results uncovered the stress gateway reflex acting as a direct link between micro-inflammation at a particular site in the brain and gastrointestinal homeostasis (**Figure 5**) [38].

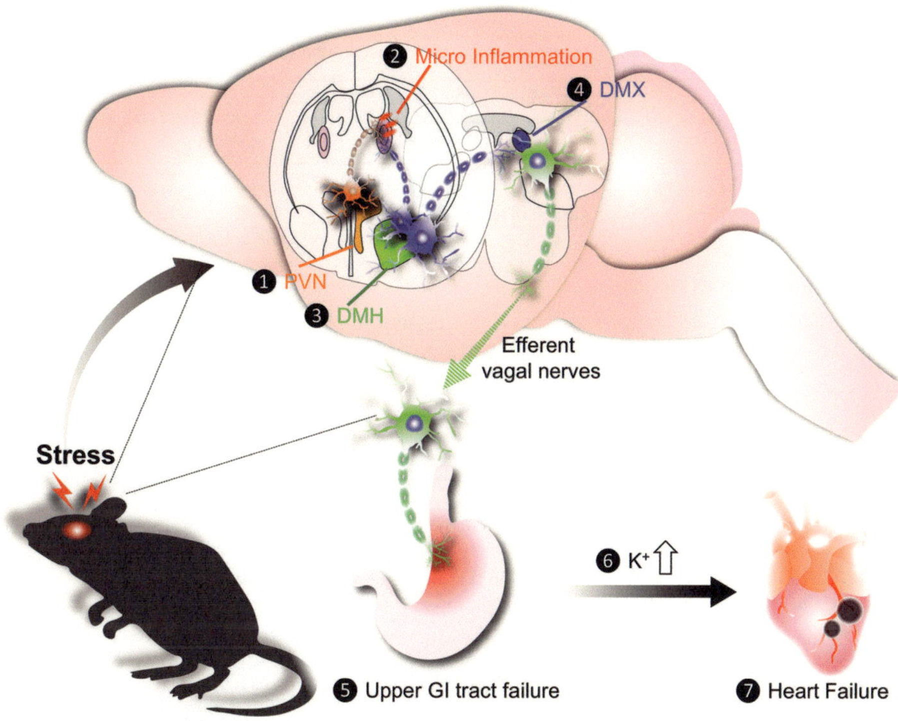

Figure 5. Stress gateway reflex. Chronic stress induces activation of the paraventricular nucleus (PVN) (1). We identified neural connections from the PVN to the specific vessels of the boundary area of the third ventricle, thalamus and dentate gyrus, which induces micro-inflammation around the specific vessels (2). The resulting micro-inflammation induces activation of a neural pathway that connects to the dorsomedial nucleus of hypothalamus (DMH) (3) and dorsal motor nucleus of the vagal nerve (DMX) (4), resulting in severe upper gastrointestinal (GI) tract failure via efferent vagal nerves (5). The increase of potassium ions (K^+) in blood circulation by the upper GI tract failure (6) explains at least in part heart failure associated with cardiac myocyte necrosis (7).

Several reports have shown the concurrence of inflammatory bowel diseases and MS [61–65]. Moreover, brain micro-inflammations were observed in patients with neurodegenerative diseases such as Alzheimer's disease and Parkinson's disease [66, 67], epilepsy [68], and psychological disorders [69]. We therefore suggest that brain micro-inflammations could regulate the homeostasis of organ functions including the brain itself by acting as a switch to stimulate resting neural pathways, which could account for the comorbidities observed in many diseases.

2.5. Cholinergic anti-inflammatory pathway

The Tracey laboratory has demonstrated using a mouse model of sepsis that the activation of vagal nerves, which mainly consist of parasympathetic nerves, suppresses an inflammatory response [70]. They revealed that lipopolysaccharide induces norepinephrine release in the spleen via vagal and splenic nerves. The norepinephrine stimulates a novel subset of CD4+ T cells that express β1/2 adrenergic receptor and produce acetylcholine. The resulting acetylcholine then acts on activated macrophages expressing $\alpha 7$ nicotinic receptor to suppress the

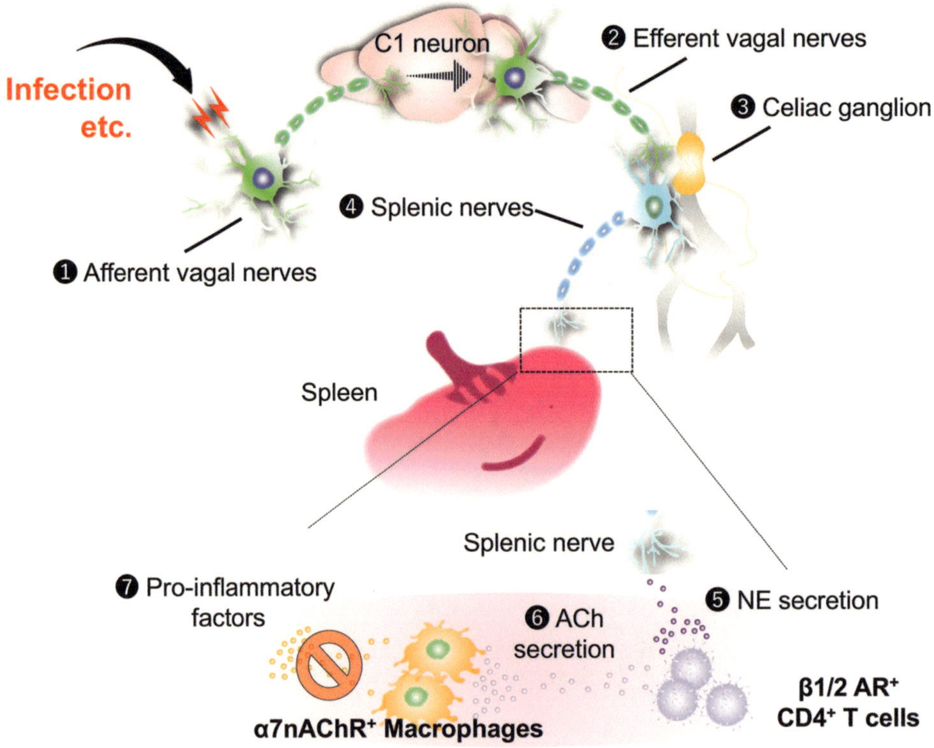

Figure 6. Inflammatory reflex. Afferent (1) and efferent vagal nerve (2) activation by infection, and so on induces neural activation in the celiac ganglion (3), followed by the production of norepinephrine (NE) by the splenic nerves (4). NE stimulates the release of acetylcholine (ACh) from a subset of CD4+ T cells expressing β1/2 adrenaline receptor (AR) (5). Then, ACh (6) acts on macrophages expressing α7 nicotinic acetylcholine receptor (α7nAChR) to suppress the expression of pro-inflammatory factors such as HMGB1 and TNFα. It is suggested that C1 neurons in the brain mediate the cholinergic anti-inflammatory effect.

expression of pro-inflammatory factors such as TNFα and HMGB1 (**Figure 6**) [71–73]. The stimulation of macrophages with nicotine inhibits NF-κB activation, but not MAP kinase activation in response to endotoxin [72]. This cholinergic anti-inflammatory pathway is called the "inflammatory reflex" [27–29, 31, 32, 34, 74–76]. In addition to infection, this pathway exerts anti-inflammatory effects in various disease models including renal ischemia–reperfusion injury, acute kidney injury, pressure overload-induced cardiac hypertrophy and neointimal hyperplasia [77–81]. The direct stimulation of C1 neurons in the medullary reticular formation induced this anti-inflammatory effect in mice [79], suggesting the involvement of C1 neurons between afferent and efferent vagal nerves. Activation of the cholinergic anti-inflammatory pathway can be induced by acupuncture and ultrasound, thus inhibiting the pathology of animal models [82, 83]. These findings establish a scientific basis for acupuncture and physical therapy. In addition, activation of the inflammatory reflex through vagal nerve stimulation by an implantable device has been tested in humans for chronic inflammatory diseases including rheumatoid arthritis with promising results [84, 85].

2.6. Future directions

Accumulating evidence has demonstrated the significant effects of regional neuro-immune interactions on organ homeostasis, particularly during inflammation and diseases. The gateway reflex and inflammatory reflex can be induced by various stimulations to activate these interactions. Stimulation of the vagal nerves by an implantable device has already shown an anti-inflammatory effect in humans [84, 85]. The stimulation of the neurons responsible for the gateway reflex and inflammatory reflex at the body surface, as is the case with acupuncture, could lead to therapies that are less invasive and less costly, although elucidation of the precise neural circuits activated by the reflexes is required. Neural mapping is still challenging, but recent techniques and tools including the tissue clearing method CUBIC [86–89], optogenetics and chemogenetics [90, 91], and transgenic mice reporting neural circuitry and activations [92–94] will greatly contribute. Because neural circuits run throughout the body and because nonimmune cells as well as immune cells [18, 19, 71, 95–97] are able to secrete and respond to neurotransmitters, specific regional neuro-immune interactions such as the gateway reflex and inflammatory reflex bear potential as a significant therapeutic strategy to recover organ homeostasis.

Acknowledgements

We appreciate the outstanding experimental contributions provided by Dr. Y. Arima and the excellent technical assistances by Ms. Ezawa and Ms. Nakayama for our work described here. We also thank Ms. Fukumoto for her excellent secretarial assistance, and Dr. P. Karagiannis (CiRA, Kyoto University, Kyoto, Japan) for carefully reading the manuscript. Our work described here was supported by KAKENHI (D. K. and M. M.), Takeda Science Foundation (M. M.), Institute for Fermentation Osaka (M. M.), Mitsubishi Foundation (M. M.), Mochida Memorial Foundation for Medical and Pharmaceutical Research (D. K.), Suzuken Memorial Foundation (D. K.), and Tokyo Medical Research Foundation (M. M.).

Conflict of interest

The authors have no conflict of interest for the article.

Author details

Daisuke Kamimura, Yuki Tanaka, Takuto Ohki and Masaaki Murakami*

*Address all correspondence to: murakami@igm.hokudai.ac.jp

Molecular Psychoimmunology, Institute for Genetic Medicine, Graduate School of Medicine, Hokkaido University, Sapporo, Hokkaido, Japan

References

[1] Petersen N, Lambrecht G, Scott J, Hirsch N, Stokes M, Mester J. Postflight reconditioning for European astronauts—A case report of recovery after six months in space. Musculoskeletal Science & Practice. 2017;**27**(Suppl 1):S23-S31. DOI: 10.1016/j.msksp.2016.12.010

[2] Sibonga JD. Spaceflight-induced bone loss: Is there an osteoporosis risk? Current Osteoporosis Reports. 2013;**11**(2):92-98. DOI: 10.1007/s11914-013-0136-5

[3] De Bosscher K, Vanden Berghe W, Haegeman G. Mechanisms of anti-inflammatory action and of immunosuppression by glucocorticoids: Negative interference of activated glucocorticoid receptor with transcription factors. Journal of Neuroimmunology. 2000;**109**(1):16-22

[4] Bellavance MA, Rivest S. The HPA-immune axis and the immunomodulatory actions of glucocorticoids in the brain. Frontiers in Immunology. 2014;**5**(136). DOI: 10.3389/fimmu.2014.00136

[5] Ransohoff RM, Kivisakk P, Kidd G. Three or more routes for leukocyte migration into the central nervous system. Nature Reviews Immunology. 2003;**3**(7):569-581. DOI: 10.1038/nri1130

[6] Steed E, Balda MS, Matter K. Dynamics and functions of tight junctions. Trends in Cell Biology. 2010;**20**(3):142-149. S0962-8924(09)00297-9 [pii]. DOI: 10.1016/j.tcb.2009.12.002

[7] Zlokovic BV. The blood-brain barrier in health and chronic neurodegenerative disorders. Neuron. 2008;**57**(2):178-201. S0896-6273(08)00034-2 [pii]. DOI: 10.1016/j.neuron.2008.01.003

[8] Tsao N, Hsu HP, Wu CM, Liu CC, Lei HY. Tumour necrosis factor-alpha causes an increase in blood-brain barrier permeability during sepsis. Journal of Medical Microbiology. 2001;**50**(9):812-821

[9] Argaw AT, Zhang Y, Snyder BJ, Zhao ML, Kopp N, Lee SC, et al. IL-1beta regulates blood-brain barrier permeability via reactivation of the hypoxia-angiogenesis program. Journal of Immunology. 2006;**177**(8):5574-5584

[10] Nagarsheth N, Wicha MS, Zou W. Chemokines in the cancer microenvironment and their relevance in cancer immunotherapy. Nature Reviews Immunology. 2017;**17**(9):559-572. DOI: 10.1038/nri.2017.49

[11] Keegan BM, Noseworthy JH. Multiple sclerosis. Annual Review of Medicine. 2002;**53**:285-302. DOI: 10.1146/annurev.med.53.082901.103909

[12] McFarland HF, Martin R. Multiple sclerosis: A complicated picture of autoimmunity. Nature Immunology. 2007;**8**(9):913-919. DOI: 10.1038/ni1507

[13] International Multiple Sclerosis Genetics C, Wellcome Trust Case Control C, Sawcer S, Hellenthal G, Pirinen M, Spencer CC, et al. Genetic risk and a primary role for cell-mediated immune mechanisms in multiple sclerosis. Nature. 2011;**476**(7359):214-219. DOI: 10.1038/nature10251

[14] Australia, New Zealand Multiple Sclerosis Genetics C. Genome-wide association study identifies new multiple sclerosis susceptibility loci on chromosomes 12 and 20. Nature Genetics 2009;**41**(7):824-828. 10.1038/ng.396

[15] Moutsianas L, Jostins L, Beecham AH, Dilthey AT, Xifara DK, Ban M, et al. Class II HLA interactions modulate genetic risk for multiple sclerosis. Nature Genetics. 2015;**47**(10): 1107-1113. DOI: 10.1038/ng.3395

[16] Steinman L, Zamvil SS. Virtues and pitfalls of EAE for the development of therapies for multiple sclerosis. Trends in Immunology. 2005;**26**(11):565-571. DOI: 10.1016/j.it.2005.08.014

[17] Rangachari M, Kuchroo VK. Using EAE to better understand principles of immune function and autoimmune pathology. Journal of Autoimmunity. 2013;**45**:31-39. DOI: 10.1016/j.jaut.2013.06.008

[18] Arima Y, Harada M, Kamimura D, Park JH, Kawano F, Yull FE, et al. Regional neural activation defines a gateway for autoreactive T cells to cross the blood-brain barrier. Cell. 2012;(3):**148**:447-457. S0092-8674(12)00088-8 [pii]. DOI: 10.1016/j.cell.2012.01.022

[19] Arima Y, Kamimura D, Atsumi T, Harada M, Kawamoto T, Nishikawa N, et al. A pain-mediated neural signal induces relapse in murine autoimmune encephalomyelitis, a multiple sclerosis model. eLife. 2015;**4**:e08733. DOI: 10.7554/eLife.08733

[20] Tanaka Y, Arima Y, Higuchi K, Ohki T, Elfeky M, Ota M, et al. EAE induction by passive transfer of MOG-specific CD4+ T cells. Bio-protocol. 2017;**7**(13):7/5/2017. DOI: 10.21769/BioProtoc.2370

[21] Komiyama Y, Nakae S, Matsuki T, Nambu A, Ishigame H, Kakuta S, et al. IL-17 plays an important role in the development of experimental autoimmune encephalomyelitis. Journal of Immunology. 2006;**177**(1):566-573

[22] Korn T, Bettelli E, Oukka M, Kuchroo VK. IL-17 and Th17 cells. Annual Review of Immunology. 2009;**27**:485-517. DOI: 10.1146/annurev.immunol.021908.132710

[23] Ohira Y, Kawano F, Stevens JL, Wang XD, Ishihara A. Load-dependent regulation of neuromuscular system. Journal of Gravitational Physiology. 2004;**11**:127-128

[24] Shen J, Wang HY, Chen JY, Liang BL. Morphologic analysis of normal human lumbar dorsal root ganglion by 3D MR imaging. American Journal of Neuroradiology. 2006;**27**(10):2098-2103

[25] Morey-Holton ER, Globus RK. Hindlimb unloading rodent model: Technical aspects. Journal of Applied Physiology. 2002;**92**(4):1367-1377. DOI: 10.1152/japplphysiol.00969.2001

[26] Ohira T, Kawano F, Ohira T, Goto K, Ohira Y. Responses of skeletal muscles to gravitational unloading and/or reloading. The Journal of Physiological Sciences. 2015;**65**(4): 293-310. DOI: 10.1007/s12576-015-0375-6

[27] Tracey KJ. Immune cells exploit a neural circuit to enter the CNS. Cell. 2012;**148**:392-394

[28] Andersson U, Tracey KJ. Neural reflexes in inflammation and immunity. The Journal of Experimental Medicine. 2012;**209**(6):1057-1068. DOI: 10.1084/jem.20120571

[29] Deutschman CS, Tracey KJ. Sepsis: Current dogma and new perspectives. Immunity. 2014;**40**(4):463-475. DOI: 10.1016/j.immuni.2014.04.001

[30] Sabharwal L, Kamimura D, Meng J, Bando H, Ogura H, Nakayama C, et al. The gateway reflex, which is mediated by the inflammation amplifier, directs pathogenic immune cells into the CNS. Journal of Biochemistry. 2014;**156**:299-304

[31] Tracey KJ. Reflexes in immunity. Cell. 2016;**164**(3):343-344. DOI: 10.1016/j.cell.2016.01.018

[32] Pavlov VA, Tracey KJ. Neural regulation of immunity: Molecular mechanisms and clinical translation. Nature Neuroscience. 2017;**20**(2):156-166. DOI: 10.1038/nn.4477

[33] Ohki T, Kamimura D, Arima Y, Murakami M. Gateway reflex, a new paradigm of neuro-immune interaction. Clinical & Experimental Neuroimmunology. 2017;**8**:23-32. DOI: 10.1111/cen3.12378

[34] Chavan SS, Pavlov VA, Tracey KJ. Mechanisms and therapeutic relevance of neuro-immune communication. Immunity. 2017;**46**(6):927-942. DOI: 10.1016/j.immuni.2017.06.008

[35] Ogura H, Murakami M, Okuyama Y, Tsuruoka M, Kitabayashi C, Kanamoto M, et al. Interleukin-17 promotes autoimmunity by triggering a positive-feedback loop via interleukin-6 induction. Immunity. 2008;**29**:628-636

[36] Atsumi T, Singh R, Sabharwal L, Bando H, Jie M, Arima Y, et al. Inflammation amplifier, a new paradigm in cancer biology. Cancer Research. 2014;**74**:8-14

[37] Nakagawa I, Kamimura D, Atsumi T, Arima Y, Murakami M. Role of inflammation amplifier-induced growth factor expression in the development of inflammatory diseases. Critical Reviews in Immunology. 2015;**35**(5):365-378

[38] Arima Y, Ohki T, Nishikawa N, Higuchi K, Ota M, Tanaka Y, et al. Brain micro-inflammation at specific vessels dysregulates organ-homeostasis via the activation of a new neural circuit. eLife. 2017;**6**:e25517. DOI: 10.7554/eLife.25517

[39] Murakami M, Okuyama Y, Ogura H, Asano S, Arima Y, Tsuruoka M, et al. Local microbleeding facilitates IL-6- and IL-17-dependent arthritis in the absence of tissue antigen recognition by activated T cells. The Journal of Experimental Medicine. 2011;**208**(1): 103-114. DOI: 10.1084/jem.20100900

[40] Lee J, Nakagiri T, Oto T, Harada M, Morii E, Shintani Y, et al. IL-6 amplifier, NF-kappaB-triggered positive feedback for IL-6 signaling, in grafts is involved in allogeneic rejection responses. Journal of Immunology. 2012;**189**(4):1928-1936. DOI: 10.4049/jimmunol.1103613

[41] Murakami M, Harada M, Kamimura D, Ogura H, Okuyama Y, Kumai N, et al. Disease-association analysis of an inflammation-related feedback loop. Cell Reports. 2013;**3**(3): 946-959. DOI: 10.1016/j.celrep.2013.01.028

[42] Lee J, Nakagiri T, Kamimura D, Harada M, Oto T, Susaki Y, et al. IL-6 amplifier activation in epithelial regions of bronchi after allogeneic lung transplantation. International Immunology. 2013;**25**(5):319-332. DOI: 10.1093/intimm/dxs158

[43] Harada M, Kamimura D, Arima Y, Kohsaka H, Nakatsuji Y, Nishida M, et al. Temporal expression of growth factors triggered by epiregulin regulates inflammation development. Journal of Immunology. 2015;**194**(3):1039-1046. DOI: 10.4049/jimmunol.1400562

[44] Meng J, Jiang JJ, Atsumi T, Bando H, Okuyama Y, Sabharwal L, et al. Breakpoint cluster region-mediated inflammation is dependent on casein kinase II. Journal of Immunology. 2016;**197**(8):3111-3119. DOI: 10.4049/jimmunol.1601082

[45] Atsumi T, Suzuki H, Jiang JJ, Okuyama Y, Nakagawa I, Ota M, et al. Rbm10 regulates inflammation development via alternative splicing of Dnmt3b. International Immunology. 2017;**29**(12):581-591. DOI: 10.1093/intimm/dxx067

[46] Kipnis J, Cohen H, Cardon M, Ziv Y, Schwartz M. T cell deficiency leads to cognitive dysfunction: Implications for therapeutic vaccination for schizophrenia and other psychiatric conditions. Proceedings of the National Academy of Sciences of the United States of America. 2004;**101**(21):8180-8185. DOI: 10.1073/pnas.0402268101

[47] Wolf SA, Steiner B, Akpinarli A, Kammertoens T, Nassenstein C, Braun A, et al. CD4-positive T lymphocytes provide a neuroimmunological link in the control of adult hippocampal neurogenesis. Journal of Immunology. 2009;**182**(7):3979-3984. DOI: 10.4049/jimmunol.0801218

[48] Derecki NC, Cardani AN, Yang CH, Quinnies KM, Crihfield A, Lynch KR, et al. Regulation of learning and memory by meningeal immunity: A key role for IL-4. The Journal of Experimental Medicine. 2010;**207**(5):1067-1080. DOI: 10.1084/jem.20091419

[49] Morales-Lazaro SL, Simon SA, Rosenbaum T. The role of endogenous molecules in modulating pain through transient receptor potential vanilloid 1 (TRPV1). The Journal of Physiology. 2013;**591**(Pt 13):3109-3121. DOI: 10.1113/jphysiol.2013.251751

[50] Bennett DL, Woods CG. Painful and painless channelopathies. The Lancet Neurology. 2014;**13**(6):587-599. DOI: 10.1016/S1474-4422(14)70024-9

[51] Feinstein A, Magalhaes S, Richard JF, Audet B, Moore C. The link between multiple sclerosis and depression. Nature Reviews Neurology. 2014;**10**(9):507-517. DOI: 10.1038/nrneurol.2014.139

[52] Ehde DM, Gibbons LE, Chwastiak L, Bombardier CH, Sullivan MD, Kraft GH. Chronic pain in a large community sample of persons with multiple sclerosis. Multiple Sclerosis. 2003;**9**:605-611

[53] Ehde DM, Osborne TL, Hanley MA, Jensen MP, Kraft G H. The scope and nature of pain in persons with multiple sclerosis. Multiple Sclerosis. 2006;**12**:629-638

[54] O'Connor AB, Schwid SR, Herrmann DN, Markman JD, Dworkin RH. Pain associated with multiple sclerosis: Systematic review and proposed classification. Pain. 2008;**137**(1):96-111. DOI: 10.1016/j.pain.2007.08.024

[55] Lu J, Kurejova M, Wirotanseng LN, Linker RA, Kuner R, Tappe-Theodor A. Pain in experimental autoimmune encephalitis: A comparative study between different mouse models. Journal of Neuroinflammation. 2012;**9**:233. DOI: 10.1186/1742-2094-9-233

[56] Khan N, Smith MT. Multiple sclerosis-induced neuropathic pain: Pharmacological man agement and pathophysiological insights from rodent EAE models. Inflammopharmacology. 2014;**22**(1):1-22. DOI: 10.1007/s10787-013-0195-3

[57] Thygesen TH, Baad-Hansen L, Svensson P. Sensory action potentials of the maxillary nerve: A methodologic study with clinical implications. Journal of Oral and Maxillofacial Surgery. 2009;**67**:537-542

[58] Ulrich-Lai YM, Herman JP. Neural regulation of endocrine and autonomic stress responses. Nature Reviews Neuroscience. 2009;**10**(6):397-409. DOI: 10.1038/nrn2647

[59] Burnstock G. Historical review: ATP as a neurotransmitter. Trends in Pharmacological Sciences. 2006;**27**(3):166-176. DOI: 10.1016/j.tips.2006.01.005

[60] Fiebich BL, Akter S, Akundi RS. The two-hit hypothesis for neuroinflammation: Role of exogenous ATP in modulating inflammation in the brain. Frontiers in Cellular Neuroscience. 2014;**8**:260. DOI: 10.3389/fncel.2014.00260

[61] Rang EH, Brooke BN, Hermon-Taylor J. Association of ulcerative colitis with multiple sclerosis. Lancet. 1982;**2**(8297):555

[62] Sadovnick AD, Paty DW, Yannakoulias G. Concurrence of multiple sclerosis and inflammatory bowel disease. The New England Journal of Medicine. 1989;**321**(11):762-763

[63] Gupta G, Gelfand JM, Lewis JD. Increased risk for demyelinating diseases in patients with inflammatory bowel disease. Gastroenterology. 2005;**129**(3):819-826. DOI: 10.1053/j.gastro.2005.06.022

[64] Kimura K, Hunter SF, Thollander MS, Loftus EV Jr, Melton LJ 3rd, O'Brien PC, et al. Concurrence of inflammatory bowel disease and multiple sclerosis. Mayo Clinic Proceedings. 2000;**75**(8):802-806. DOI: 10.4065/75.8.802

[65] Pokorny CS, Beran RG, Pokorny MJ. Association between ulcerative colitis and multiple sclerosis. Internal Medicine Journal. 2007;**37**(10):721-724. DOI: 10.1111/j.1445-5994.2007.01452.x

[66] Togo T, Akiyama H, Iseki E, Kondo H, Ikeda K, Kato M, et al. Occurrence of T cells in the brain of Alzheimer's disease and other neurological diseases. Journal of Neuroimmunology. 2002;**124**(1-2):83-92

[67] Appel SH, Beers DR, Henkel JS. T cell-microglial dialogue in Parkinson's disease and amyotrophic lateral sclerosis: Are we listening? Trends in Immunology. 2010;**31**:7-17

[68] Vezzani A, French J, Bartfai T, Baram TZ. The role of inflammation in epilepsy. Nature Reviews. Neurology. 2011;**7**:31-40

[69] Najjar S, Pearlman DM, Alper K, Najjar A, Devinsky O. Neuroinflammation and psychiatric illness. Journal of Neuroinflammation. 2013;**10**:43. DOI: 10.1186/1742-2094-10-43

[70] Borovikova LV, Ivanova S, Zhang M, Yang H, Botchkina GI, Watkins LR, et al. Vagus nerve stimulation attenuates the systemic inflammatory response to endotoxin. Nature. 2000;**405**:458-462

[71] Wang H, Yu M, Ochani M, Amella CA, Tanovic M, Susarla S, et al. Nicotinic acetylcholine receptor alpha7 subunit is an essential regulator of inflammation. Nature. 2003;**421**(6921):384-388. DOI: 10.1038/nature01339

[72] Wang H, Liao H, Ochani M, Justiniani M, Lin X, Yang L, et al. Cholinergic agonists inhibit HMGB1 release and improve survival in experimental sepsis. Nature Medicine. 2004;**10**(11):1216-1221. DOI: 10.1038/nm1124

[73] Rosas-Ballina M, Olofsson PS, Ochani M, Valdes-Ferrer SI, Levine YA, Reardon C, et al. Acetylcholine-synthesizing T cells relay neural signals in a vagus nerve circuit. Science. 2011;**334**(6052):98-101. DOI: 10.1126/science.1209985

[74] Andersson U, Tracey KJ. Reflex principles of immunological homeostasis. Annual Review of Immunology. 2012;**30**:313-335. DOI: 10.1146/annurev-immunol-020711-075015

[75] Olofsson PS, Tracey KJ. Bioelectronic medicine: Technology targeting molecular mechanisms for therapy. Journal of Internal Medicine. 2017;**282**(1):3-4. DOI: 10.1111/joim.12624

[76] Tanaka Y, Arima Y, Kamimura D, Murakami M. The gateway reflex, a novel neuroimmune interaction for the regulation of regional vessels. Frontiers in Immunology. 2017;**8**:1321. DOI: 10.3389/fimmu.2017.01321

[77] Mao Y, Tokudome T, Kishimoto I, Otani K, Nishimura H, Yamaguchi O, et al. Endogenous ghrelin attenuates pressure overload-induced cardiac hypertrophy via a cholinergic anti-inflammatory pathway. Hypertension. 2015;**65**(6):1238-1244. DOI: 10.1161/HYPERTENSIONAHA.114.04864

[78] Inoue T, Abe C, Sung SS, Moscalu S, Jankowski J, Huang L, et al. Vagus nerve stimulation mediates protection from kidney ischemia-reperfusion injury through alpha7nAChR+ splenocytes. The Journal of Clinical Investigation. 2016;**126**(5):1939-1952. DOI: 10.1172/JCI83658

[79] Abe C, Inoue T, Inglis MA, Viar KE, Huang LP, Ye H, et al. C1 neurons mediate a stress-induced anti-inflammatory reflex in mice. Nature Neuroscience. 2017;**20**(5):700-707. DOI: 10.1038/nn.4526

[80] Chatterjee PK, Yeboah MM, Solanki MH, Kumar G, Xue X, Pavlov VA, et al. Activation of the cholinergic anti-inflammatory pathway by GTS-21 attenuates cisplatin-induced acute kidney injury in mice. PLoS One. 2017;**12**(11):e0188797. DOI: 10.1371/journal.pone.0188797

[81] Li DJ, Fu H, Tong J, Li YH, Qu LF, Wang P, et al. Cholinergic anti-inflammatory pathway inhibits neointimal hyperplasia by suppressing inflammation and oxidative stress. Redox Biology. 2018;**15**:22-33. DOI: 10.1016/j.redox.2017.11.013

[82] Torres-Rosas R, Yehia G, Pena G, Mishra P, del Rocio Thompson-Bonilla M, Moreno-Eutimio MA, et al. Dopamine mediates vagal modulation of the immune system by electroacupuncture. Nature Medicine. 2014;**20**(3):291-295. DOI: 10.1038/nm.3479

[83] Gigliotti JC, Huang L, Bajwa A, Ye H, Mace EH, Hossack JA, et al. Ultrasound modulates the splenic neuroimmune axis in attenuating AKI. Journal of the American Society of Nephrology. 2015;**26**(10):2470-2481. DOI: 10.1681/ASN.2014080769

[84] Koopman FA, Chavan SS, Miljko S, Grazio S, Sokolovic S, Schuurman PR, et al. Vagus nerve stimulation inhibits cytokine production and attenuates disease severity in rheumatoid arthritis. Proceedings of the National Academy of Sciences of the United States of America. 2016;**113**(29):8284-8289. DOI: 10.1073/pnas.1605635113

[85] Bonaz B, Sinniger V, Pellissier S. Vagus nerve stimulation: A new promising therapeutic tool in inflammatory bowel disease. Journal of Internal Medicine. 2017;**282**(1):46-63. DOI: 10.1111/joim.12611

[86] Susaki EA, Tainaka K, Perrin D, Kishino F, Tawara T, Watanabe TM, et al. Whole-brain imaging with single-cell resolution using chemical cocktails and computational analysis. Cell. 2014;**157**(3):726-739. DOI: 10.1016/j.cell.2014.03.042

[87] Tainaka K, Kubota SI, Suyama TQ, Susaki EA, Perrin D, Ukai-Tadenuma M, et al. Whole-body imaging with single-cell resolution by tissue decolorization. Cell. 2014;**159**(4): 911-924. DOI: 10.1016/j.cell.2014.10.034

[88] Tainaka K, Kuno A, Kubota SI, Murakami T, Ueda HR. Chemical principles in tissue clearing and staining protocols for whole-body cell profiling. Annual Review of Cell and Developmental Biology. 2016;**32**:713-741. DOI: 10.1146/annurev-cellbio-111315-125001

[89] Murakami TC, Mano T, Saikawa S, Horiguchi SA, Shigeta D, Baba K, et al. A three-dimensional single-cell-resolution whole-brain atlas using CUBIC-X expansion microscopy and tissue clearing. Nature Neuroscience. 2018;**21**:625-637. DOI: 10.1038/s41593-018-0109-1

[90] Kim CK, Adhikari A, Deisseroth K. Integration of optogenetics with complementary methodologies in systems neuroscience. Nature Reviews Neuroscience. 2017;**18**(4):222-235. DOI: 10.1038/nrn.2017.15

[91] Campbell EJ, Marchant NJ. The use of chemogenetics in behavioural neuroscience: Receptor variants, targeting approaches, and caveats. British Journal of Pharmacology. 2018;**175**:994-1003. DOI: 10.1111/bph.14146

[92] Livet J, Weissman TA, Kang H, Draft RW, Lu J, Bennis RA, et al. Transgenic strategies for combinatorial expression of fluorescent proteins in the nervous system. Nature. 2007;**450**(7166):56-62. DOI: 10.1038/nature06293

[93] Eguchi M, Yamaguchi S. In vivo and in vitro visualization of gene expression dynamics over extensive areas of the brain. NeuroImage. 2009;**44**(4):1274-1283. DOI: 10.1016/j.neuroimage.2008.10.046

[94] Kebschull JM, Garcia da Silva P, Reid AP, Peikon ID, Albeanu DF, Zador AM. High-throughput mapping of single-neuron projections by sequencing of barcoded RNA. Neuron. 2016;**91**(5):975-987. DOI: 10.1016/j.neuron.2016.07.036

[95] Levite M. Neurotransmitters activate T-cells and elicit crucial functions via neurotransmitter receptors. Current Opinion in Pharmacology. 2008;**8**(4):460-471. DOI: 10.1016/j.coph.2008.05.001

[96] Selmeczy Z, Vizi ES, Csoka B, Pacher P, Hasko G. Role of nonsynaptic communication in regulating the immune response. Neurochemistry International. 2008;**52**(1-2):52-59. DOI: 10.1016/j.neuint.2007.06.001

[97] Camacho-Arroyo I, Lopez-Griego L, Morales-Montor J. The role of cytokines in the regulation of neurotransmission. Neuroimmunomodulation. 2009;**16**(1):1-12. DOI: 10.1159/000179661

Chapter 5

Platelets: From Formation to Function

Laura Twomey, Robert G. Wallace,
Philip M. Cummins, Bernard Degryse,
Sinead Sheridan, Michael Harrison, Niall Moyna,
Gerardene Meade-Murphy, Nastassia Navasiolava,
Marc-Antoine Custaud and Ronan P. Murphy

Additional information is available at the end of the chapter

http://dx.doi.org/10.5772/intechopen.80924

Abstract

Platelets are small, anucleate cells that travel as resting discoid fragments in the circulation. Their average circulating life span is 8–9 days, and their formation is an elegant and finely orchestrated series of cellular processes known as *megakaryocytopoiesis* and *thrombopoiesis*. This involves the commitment of haematopoietic stem cells, proliferation, terminal differentiation of megakaryocytic progenitors and maturation of megakaryocytes to produce functional platelets. This complex process occurs in specialised endosteal and vascular niches in the bone marrow where megakaryocytes form proplatelet projections, releasing platelets into the circulation. Upon contact with an injured blood vessel, they prevent blood loss through processes of adhesion, activation and aggregation. Platelets play a central role in cardiovascular disease (CVD), both in the development of atherosclerosis and as the cellular mediator in the development of thrombosis. Platelets have diverse roles not limited to thrombosis/haemostasis, also being involved in many vascular inflammatory conditions. Depending on the physiological context, platelet functions may be protective or contribute to adverse thrombotic and inflammatory outcomes. In this chapter, we will discuss platelets in context of their formation and function. Because of their multifaceted role in maintaining physiological homeostasis, current and development of platelet function testing platforms will be discussed.

Keywords: platelets, megakaryocytes, megakaryocytopoiesis and thrombopoiesis, thrombosis and haemostasis, cardiovascular disease, platelet function testing

1. Introduction

Efforts in coping with the socio-economic and health burden of CVD require further understanding of its aetiology and the risk factors behind it in order to develop cost-effective preventive strategies (primordial, primary, secondary and tertiary) to prevent and manage it. CVD risk factors can be classed as modifiable or non-modifiable. Modifiable risk factors included smoking, diabetes, unhealthy diet, cholesterol, physical inactivity (sedentary lifestyle and low cardiorespiratory fitness), overweight and obesity. Risk factors for CVD track from childhood into adulthood and are strong predictors of subclinical disease in early adulthood [1]. Up to 80% of CVD may be prevented if modifiable risk factors are evaded [2].

The main functions of blood are to supply oxygen and nutrients to tissues and cells, removal of waste, and regulation of pH and body temperature [3]. An average adult has approximately five litres of blood, accounting for about 7% of their body weight. Blood is composed of approximately 55% plasma; a pale yellow fluid mainly consisting of water, proteins, sugars and fat particles, and 45% blood cells. Blood cells include erythrocytes, leukocytes and platelets. Each of these cells are derived from a haematopoietic stem cell (HSC), which reside in the bone marrow and sit at the peak of a developmental hierarchy, with unique ability to self-renew and give rise to cells of all of the blood lineages [4]. In adults, nearly a trillion new blood cells are produced daily to sustain steady state in circulation [5]. In the classic model of haematopoiesis (the production of blood cells), an important bifurcation occurs between the lymphoid and myeloid branches, which then further divide into a number of progenitor cells (**Figure 1**). While the main function of red blood cells (RBCs) and white blood cells (WBCs) is oxygen transport and defence, respectively, this chapter will focus on platelets, the final product of one of the myeloid

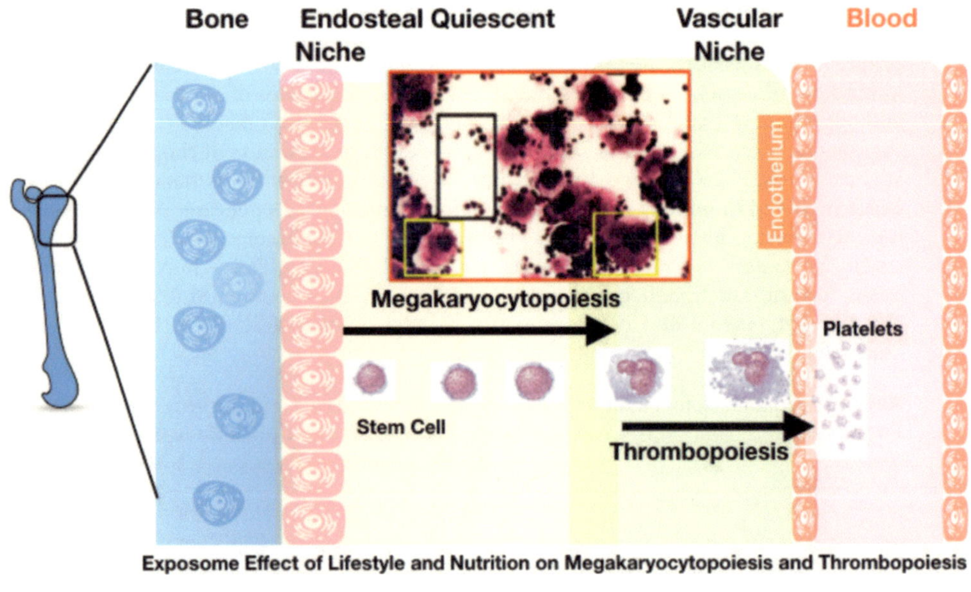

Figure 1. Pictorial representation of megakaryocytopoiesis and thrombopoiesis, and the potential for exposome and epigenome modulation of platelet phenotype and function during this dynamic cellular differentiation process.

cell lines. In addition, the principle hypothesis and paradigm discussed will be that the modifiable lifestyle factors of physical activity and inactivity can impact on the processes of both *megakaryocytopoiesis and thrombopoiesis, via* epigenetic mechanisms. We propose that physical activity/inactivity can modulate and program platelet phenotype and therefore function. Through recent studies, including our own (manuscripts in preparation), it is becoming increasingly evident that lifestyle factors such as physical (in)-activity and high BMI do impact on platelet function. Thus, various research studies have collectively demonstrated that platelets are indeed reflective of physiological and lifestyle changes, making them sensitive biomarkers of human health. Platelets represent a tangible link to physiological and pathological changes within the body. Future investigations will undoubtedly contribute to a greater mechanistic understanding of the relationship between cardiovascular health, lifestyle factors and platelet biology.

2. Platelets

2.1. Platelet production

Gulio Bizzozero first described platelets as 'spherules piastrine' (little plates) as small cell fragments that clumped together at an injured blood vessel site. He also showed that these blood elements did not have a nucleus [6]. Circulating anucleate platelets are now described as dynamic specialised cells, formed in an elaborate style from their precursor cell, the megakaryocyte. Normal platelet counts ranges between 150 and 450 × 10^3 per microliter of blood, constituting the second most abundant cell type in blood after red blood cells. The average adult produces 10^{11} platelets per day to preserve this count. Platelets travel as resting (quiescent) discoid fragments in the circulation, while an elaborate internal cytoskeleton allows shape changes to occur upon contact with an injured blood vessel.

The size of a mature platelet is approximately 2–4 μm, making them the smallest cells in circulation, while their average thickness is 0.5 μm [7] and their volume about 7 μm^3. Their small size facilitates in their role as 'guardians of the vasculature', as under laminar flow environments, platelets are pushed to the periphery by larger white and red blood cells. Consequently, they remain in close proximity to the blood vessel wall where they can quickly respond to any vascular damage [8]. This enables platelets to perform their main physiological function to prevent blood loss in primary haemostasis by the formation of a 'platelet plug' [9].

Platelets are formed and released into the bloodstream from megakaryocytes (MKs), which reside in the bone marrow [10, 11]. Their production is arguably the most elegant and distinct developmental process in eukaryotes [12]. While accounting for only 0.01% of nucleated bone marrow cells, MKs are also the largest cells, measuring between 50 and 100 μm [13]. Both MK and platelet production, termed megakaryocytopoiesis and thrombopoiesis, are regulated by multiple cytokines, with thrombopoietin (TPO), a hormone produced by the liver and kidneys, being the key regulator. In response to TPO, HSCs differentiate into MKs by differential expression of various transcription factors. This maturation is characterised by a growth in MK size and DNA ploidy levels (endomitosis), enabling the accumulation of RNA, protein and organelles in the MK for packaging into platelets. MKs then migrate to the sinusoidal blood vessels in the vascular niche where numerous long processes called pro-platelets are formed. MKs can extend as many as 20 pro-platelets which branch repeatedly over time. Platelets form at the tips of pro-platelets, receiving organelles, genetic material and granule contents that are transported from the MK cell body. The final point of platelet production occurs in circulation whereby anucleate fragments of pro-platelets bud into pre-platelets [14] and barbell-shaped platelets [15] that are subsequently converted into single platelets in a microtubule-driven process [16] aided by the shear forces within the bloodstream [11]. Platelets then have an average lifespan of 8–10 days, after which they are cleared via phagocytic cells such as macrophages in the spleen. Apoptosis (programmed cell death) is also well recognised in the anucleate platelet [17]. The constant number of platelets in circulation is a consequence of a homeostatic balance between their production and destruction/clearance.

2.2. Platelet structure

2.2.1. Internal

Platelets are unique in their structural composition and, while anucleate, contain a large variety of cellular organelles, granules and mitochondria. Granules are generally secretory vesicles that release their contents either to the platelet surface or to extracellular fluid by endocytosis. Over 300 proteins from platelet granules have been identified in the platelet releasate following activation [18, 19]. Three types of platelet granules have been identified: α-granules, dense granules, and lysosomal granules all of which derive their cargo from MKs. α-granules are the largest and most numerous (50–80 per platelet) encompassing roughly 10% of the platelet volume [20]. They harbour a vast assortment of proteins important for primary haemostasis including integrins (αIIbβ3) immunoglobulin family receptors (e.g. GPVI, PECAM), leucine-rich repeat family receptors (e.g., GPIb-IX-V complex), tetraspanins (e.g., CD9) and other adhesive proteins such as von Willebrand Factor (vWF), fibrinogen, and coagulation factors (Factors V, XI) that participate in secondary haemostasis. While it was previously assumed that platelet α-granules were homogenous populations, [21] suggested that platelets have distinct subpopulations of alpha granules which differentially release their cargo in a context-dependent manner. Dense granules are smaller in size and number (3–8 per platelet) storing high concentrations of non-protein molecules that potentiate platelet activation such as adenosine diphosphate (ADP), adenosine triphosphate (ATP), calcium, histamine, polyphosphate and serotonin [22, 23]. Lysosomal granules are sparse and harbour enzymes such as acid hydrolases and proteases). They function in the digestion of cytosolic components. Secretion of the lysosomal content has key extracellular functions including receptor cleavage, fibrinolysis and degradation of extracellular matrix (ECM) [24]. A recent report has described a possible new type of secretory granule termed a T-granule, after their tubular morphology [11]. These novel electron-dense granules have been proposed to function in toll-like receptor (TLR) organisation and signalling. Platelet granule deficiencies or defects such as the Grey Platelet Syndrome (α-granule deficiency) or Hermansky-Pudlak Syndrome (dense granule deficiency) can cause mild to severe bleeding disorders [25].

Platelets contain functional mitochondria, which despite being few in number, have higher rates of ATP turnover than resting mammalian muscle, suggesting they are very metabolically active [26]. The traditional role of mitochondria in the platelet is to the supply of energy in the form of ATP for primary platelet functions. However, novel functions for mitochondria continue to emerge. Dual activation of platelets with collagen and thrombin results in a subtype of platelets known as collagen and thrombin activated (COAT) platelets. COAT platelets display striking alterations in function and structure to typical "activated platelets" by exhibiting a myriad of features such as phosphatidylserine exposure due to cytoskeletal reorganisation, high microparticle release, and increased levels of fibrinogen on the platelet surface [27]. Mitochondrial membrane potential ($\Delta\Psi_m$) is reduced in (COAT) platelets and decreases in parallel with elevated mitochondrial ROS levels that are necessary for facilitating platelet PS exposure upon activation [28]. Mitochondria are involved in the process of platelet apoptosis [29] and can be released from platelets as potential inflammatory mediators [30].

2.2.2. *Surface receptors*

Platelets express a wide variety of receptors on their membrane, which are fundamental to platelet function and downstream signalling [9]. Major receptors include integrins, leucine-rich repeat receptors (Glycoprotein GPIb/IX/V, Toll-like receptors), C-type lectin receptors (P-Selectin, CLEC-2), tyrosine kinase receptors (Ephrins and Eph kinases), proteins belonging to the immunoglobulin superfamily (GPVI, FcγRIIA) and other receptors shared with vascular cells (TNF receptor type, CD63, CD36, PSGL-1).

Integrins are type I transmembrane cell adhesion receptors [31] consisting of a short intracellular and larger extracellular domain. All integrins contain an α subunit and a β subunit, capable of bi-directional signalling. During signal transduction, they transmit information concerning the chemical and mechanical status of the ECM to the cell [32, 33]. Platelets express five integrin receptors—each with affinity for specific ligands; αIIbβ3 (fibrinogen), α2β1(collagen), α5β1(fibronectin), αVβ3 (vitronectin) and α6β1 (laminin), all of which share related signal transduction processes [34]. On the cytoplasmic face of the plasma membrane, integrins regulate cytoskeletal dynamics and signalling complexes. On the extracellular side, integrins bind with high affinity to either ECM ligands or counter receptors on adjacent cell surfaces [35]. αIIbβ3 is the most abundant platelet integrin, with copy numbers of ~50,000 per platelet, and is present in both alpha granules and the platelet surface. Platelets express a number of G-protein coupled receptors (GPCRs) which constitute a large family of receptors that can identify molecules outside the cell and initiate signal transduction pathways and ultimately cell function. The main GPCRs present on platelets include thrombin receptors called protease-activated receptors (PARs) (PAR1 and PAR4), ADP receptors ($P2Y_1$, $P2Y_{12}$), of which approximately 150 $P2Y_1$ receptors are present on the platelet [36], thromboxane receptors (TPα and TPβ) and glycoprotein receptors [37].

2.3. Platelet function

2.3.1. *Platelet function in primary haemostasis*

Haemostasis can be subdivided into primary haemostasis, secondary haemostasis and fibrinolysis (**Figure 2**) [38]. Platelets prevent blood loss in primary haemostasis, the physiological process which halts bleeding at an injured blood vessel, while maintaining normal blood flow elsewhere in circulation, by the formation of a 'platelet plug' [11]. Secondary haemostasis refers to the deposition of insoluble fibrin that is generated by the coagulation cascade.

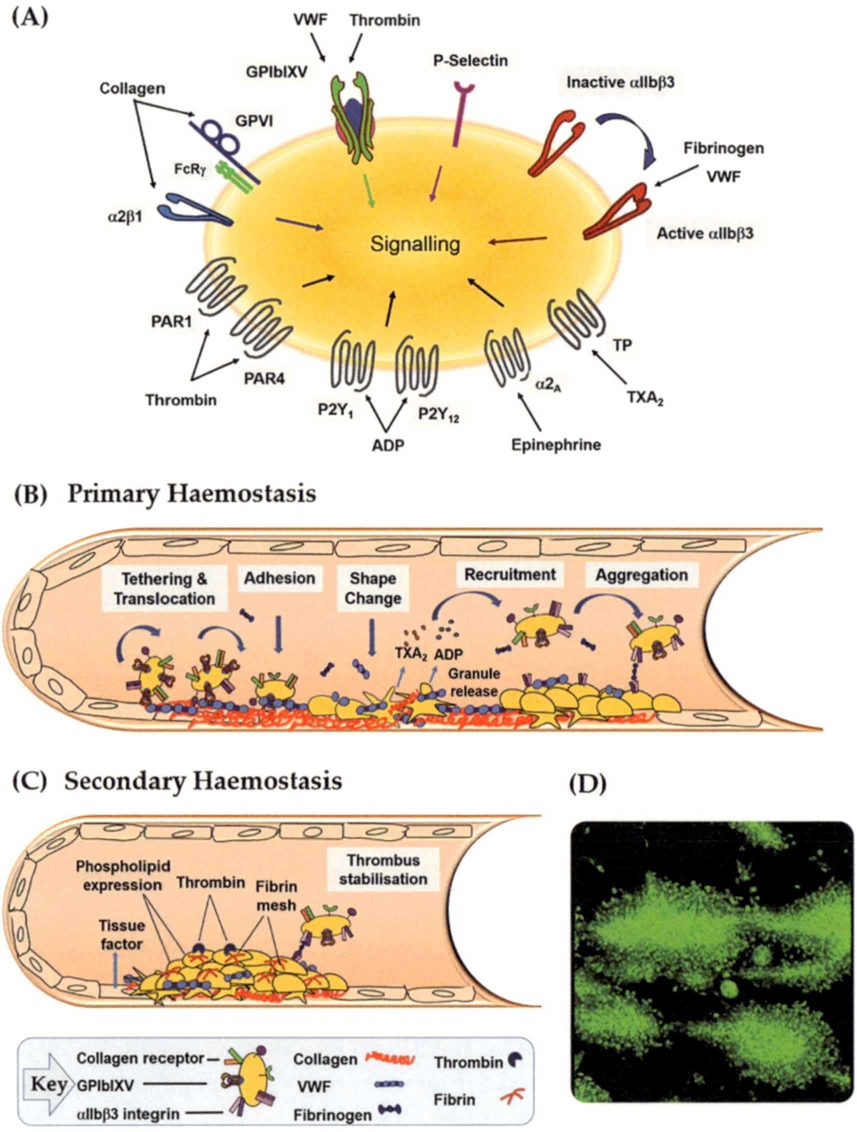

Figure 2. (A) Diagram showing key platelet membrane receptors involved in platelet function in haemostasis and thrombosis. (B) Distinct phases of platelet function in primary haemostasis: Initial platelet tethering and adhesion at a site of vascular damage, firm adhesion and shape change, activation, granule secretion and further recruitment of platelets leading to platelet aggregation. (C) Secondary haemostasis: Formation, deposition and cross-linking of insoluble fibrin, generated by the coagulation cascade, to stabilise the primary platelet plug. (D) Image of platelet aggregates (fluorescently-labelled, green) on collagen fibrils following *in vitro* blood flow at arterial shear (1800 s^{-1}) using a parallel-plate blood perfusion chamber (*unpublished data—Gerardene Meade-Murphy*).

Finally, fibrinolysis results in the breakdown of blood clots during wound healing involving the interplay of a number of enzymes [38]. A healthy endothelium provides a non-adhesive surface for platelets. However, in areas of vascular injury, the sub-endothelium is exposed and platelets may adhere quickly to different extracellular matrix components, and then form a platelet plug. This process is achieved through three distinct processes—platelet adhesion, platelet activation and secretion, and platelet aggregation [39].

2.3.2. *Platelet adhesion*

Platelet adhesion entails a collaborative effort of various platelet receptors, fundamentally leading to platelet activation and aggregation. ECM elements that platelets adhere to include proteins such as collagen, vWF, fibronectin, laminin and fibrinogen among others [40, 41]. Among these subendothelial substrates, the thrombogenic fibrillar collagens type I and III are the most powerful intermediaries of platelet adhesion due to their robust activating potential and affinity for vWF [42]. Following vascular damage, initial platelet 'tethering' is mediated by the interaction between the A1 domain of vWF deposited in the subendothelial matrix of the damaged vessel wall, and the GPIbα in the platelet receptor GPIb-IX-V. This interaction is particularly important at high shear rates supporting platelet translocation (i.e. decelerating platelets and keeping them in close contact with the endothelium) over the subendothelium, but not stable adhesion [43]. This interaction allows engagement of other platelet receptors. vWF/GPIb-IX interface also induces platelet activation signalling events, resulting in integrin activation [44].

Following platelet tethering, platelet collagen receptors, GPVI and α2β1 interact with exposed collagen and promote platelet adhesion and activation. GPVI is non-covalently coupled to the Fc Receptor chain (FcRγ) [45, 46] and has been acknowledged as the major signalling receptor for collagen. FcRγ has an immunoreceptor tyrosine-based activation motif (ITAM) on its cytoplasmic sphere. After collagen binding to GPVI, the ITAM motif on the GPVI/FcRγ complex is phosphorylated, resulting in activation of the Syk kinase pathways that phosphorylate downstream targets, ultimately resulting in increased cytosolic Ca^{2+} and subsequent platelet shape change, granule secretion and integrin activation. GPVI has a low affinity for collagen, rendering it unable to mediate stable adhesion alone. The α2β1 integrin then maintains stable adhesion to collagen. α2β1 stimulates downstream steps indirectly by reinforcing GPVI-collagen interactions [41, 47] and by direct signalling leading to activation of αIIbβ3. Platelet Ca^{2+} signalling is markedly dissimilar between GPVI and α2β1 suggesting that the alliance regarding GPVI and α2β1 supports optimal platelet adhesion.

The final step of platelet adhesion occurs via binding of platelets to other ECM components such as fibronectin, laminin and immobilised vWF. Platelets bind to fibronectin via the α5β1 receptor and αIIbβ3, whilst adhesion to laminin is mediated by their α6β1 receptor. Stable binding of platelets elicits activation pathways involving tyrosine kinases and signal transduction GPCR receptor signalling, cumulatively resulting in elevated cytosolic Ca^{2+} levels, cytoskeletal reorganisation and integrin activation.

2.3.3. *Platelet activation and secretion*

Once platelet adhesion has occurred at the site of vessel wall damage, platelet activation needs to be maintained for haemostasis to continue. Essential for the amplification of platelet activation is the production and release of soluble agonists at the site of damage [26], which act in an autocrine and paracrine manner to amplify platelet activation and recruit further circulating platelets. These agonists consist of TxA_2, ADP, epinephrine and thrombin. ADP is secreted from platelet dense granules and binds to its relevant receptors, $P2Y_{12}$ and $P2Y_1$ on the platelet surface [48]. ADP is also released from red blood cells at the site of vascular damage [37]. Binding of ADP initiates a full complement of activation events such as elevation of intracellular platelet Ca^{2+}, TxA_2 synthesis, protein phosphorylation, shape change, granule release, and most importantly, activation of αIIbβ3 [49]. $P2Y_{12}$ is also the target of a class of antiplatelet drugs called thienopyridines (ticlopidine, clopidogrel, prasugrel), widely used in the prevention of vascular events in patients with CVD.

TxA_2 is a potent platelet agonist synthesised from arachidonic acid through the COX pathway and TxA_2 synthase enzymes. It subsequently binds to TPα and TPβ receptors that differ in their cytoplasmic tails, causing vasoconstriction, shape change, protein phosphorylation, secretion and platelet aggregation [50, 51]. Indeed, high levels of TxA_2 have been implicated in CVD, whilst inhibition of TxA_2 synthesis through aspirin-mediated COX inhibition is a major anti-platelet target.

The agonist thrombin rapidly accumulates at sites of vascular damage and has major functions in promoting and stabilising thrombus formation. Platelets release factors that support the activation of prothrombin, which after a complex series of sequential events in the coagulation cascade, results in the generation of thrombin [52]. The increase in cytosolic Ca^{2+} after platelet activation results in platelet phosphatidylserine (PS) exposure on the activated platelet membrane providing a procoagulant surface for thrombin to interact with its PAR1 and PAR4 (G protein-coupled) receptors.

Uniquely, thrombin activates its PAR-receptors by cleaving an N-terminal part at a consensus site. Cleavage exposes a new binding site that acts as a ligand to activate the receptor. Thrombin is the most powerful platelet activator, initiating an entire complement of platelet responses (shape change, granule secretion, TxA_2 synthesis, aggregation etc.) [53]. Thrombin can activate platelets at extremely low concentrations and within seconds, increases cytosolic level of Ca^{2+}, eliciting downstream signalling events. Unlike ADP, TxA_2 or thrombin, the catecholamine epinephrine is a weak agonist unable to cause shape change alone. However, it works collectively with the other agonists increasing their potential to activate. Epinephrine's mode of action is to inhibit cAMP formation by the platelet α_2A-adrenergic receptor [54].

2.3.4. Platelet aggregation

The ultimate step in primary haemostasis is platelet aggregation, caused by crosslinking of αIIbβ3 on adjacent platelets by fibrinogen (**Figure 2**). While platelet aggregation is a complex process involving different receptors (αIIbβ3 and GPIbα) and ligands (fibrinogen, fibronectin, and vWF), the main process involves the integrin αIIbβ3. On resting platelets, integrin αIIbβ3 has a low affinity for its ligands fibrinogen and vWF, which dramatically increases upon platelet activation. Binding of the main agonists to their respective receptors induce intracellular signals that disrupt the complex between the cytoplasmic tails of αIIbβ3. This ultimately leads to a conformational change in its extracellular globular head domains from a low affinity resting state to a high affinity activated state in order to bind extracellular ligands such as fibrinogen and vWF. Irreversible activation of αIIbβ3 is a prerequisite for the development of irreversible platelet aggregates. Due to the symmetrical nature of fibrinogen, platelets can be 'bridged' and platelet aggregates are formed [40]. While αIIbβ3 is the major player in platelet aggregation, couplings between other platelet receptors and their ligands could be incorporated in aggregation. Some of these comprise CD40 ligand [55] interaction with αIIbβ3, the vWF-GPIb complex [56], and an involvement of fibronectin in stabilising platelet aggregation [57, 58]. Cadherin-6 was recently acknowledged as a new counter-receptor for αIIbβ3, involved in platelet aggregation [59].

Bi-directional 'inside out' and 'outside in' signals are transferred by both integrin subunits, mediating receptor conformation and platelet function [44]. After platelet stimulation with ADP, the signal from the ADP receptors is conducted to the intracellular domain of the cytoplasmic tail of αIIbβ3 and subsequently transmitted through a series of events to the extracellular domain (inside-out) causing a conformational change in the extracellular domain that binds to its ligand. Inside-out signalling necessitates binding of talin and kindlins to the cytoplasmic domain of β3. After ligand binding, a signal is sent to the cell (outside-in) to control platelet function such as filopodia and lamellipodia extension to dull platelet spreading [34, 44]. Furthermore, αIIbβ3 outside-in signals can also act as a break to curb excessive platelet activation by activated SHIP-1 [60].

2.3.5. *Bio rheological factors in platelet aggregation*

A significant factor influencing platelet aggregation is the distinct shear environment experienced within the vascular system. Platelets are subjected to fluctuating haemodynamic conditions *in vivo* such as shear stress and shear rate. The latter refers to the rate of increase of blood flow velocity, whereas the former denotes the force per unit area on the vessel wall [61]. Shear rates experienced by platelets range from slow flow in veins (shear rate 10 s^{-1} to 500 s^{-1}) to small arteries (approx. 2000 s^{-1}) to diseased or pathological arteries, where extremely high shear rates (up to 40,000) have been described [61, 62]. Increasing shear rate activates platelets itself. At low shear rates (<1000 s^{-1}), platelet aggregation is primarily facilitated by αIIbβ3-fibrinogen interactions. At shear rates typically over 5000 s^{-1} (but between 1000 and 10,000), a two-step sequential process occurs. The first depends on the adhesive properties of GP1bα and αIIbβ3 and is facilitated by the formation of reversible platelet aggregates. The second relies on the generation of platelet agonists and involves the irreversible activation of αIIbβ3 to form stable aggregates [63]. Accordingly, both fibrinogen and vWF, and receptors GPIbα and αIIbβ3, have distinctive but complementary roles in platelet aggregation subject to the haemodynamic environment.

2.3.6. *Signal transduction during platelet function*

The role of platelets in haemostasis is reliant on the equilibrium between activatory and inhibitory signals [64]. Inhibitory signals from the vasculature prevent platelet activation in healthy vessels. Activatory signals present at an injured blood vessel initiate platelet activation, and are managed by endogenous negative signalling regulators. Negative regulators and pathways include the ITIM containing receptors that are postulated to reduce activation of PLC, PI3K and integrin αIIbβ3 [64]. The Wnt-β-catenin pathway has recently gained attention as negative regulator of platelet function [65, 66]. Wnt3a is one of these glycoproteins, which is secreted from activated platelets. It has been suggested to activate the canonical Wnt-β-catenin pathway as constituents of this pathway have been identified in platelets. The regulation of small GTPases such as Rap1, Rac1, RhoA and Cdc42 has been suggested as players in this inhibition of platelet function [65]. These pathways are less well characterised than the classical activation pathways [67]. The primary platelet inhibiting signals produced by healthy ECs are nitric oxide and prostacyclin. Both NO and prostacyclin relax blood vessels and prevent platelet activation [68]. NO is synthesised from several cells including platelets, ECs and RBCs and plays vital roles in maintaining platelets in a resting state [69–71]. Prostacyclin (PGI_2) is a physiological anti-aggregating agent produced constitutively by ECs as a result of arachidonic acid metabolism by cyclooxygenase (COX) enzymes.

Following endothelial damage, endogenous inhibitory signals are overcome and platelets react rapidly to limit blood loss. The activating stimuli, such as collagen, vWF, ADP, TxA_2 and thrombin, which induce platelet adhesion, activation and aggregation ultimately regulate a central set of signalling mediators that support activation. Three principle mediator families of platelet activation are phospholipase C (PLC), protein kinase C (PKC) and phosphatidylinositide-3-kinase (PI3K), and underlie two crucial events in platelet activation—secretion of amplifying mediators and activation of integrin αIIbβ3.

2.4. Platelet function beyond haemostasis

Platelets are recognised as pivotal players in numerous other processes ranging from inflammation and atherosclerosis, fighting microbial infection and tumour growth and metastasis. Platelets are equipped to influence inflammation, the innate immune response and infection, by acting as sentinels in pathogen detection. They express a collection of pattern recognition receptors called toll-like receptors (TLRs) that identify molecular motifs called pathogen associated molecular patterns (PAMPs) and initiate immune responses [72]. Platelets express functional TLR 1–9, whilst TLR2 stimulation in platelets by bacteria through the activation of the PI3K signalling pathway induces a pro-inflammatory response [20]. Platelet TLR9 activation has been associated with thrombosis and oxidative stress, and is found within the T-granules in platelets [11].

Platelets interact with leukocytes, monocytes and granulocytes through different receptor-ligand interactions (P-Selectin and PSGL-1 interaction) enabling intercellular communication [73]. The capacity of platelets to store and release copious inflammatory cytokines and chemokines is intimately associated to their role in inflammation. Platelets release microparticles linked to inflammatory pathways and which are associated with inflammatory diseases such as rheumatoid arthritis [74]. Platelets play key roles in infection and immune response to bacterial and viral infections. They are the first cell type to arrive at areas of vascular infection [75]. Thrombocytopenia (low platelet count) is a well-established manifestation of sepsis and studies suggests platelets play a functional role in the pathogenesis of sepsis and multi-organ failure [76].

Platelets have significant roles in the pathogenesis of metastasis [77]. Tumour cells can aggregate platelets *in vitro* and it has been proposed that platelet adhesion to metastatic cells can act as a 'cloak' around circulating tumour cells therefore acting as a shield for immune clearance. This phenomenon of platelet cloaking has resulted in pro-survival, pro-angiogenic and epithelial mesenchymal transition (EMT) in cancer cells [78]. Platelets also release growth factors such as VEGF and PDGF that can expedite tumour growth [79]. Platelets are deeply implicated in wound healing and bone health, and indeed the use of platelet rich plasma (PRP) therapy (i.e. rich in growth factors and bioactive substances) is effective in osteoarthritis [10] and in muscle damage such as rotator cuff tendinopathy [80].

2.5. Assessing function: platelet function tests (PFT)

The different functions of platelets may be reliably detected with a wide spectrum of tests (**Table 1**). These can be utilised to identify inherited or acquired platelet dysfunction, monitor antiplatelet therapy, manage various aspects of platelet banking and transfusion, and to aid in the understanding of platelet physiology in basic research. PFTs are centred around principles of platelet function such as platelet adhesion and aggregation, platelet function under shear conditions, and measurement of the platelet releasate [81]. Platelet function testing began with the evaluation of the bleeding time (the time taken for platelets to occlude an *in vivo* wound), using the Duke procedure [82], before the development of light transmission aggregometry (LTA) revolutionised the study of platelet function. Considered the historical gold standard, LTA is a relatively easy technique that involves stirring a suspension of platelet rich plasma in a cuvette in the presence of a platelet agonist (such as ADP or collagen). The cuvette is placed between a light source and photocell. Agonist addition causes *in vitro* platelet aggregation and changes in light absorbance, which is detected by the photocell [83].

Investigation of platelet function in the environmental *milieu* of whole blood under conditions that take into account most of the physiological parameters that influence platelet adhesion and aggregation (red blood cells, white blood cells, plasma) is important. The PFA-100, The Impact-R Cone and Plate analyser and the global thrombosis test are examples of such assays. The PFA-100 assesses platelet aggregation under high shear where platelets are activated in whole blood by an amalgamation of high shear stress (5000–6000 s^{-1}) and agonists (e.g. collagen and ADP), resulting in closure of an aperture [84]. The Impact-R Cone and Plate analyser is a point-of-care (POC) device which measures global platelet function by testing platelet adhesion and aggregation in whole blood, under arterial shear conditions [85, 86]. In this assay, platelet adhesion is dependent on plasma proteins vWF, fibrinogen and RBCs and WBCs. The addition of platelet agonists such as arachidonic acid (AA) and ADP in the system has enabled the evaluation of dual anti-platelet therapy [87–89]. The system is effective in the assessment of platelet function disorders in adults, [90] children [91] and new-borns [92].

2.6. Platelet indices

Platelet indices are useful as inexpensive non-invasive biomarkers for assessing platelet activation [93]. Platelet indices are straightforwardly measured by semi-automated counters in complete blood counts (CBC) and usually include four factors; platelet count (PLT), mean

Method	Sample	Principle
Platelet aggregation assays		
Light transmission aggregometry (LTA)	Citrated PRP	Measurement of light transmission in response to agonist-induced platelet activation
Impedance Aggregation	Citrated WB	Monitors changes in electrical impedance in relation to agonist-induced platelet aggregation. Whole blood specimen diluted 1:1 with physiologic saline, prior to testing
VerifyNow	Citrated WB	Turbidimetric optical detection of platelet aggregation in whole blood in response to agonists/inhibitors
Plateletworks	Citrated WB	Platelet counting pre- and post-activation
Assays measuring release reactions		
Lumi-aggregometry	Citrated WB	Combination of LTA/WBA with nucleotide release
Soluble platelet release markers (Txb2, PF4, BTF, scD40L)	Urine, serum, citrated plasma	Typically measured by ligand binding ELISA immunoassays
Shear-based assays		
PFA-100/200	Citrated WB	High-shear platelet adhesion and aggregation during formation of a platelet plug
Impact R cone and plate analyser	Citrated WB	Shear-induced platelet adhesion-aggregation upon specific surface
Global thrombosis test	Native WB	High-shear platelet plug formation—measurement of time cessation of WB flow
Platelet activation based assays		
VASP phosphorylation	Citrated WB	Flow cytometry or ELISA measurement of VASP phosphorylation
Flow cytometry	Citrated WB, PRP, washed platelets	Measurement of platelet glycoproteins, and activation markers; platelet leukocyte aggregates, platelet microvesicles

Typical platelet function tests range from assessment of their primary haemostatic function including measurement of granule secretion using lumi-aggregometry to whole blood shear based assays, which measure platelet function under flow conditions. Platelet function can be tested in washed platelets, whole blood or platelet rich plasma.

Table 1. Platelet function tests.

platelet volume (MPV), platelet distribution width (PDW), plateletcrit (PCT), and depending on the analyser, platelet large cell ratio (PLCR). PLT is a universal indicator of haemostasis in a clinical setting and is utilised as a sensitive biomarker for a range of diseases. High PLT, even within the physiological range of 150–450 μl, is associated with a greater risk of thrombosis and CVD suggesting that enhanced PLT encourage platelet hyperactivity and a pro-inflammatory state [94]. However, the consequence of high platelet numbers that are still within physiologic ranges remains unclear [95].

The indices MPV, PDW and PLCR are quantitative measures of the variability in platelet size. MPV reflects the average platelet size while PDW reflects the volume variability in platelet size

[94, 96]. The volume of circulating platelets is heterogeneous with subsequent functional differences. Some authors suggest that larger platelets are metabolically more active than smaller platelets, that they have faster rates of aggregation and release higher quantities of pro-thrombotic elements such as TxA_2 and ADP [97]. MPV and PDW levels can be altered in several diseases including T2DM [98], CVD and atherosclerosis [93] (Berger et al., 2010), and in this regard they have been suggested as markers of subclinical platelet activation. PLCR and PCT may serve as sensitive biomarkers of platelet health [99]. PLCR indicates the percentage of large platelets present in blood [93]. PLCR is significantly higher in subjects with dyslipidaemia compared to healthy subjects [100] and higher in children with T2DM compared to healthy children. Moreover, Rechcinski et al., have hypothesised that PLCR has the potential to be a prognostic biomarker [101]. Importantly, thrombogenicity of large platelets may put individuals at higher risk of acute cardiovascular events. PCT is the volume of blood occupied by platelets as a percentage, similar to the erythrocyte measurement of haematocrit (HCT). PCT reflects total platelet mass and is calculated as PLT × MPV/10^7, providing comprehensive information about platelet activity. PCT has been proposed as a novel predictor of cardiovascular risk and higher PCT is associated with the risk of re-infarction and long-term mortality in CVD patients [102]. However, the clinical significance, reference values and efficacy of some of these parameters are still under exploration.

In peripheral blood, there is ample interplay between RBCs, WBCs and platelets [103] and altered levels of blood cells and their morphology have been associated with CVD [104]. Platelet adhesion and aggregate size, is influenced by platelet indices, RBC and WBC [86]. RBCs encourage platelets towards the vessel wall [105], which can affect platelet adhesion and aggregation. In this context, it is important to investigate the associations between the various indices of each blood cell to interpret the multicellular contribution to both thrombogenesis and CVD risk.

3. Conclusion

Novel techniques continue to emerge and develop the knowledge surrounding the platelet function regulation. The modern "omics" revolution enables simultaneous quantification of hundreds of molecules (e.g. protein or mRNA) from a single sample and their signatures may be reflective of platelet function changes. The amalgamation of transcriptomic and proteomic data and subsequent bioinformatic analysis will lead to a more complete characterisation of platelet function in response to environmental stimuli [106, 107]. Epigenetics and its ancillary elements, including platelet secreted microvesicles (MVs), and microRNA (miRNA), and regulation of the platelet mitochondrial genome are new avenues of investigation and testing in platelet research [108–110].

Acknowledgements

This work was supported through the Irish Higher Education Authority Programme for Research in Third Level Institutes (HEA-PRTLI Cycle 4: T3 Targeted Therapeutics & Theranostics), the Health Research Board of Ireland (HRB-RP/2005/184, HRA/2009/122, HRA/2009/122/R),

Science Foundation Ireland (SFI-04/BR/B0577, 11/TIDA/B1927) to Dr. R.P. Murphy, and the International Visitor Award, Dublin City University, Dublin, Ireland to Drs. B. Degryse and R.P. Murphy. R.P. Murphy is a management committee member of the EU H2020 COST action-CM1406, EpiChemBio. The ground based protocol and studies of microgravity (3-Day Dry Immersion) were undertaken in MEDES, The Institute for Space Medicine and Physiology, *Toulouse* University Hospital, France and was supported and funded by the Centre National d'Etudes Spatiales (*CNES*—National Centre for Space Studies) to M-A.C., R.P.M., N.N., L.T.

Conflict of interest

We wish to confirm that there are no known conflicts of interest associated with this publication and there has been no significant financial support for this work that could have influenced its outcome.

Author details

Laura Twomey[1], Robert G. Wallace[1], Philip M. Cummins[2], Bernard Degryse[1], Sinead Sheridan[3], Michael Harrison[4], Niall Moyna[1], Gerardene Meade-Murphy[5], Nastassia Navasiolava[6], Marc-Antoine Custaud[6] and Ronan P. Murphy[1*]

*Address all correspondence to: ronan.murphy@dcu.ie

1 School of Health and Human Performance, Dublin City University, Dublin, Ireland

2 School of Biotechnology, Dublin City University, Dublin, Ireland

3 School of Public Health, Li Ka Shing Faculty of Medicine, The University of Hong Kong, Hong Kong

4 School of Health Sciences—Health Sport and Exercise Science, Waterford Institute of Technology, Ireland

5 Department of Pharmacology and Therapeutics, University College Cork, Cork, Ireland

6 Centre de Recherche Clinique du CHU d'Angers, Angers, France

References

[1] Juhola J, Magnussen CG, Viikari JS, Kahonen M, Hutri-Kahonen N, Jula A, et al. Tracking of serum lipid levels, blood pressure, and body mass index from childhood to adulthood: The cardiovascular risk in young Finns study. The Journal of Pediatrics. 2011;**159**(4):584-590

[2] McNeal CJ, Dajani T, Wilson D, Cassidy-Bushrow AE, Dickerson JB, Ory M. Hypercholesterolemia in youth: Opportunities and obstacles to prevent premature atherosclerotic cardiovascular disease. Current Atherosclerosis Reports. 2010;**12**(1):20-28

[3] Ruggeri ZM. Platelet adhesion under flow. Microcirculation (New York, NY: 1994). 2009;**16**(1):58-83

[4] Woolthuis CM, Park CY. Hematopoietic stem/progenitor cell commitment to the megakaryocyte lineage. Blood. 2016;**127**(10):1242-1248

[5] Huang H, Cantor AB. Common features of megakaryocytes and hematopoietic stem cells: What's the connection? Journal of Cellular Biochemistry. 2009;**107**(5):857-864

[6] Roy GC. Micro-organisms or granules of Bizzozero in the blood. The Indian Medical Gazette. 1882;**17**(4):110

[7] Xu XR, Zhang D, Oswald BE, Carrim N, Wang X, Hou Y, et al. Platelets are versatile cells: New discoveries in hemostasis, thrombosis, immune responses, tumor metastasis and beyond. Critical Reviews in Clinical Laboratory Sciences. 2016;**53**(6):409-430

[8] McFadyen JD, Kaplan ZS. Platelets are not just for clots. Transfusion Medicine Reviews. 2015;**29**(2):110-119

[9] Cimmino G, Golino P. Platelet biology and receptor pathways. Journal of Cardiovascular Translational Research. 2013;**6**(3):299-309

[10] Patel S, Dhillon MS, Aggarwal S, Marwaha N, Jain A. Treatment with platelet-rich plasma is more effective than placebo for knee osteoarthritis: A prospective, double-blind, randomized trial. The American Journal of Sports Medicine. 2013;**41**(2):356-364

[11] Thon JN, Peters CG, Machlus KR, Aslam R, Rowley J, Macleod H, et al. T granules in human platelets function in TLR9 organization and signaling. The Journal of Cell Biology. 2012;**198**(4):561-574

[12] Simon LM, Edelstein LC, Nagalla S, Woodley AB, Chen ES, Kong X, et al. Human platelet microRNA-mRNA networks associated with age and gender revealed by integrated plateletomics. Blood. 2014;**123**(16):e37-e45

[13] Machlus KR, Italiano JEJ. The incredible journey: From megakaryocyte development to platelet formation. The Journal of Cell Biology. 2013;**201**(6):785-796

[14] Thon JN, Italiano JE. Platelet formation. Seminars in Hematology. 2010;**47**(3):220-226

[15] Schwertz H, Weyrich AS. Platelet precursors display bipolar behavior. The Journal of Cell Biology. 2010;**191**(4):699-700

[16] Poulter NS, Thomas SG. Cytoskeletal regulation of platelet formation: Coordination of F-actin and microtubules. The International Journal of Biochemistry & Cell Biology. 2015;**66**:69-74

[17] Gyulkhandanyan AV, Mutlu A, Freedman J, Leytin V. Selective triggering of platelet apoptosis, platelet activation or both. British Journal of Haematology. 2013;**161**(2):245-254

[18] Coppinger JA, Cagney G, Toomey S, Kislinger T, Belton O, McRedmond JP, et al. Characterization of the proteins released from activated platelets leads to localization of novel platelet proteins in human atherosclerotic lesions. Blood. 2004;**103**(6):2096-2104

[19] Wijten P, van Holten T, Woo LL, Bleijerveld OB, Roest M, Heck AJ, et al. High precision platelet releasate definition by quantitative reversed protein profiling—Brief report. Arteriosclerosis, Thrombosis, and Vascular Biology. 2013;**33**(7):1635-1638

[20] Blair P, Flaumenhaft R. Platelet alpha-granules: Basic biology and clinical correlates. Blood Reviews. 2009;**23**(4):177-189

[21] Italiano JE Jr, Battinelli EM. Selective sorting of alpha-granule proteins. Journal of Thrombosis and Haemostasis. 2009;**7**(Suppl 1):173-176

[22] Morrell CN, Aggrey AA, Chapman LM, Modjeski KL. Emerging roles for platelets as immune and inflammatory cells. Blood. 2014;**123**(18):2759-2767

[23] Ruiz FA, Lea CR, Oldfield E, Docampo R. Human platelet dense granules contain polyphosphate and are similar to acidocalcisomes of bacteria and unicellular eukaryotes. The Journal of Biological Chemistry. 2004;**279**(43):44250-44257

[24] Heijnen H, van der Sluijs P. Platelet secretory behaviour: as diverse as the granules...or not? Journal of Thrombosis and Haemostasis. 2015;**13**(12):2141-2151

[25] Nurden AT, Nurden P. The Gray Platelet syndrome: Clinical spectrum of the disease. Blood Reviews. 2007;**21**(1):21-36

[26] Zharikov S, Shiva S. Platelet mitochondrial function: From regulation of thrombosis to biomarker of disease. Biochemical Society Transactions. 2013;**41**(1):118-123

[27] Jobe SM, Wilson KM, Leo L, Raimondi A, Molkentin JD, Lentz SR, et al. Critical role for the mitochondrial permeability transition pore and cyclophilin D in platelet activation and thrombosis. Blood. 2008;**111**(3):1257-1265

[28] Choo HJ, Saafir TB, Mkumba L, Wagner MB, Jobe SM. Mitochondrial calcium and reactive oxygen species regulate agonist-initiated platelet phosphatidylserine exposure. Arteriosclerosis, Thrombosis, and Vascular Biology. 2012;**32**(12):2946-2955

[29] Hayashi T, Tanaka S, Hori Y, Hirayama F, Sato EF, Inoue M. Role of mitochondria in the maintenance of platelet function during in vitro storage. Transfusion Medicine (Oxford, England). 2011;**21**(3):166-174

[30] Boudreau LH, Duchez AC, Cloutier N, Soulet D, Martin N, Bollinger J, et al. Platelets release mitochondria serving as substrate for bactericidal group IIA-secreted phospholipase A2 to promote inflammation. Blood. 2014;**124**(14):2173-2183

[31] Hynes RO, Zhao Q. The evolution of cell adhesion. The Journal of Cell Biology. 2000;**150**(2):F89-F96

[32] Cosemans JM, Iserbyt BF, Deckmyn H, Heemskerk JW. Multiple ways to switch platelet integrins on and off. Journal of Thrombosis and Haemostasis. 2008;**6**(8):1253-1261

[33] Bennett JS, Berger BW, Billings PC. The structure and function of platelet integrins. Journal of Thrombosis and Haemostasis. 2009;**7**(Suppl 1):200-205

[34] Shattil SJ, Newman PJ. Integrins: Dynamic scaffolds for adhesion and signaling in platelets. Blood. 2004;**104**(6):1606-1615

[35] Humphries JD, Byron A, Humphries MJ. Integrin ligands at a glance. Journal of Cell Science. 2006;**119**(Pt 19):3901-3903

[36] Ohlmann P, de Castro S, Brown GGJ, Gachet C, Jacobson KA, Harden TK. Quantification of recombinant and platelet P2Y(1) receptors utilizing a [(125)I]-labeled high-affinity antagonist 2-iodo-N(6)-methyl-(N)-methanocarba-2′-deoxyadenosine-3′,5′-bisphosphate ([(125)I]MRS2500). Pharmacological Research. 2010;**62**(4):344-351

[37] Rivera J, Lozano ML, Navarro-Nunez L, Vicente V. Platelet receptors and signaling in the dynamics of thrombus formation. Haematologica. 2009;**94**(5):700-711

[38] Gale AJ. Continuing education course #2: Current understanding of hemostasis. Toxicologic Pathology. 2011;**39**(1):273-280

[39] Clemetson KJ. Platelets and primary haemostasis. Thrombosis Research. 2012;**129**(3): 220-224

[40] Ruggeri ZM, Mendolicchio GL. Adhesion mechanisms in platelet function. Circulation Research. 2007;**100**(12):1673-1685

[41] Broos K, Feys HB, De Meyer SF, Vanhoorelbeke K, Deckmyn H. Platelets at work in primary hemostasis. Blood Reviews. 2011;**25**(4):155-167

[42] Farndale RW, Siljander PR, Onley DJ, Sundaresan P, Knight CG, Barnes MJ. Collagen-platelet interactions: Recognition and signalling. Biochemical Society Symposium. 2003;**70**:81-94

[43] Cranmer SL, Ashworth KJ, Yao Y, Berndt MC, Ruggeri ZM, Andrews RK, et al. High shear-dependent loss of membrane integrity and defective platelet adhesion following disruption of the GPIbalpha-filamin interaction. Blood. 2011;**117**(9):2718-2727

[44] Li Z, Delaney MK, O'Brien KA, Du X. Signaling during platelet adhesion and activation. Arteriosclerosis, Thrombosis, and Vascular Biology. 2010;**30**(12):2341-2349

[45] Tsuji M, Ezumi Y, Arai M, Takayama H. A novel association of Fc receptor gamma-chain with glycoprotein VI and their co-expression as a collagen receptor in human platelets. The Journal of Biological Chemistry. 1997;**272**(38):23528-23531

[46] Jarvis GE, Atkinson BT, Snell DC, Watson SP. Distinct roles of GPVI and integrin alpha(2) beta(1) in platelet shape change and aggregation induced by different collagens. British Journal of Pharmacology. 2002;**137**(1):107-117

[47] Mazzucato M, Cozzi MR, Battiston M, Jandrot-Perrus M, Mongiat M, Marchese P, et al. Distinct spatio-temporal Ca^{2+} signaling elicited by integrin alpha2beta1 and glycoprotein VI under flow. Blood. 2009;**114**(13):2793-2801

[48] Kaplan ZS, Jackson SP. The role of platelets in atherothrombosis. Hematology American Society of Hematology Education Program. 2011;**2011**:51-61

[49] Andrews RK, Berndt MC. Platelet physiology and thrombosis. Thrombosis Research. 2004;**114**(5-6):447-453

[50] Nakahata N. Thromboxane A2: Physiology/pathophysiology, cellular signal transduction and pharmacology. Pharmacology & Therapeutics. 2008;**118**(1):18-35

[51] Angiolillo DJ, Ueno M, Goto S. Basic principles of platelet biology and clinical implications. Circulation Journal: Official Journal of the Japanese Circulation Society. 2010;**74**(4):597-607

[52] Monroe DM, Hoffman M, Roberts HR. Platelets and thrombin generation. Arteriosclerosis, Thrombosis, and Vascular Biology. 2002;**22**(9):1381-1389

[53] de Candia M, Fiorella F, Lopopolo G, Carotti A, Romano MR, Lograno MD, et al. Synthesis and biological evaluation of direct thrombin inhibitors bearing 4-(piperidin-1-yl)pyridine at the P1 position with potent anticoagulant activity. Journal of Medicinal Chemistry. 2013;**56**(21):8696-8711

[54] Horn NA, Anastase DM, Hecker KE, Baumert JH, Robitzsch T, Rossaint R. Epinephrine enhances platelet-neutrophil adhesion in whole blood in vitro. Anesthesia and Analgesia. 2005;**100**(2):520-526

[55] Andre P, Prasad KS, Denis CV, He M, Papalia JM, Hynes RO, et al. CD40L stabilizes arterial thrombi by a beta3 integrin-dependent mechanism. Nature Medicine. 2002;**8**(3):247-252

[56] Kulkarni S, Dopheide SM, Yap CL, Ravanat C, Freund M, Mangin P, et al. A revised model of platelet aggregation. The Journal of Clinical Investigation. 2000;**105**(6):783-791

[57] Ni H, Denis CV, Subbarao S, Degen JL, Sato TN, Hynes RO, et al. Persistence of platelet thrombus formation in arterioles of mice lacking both von Willebrand factor and fibrinogen. The Journal of Clinical Investigation. 2000;**106**(3):385-392

[58] Cho J, Mosher DF. Role of fibronectin assembly in platelet thrombus formation. Journal of Thrombosis and Haemostasis. 2006;**4**(7):1461-1469

[59] Dunne E, Spring CM, Reheman A, Jin W, Berndt MC, Newman DK, et al. Cadherin 6 has a functional role in platelet aggregation and thrombus formation. Arteriosclerosis, Thrombosis, and Vascular Biology. 2012;**32**(7):1724-1731

[60] Dai B, Wu P, Xue F, Yang R, Yu Z, Dai K, et al. Integrin-alphaIIbbeta3-mediated outside-in signalling activates a negative feedback pathway to suppress platelet activation. Thrombosis and Haemostasis. 2016;**116**(5):918-930

[61] Sakariassen KS, Orning L, Turitto VT. The impact of blood shear rate on arterial thrombus formation. Future Science OA. 2015;**1**(4):FSO30

[62] Jackson SP. The growing complexity of platelet aggregation. Blood. 2007;**109**(12):5087-5095

[63] Munnix IC, Cosemans JM, Auger JM, Heemskerk JW. Platelet response heterogeneity in thrombus formation. Thrombosis and Haemostasis. 2009;**102**(6):1149-1156

[64] Jones CI, Barrett NE, Moraes LA, Gibbins JM, Jackson DE. Endogenous inhibitory mechanisms and the regulation of platelet function. Methods in Molecular Biology (Clifton, NJ). 2012;**788**:341-366

[65] Steele BM, Harper MT, Macaulay IC, Morrell CN, Perez-Tamayo A, Foy M, et al. Canonical Wnt signaling negatively regulates platelet function. Proceedings of the National Academy of Sciences of the United States of America. 2009;**106**(47):19836-19841

[66] Macaulay IC, Thon JN, Tijssen MR, Steele BM, MacDonald BT, Meade G, et al. Canonical Wnt signaling in megakaryocytes regulates proplatelet formation. Blood. 2013;**121**(1): 188-196

[67] Bye A, Rosjo H, Aspenes ST, Condorelli G, Omland T, Wisloff U. Circulating microRNAs and aerobic fitness—The HUNT-study. PLoS One. 2013;**8**(2):e57496

[68] Smolenski A. Novel roles of cAMP/cGMP-dependent signaling in platelets. Journal of Thrombosis and Haemostasis. 2012;**10**(2):167-176

[69] Gkaliagkousi E, Ferro A. Nitric oxide signalling in the regulation of cardiovascular and platelet function. Frontiers in Bioscience (Landmark edition). 2011;**16**:1873-1897

[70] Gambaryan S, Tsikas D. A review and discussion of platelet nitric oxide and nitric oxide synthase: Do blood platelets produce nitric oxide from L-arginine or nitrite? Amino Acids. 2015;**47**(9):1779-1793

[71] Benz PM, Fleming I. Can erythrocytes release biologically active NO? Cell Communication and Signaling. 2016;**14**(1):22

[72] Cognasse F, Nguyen KA, Damien P, McNicol A, Pozzetto B, Hamzeh-Cognasse H, et al. The inflammatory role of platelets via their TLRs and Siglec receptors. Frontiers in Immunology. 2015;**6**:83

[73] Zago AC, Simon DI, Wang Y, Sakuma M, Chen Z, Croce K, et al. The importance of the interaction between leukocyte integrin mac-1 and platelet glycoprotein Ib-a for leukocyte recruitment by platelets and for the inflammatory response to vascular injury. Arquivos Brasileiros de Cardiologia. 2008;**90**(1):54-63

[74] Boilard E, Nigrovic PA, Larabee K, Watts GFM, Coblyn JS, Weinblatt ME, et al. Platelets amplify inflammation in arthritis via collagen-dependent microparticle production. Science. 2010;**327**(5965):580-583

[75] Metcalf Pate KA, Lyons CE, Dorsey JL, Queen SE, Adams RJ, Morrell CN, et al. TGFbeta-mediated downregulation of thrombopoietin is associated with platelet decline in

asymptomatic SIV infection. Journal of Acquired Immune Deficiency Syndromes (1999). 2014;**65**(5):510-516

[76] Guclu E, Durmaz Y, Karabay O. Effect of severe sepsis on platelet count and their indices. African Health Sciences. 2013;**13**(2):333-338

[77] Labelle M, Begum S, Hynes RO. Direct signaling between platelets and cancer cells induces an epithelial-mesenchymal-like transition and promotes metastasis. Cancer Cell. 2011;**20**(5):576-590

[78] Cooke NM, Spillane CD, Sheils O, O'Leary J, Kenny D. Aspirin and P2Y12 inhibition attenuate platelet-induced ovarian cancer cell invasion. BMC Cancer. 2015;**15**:627

[79] Erpenbeck L, Schon MP. Deadly allies: The fatal interplay between platelets and metastasizing cancer cells. Blood. 2010;**115**(17):3427-3436

[80] Kesikburun S, Tan AK, Yilmaz B, Yasar E, Yazicioglu K. Platelet-rich plasma injections in the treatment of chronic rotator cuff tendinopathy: A randomized controlled trial with 1-year follow-up. The American Journal of Sports Medicine. 2013;**41**(11):2609-2616

[81] Paniccia R, Priora R, Liotta AA, Abbate R. Platelet function tests: A comparative review. Vascular Health and Risk Management. 2015;**11**:133-148

[82] Duke WW. The relation of blood platelets to hemorrhagic disease. By W.W. Duke. Journal of the American Medical Association. 1983;**250**(9):1201-1209

[83] Born GV, Haslam RJ, Goldman M. Comparative effectiveness of adenosine analogues as inhibitors of blood-platelet aggregation and as vasodilators in man. Nature. 1965;**205**: 678-680

[84] Karger R, Reuter K, Rohlfs J, Nimsky C, Sure U, Kretschmer V. The platelet function analyzer (PFA-100) as a screening tool in neurosurgery. ISRN Hematology. 2012;**2012**:839242

[85] Varon D, Dardik R, Shenkman B, Kotev-Emeth S, Farzame N, Tamarin I, et al. A new method for quantitative analysis of whole blood platelet interaction with extracellular matrix under flow conditions. Thrombosis Research. 1997;**85**(4):283-294

[86] Peerschke EI, Silver RT, Weksler BB, Yin W, Bernhardt B, Varon D. Examination of platelet function in whole blood under dynamic flow conditions with the cone and plate(let) analyzer: Effect of erythrocytosis and thrombocytosis. American Journal of Clinical Pathology. 2007;**127**(3):422-428

[87] Savion N, Varon D. Impact—The cone and plate(let) analyzer: Testing platelet function and anti-platelet drug response. Pathophysiology of Haemostasis and Thrombosis. 2006;**35**(1-2):83-88

[88] Anand SX, Kim MC, Kamran M, Sharma SK, Kini AS, Fareed J, et al. Comparison of platelet function and morphology in patients undergoing percutaneous coronary intervention receiving bivalirudin versus unfractionated heparin versus clopidogrel pretreatment and bivalirudin. The American Journal of Cardiology. 2007;**100**(3):417-424

[89] Koshy SK, Salahuddin S, Karunakaran B, Nalakath SY, Bhaskaran J, Haridas PV, et al. Aspirin and clopidogrel resistance using the cone and plate(let) analyser in Indian patients with coronary artery disease. Heart Asia. 2014;**6**(1):159-162

[90] Goldschmidt N, Spectre G, Brill A, Zelig O, Goldfarb A, Rachmilewitz E, et al. Increased platelet adhesion under flow conditions is induced by both thalassemic platelets and red blood cells. Thrombosis and Haemostasis. 2008;**100**(5):864-870

[91] Revel-Vilk S, Varon D, Shai E, Agmon Y, Hyam E, Daas N, et al. Evaluation of children with a suspected bleeding disorder applying the impact-R [cone and plate(let) analyzer]. Journal of Thrombosis and Haemostasis. 2009;**7**(12):1990-1996

[92] Levy-Shraga Y, Maayan-Metzger A, Lubetsky A, Shenkman B, Kuint J, Martinowitz U, et al. Platelet function of newborns as tested by cone and plate(let) analyzer correlates with gestational age. Acta Haematologica. 2006;**115**(3-4):152-156

[93] Budak YU, Polat M, Huysal K. The use of platelet indices, plateletcrit, mean platelet volume and platelet distribution width in emergency non-traumatic abdominal surgery: A systematic review. Biochemia Medica. 2016;**26**(2):178-193

[94] Khandekar MM, Khurana AS, Deshmukh SD, Kakrani AL, Katdare AD, Inamdar AK. Platelet volume indices in patients with coronary artery disease and acute myocardial infarction: An Indian scenario. Journal of Clinical Pathology. 2006;**59**(2):146-149

[95] Madjid M, Fatemi O. Components of the complete blood count as risk predictors for coronary heart disease: In-depth review and update. Texas Heart Institute Journal. 2013;**40**(1):17-29

[96] Vagdatli E, Gounari E, Lazaridou E, Katsibourlia E, Tsikopoulou F, Labrianou I. Platelet distribution width: A simple, practical and specific marker of activation of coagulation. Hippokratia. 2010;**14**(1):28-32

[97] Mangalpally KK, Siqueiros-Garcia A, Vaduganathan M, Dong JF, Kleiman NS, Guthikonda S. Platelet activation patterns in platelet size sub-populations: Differential responses to aspirin in vitro. Journal of Thrombosis and Thrombolysis. 2010;**30**(3):251-262

[98] Yilmaz T, Yilmaz A. Relationship between altered platelet morphological parameters and retinopathy in patients with type 2 diabetes mellitus. Journal of Ophthalmology. 2016;**2016**:9213623

[99] Yilmaz T, Yilmaz A. Altered platelet morphological parameters in patients with retinal vein occlusion. European Review for Medical and Pharmacological Sciences. 2016;**20**(10):1934-1939

[100] Leite NR, Siqueira de Medeiros M, Mury WV, Matsuura C, Perszel MB, Noronha Filho G, et al. Platelet hyperaggregability in obesity: Is there a role for nitric oxide impairment and oxidative stress? Clinical and Experimental Pharmacology & Physiology. 2016;**43**(8):738-744

[101] Rechcinski T, Jasinska A, Forys J, Krzeminska-Pakula M, Wierzbowska-Drabik K, Plewka M, et al. Prognostic value of platelet indices after acute myocardial infarction treated with primary percutaneous coronary intervention. Cardiology Journal. 2013;**20**(5):491-498

[102] Gul M, Uyarel H, Akgul O, Akkaya E, Surgit O, Cakmak HA, et al. Long-term prognostic significance of admission plateletcrit values in patients with non-ST elevation myocardial infarction. Blood Coagulation & Fibrinolysis. 2016;**27**(6):696-701

[103] Ho CH. White blood cell and platelet counts could affect whole blood viscosity. Journal of the Chinese Medical Association. 2004;**67**(8):394-397

[104] Cetin M, Bakirci EM, Baysal E, Tasolar H, Balli M, Cakici M, et al. Increased platelet distribution width is associated with ST-segment elevation myocardial infarction and thrombolysis failure. Angiology. 2014;**65**(8):737-743

[105] Chesnutt JK, Han HC. Effect of red blood cells on platelet activation and thrombus formation in tortuous arterioles. Frontiers in Bioengineering and Biotechnology. 2013;**1**:18

[106] McRedmond JP, Park SD, Reilly DF, Coppinger JA, Maguire PB, Shields DC, et al. Integration of proteomics and genomics in platelets: A profile of platelet proteins and platelet-specific genes. Molecular & Cellular Proteomics. 2004;**3**(2):133-144

[107] Ferroni P, Riondino S, Vazzana N, Santoro N, Guadagni F, Davi G. Biomarkers of platelet activation in acute coronary syndromes. Thrombosis and Haemostasis. 2012;**108**(6):1109-1123

[108] Baccarelli AA, Byun HM. Platelet mitochondrial DNA methylation: A potential new marker of cardiovascular disease. Clinical Epigenetics. 2015;**7**:44

[109] Kaudewitz D, Skroblin P, Bender LH, Barwari T, Willeit P, Pechlaner R, et al. Association of MicroRNAs and YRNAs with platelet function. Circulation Research. 2016;**118**(3):420-432

[110] Dahiya N, Sarachana T, Kulkarni S, Wood WHIII, Zhang Y, Becker KG, et al. miR-570 interacts with mitochondrial ATPase subunit g (ATP5L) encoding mRNA in stored platelets. Platelets. 2017;**28**(1):74-81

Chapter 6

Platelets: Functional Biomarkers of Epigenetic Drift

Laura Twomey, Robert G. Wallace, Marco Mangone,
Bernard Degryse, Sinead Sheridan,
Michael Harrison, Niall Moyna,
Gerardene Meade-Murphy, Nastassia Navasiolava,
Marc-Antoine Custaud and Ronan P. Murphy

Additional information is available at the end of the chapter

http://dx.doi.org/10.5772/intechopen.83447

Abstract

Cardiovascular disease (CVD) risk factors can be classed as modifiable or non-modifiable. Physical inactivity and obesity represent major behavioural risk factors for the initiation, development and progression of CVD. Platelet dysfunction is pivotal to the aetiology of CVD, a chronic vascular inflammatory condition, which is characterised by a lag time between onset and clinical manifestation. This indicates the role of epigenetic drift, defined by stochastic patterns of gene expression not dependent on dynamic changes in coding DNA. The epigenome, a collection of chemical marks on DNA and histones, is established during embryogenesis and modified by age and lifestyle. Biogenesis and effector function of non-coding RNA, such as microRNA, play a regulatory role in gene expression and thus the epigenetic mechanism. In this chapter, we will focus on the effect of the modifiable risk factors of physical activity/inactivity and overweight/obesity on platelet function, via epigenetic changes in both *megakaryocytopoiesis* and *thrombopoiesis*. We will also discuss the role of acute exercise on platelet function and the impact of cardiorespiratory fitness (CRF) on platelet responses to acute exercise. This chapter will highlight the potential role of platelets as circulating functional biomarkers of epigenetic drift to implement, optimise and monitor CVD preventive management strategies.

Keywords: platelets, epigenetics, microRNA, lifestyle, physical activity, physical inactivity, cardiovascular disease, preventive medicine

1. Introduction

1.1. Epigenetics

Epigenetics describes modifiable changes that occur to genes, *via* chemical modifications and/or varying states of chromatin organisation and structure, which alter the gene expression without altering the DNA sequence itself. Smoking habits, obesity, ageing and physical fitness among others are examples of environmental factors that have been suggested to have a long-term influence on epigenetic changes [1]. Epigenetics may be classed as three distinct but highly interconnected processes; DNA methylation, histone modification and RNA-associated silencing (**Figure 1**). DNA methylation and histone modification alter DNA accessibility for transcriptional machinery and chromatin structure. These changes are heritable and can be passed down between generations through either mitosis or meiosis. DNA methylation involves the addition of a methyl group to the 5-position of cytosine by DNA methyl-transferases, at areas known as CpG islands. Methyl groups control gene expression by binding to promoter sites of the gene. This changes the affinity of methylation-sensitive binding proteins, and is associated with transcriptional gene silencing [2]. Whilst required for normal development, changes in DNA methylation have been linked to CVD conditions such as atherosclerosis. For example, the athero-protective oestrogen receptor genes ESR1 and ESR2, usually expressed in SMCs, are hyper-methylated in atherosclerosis [3, 4].

Unlike the platelet transcriptome and proteome, the investigation of epigenetic processes is an almost completely unexplored area in platelet biology, as analysis of these mechanisms requires DNA [2]. Although anucleate, platelets have functionally active mitochondria, with mitochondrial DNA (mtDNA) that can also be methylated, moderating the control of mitochondrial gene expression. Interestingly, Zhong and colleagues recently reported that *de novo* DNA synthesis in mitochondria, and its subsequent oxidation, plays a key role in triggering the innate immune response. Mitochondria can regulate how immune cells respond to infection and tissue damage, producing pro- or anti-inflammatory signals by regulating Krebs cycle metabolites or the production of reactive oxygen species (ROS). More and more examples are being found of mitochondrial functions being repurposed in unexpected ways to contribute to many biological processes, including inflammatory signalling [5].

Understanding epigenetic regulation of mitochondrial genes in platelets is proving crucial to understanding their implication in CVD development. Novel research by Baccarelli and Byun showed that CVD patients had significantly higher platelet mtDNA methylation than healthy individuals in MT-CO1, MT-C02, MT-CO3 and MT-TL1 genes involved in ATP synthesis [6]. These results suggest that DNA methylation in platelet mitochondria could be a potential contributor to CVD development through the regulation of platelet function.

Histone, proteins that structure DNA into units known as nucleosomes, can be modified at their amino-acid tails. Histone modifications refer to the post-translational alterations of the N-termini of these tails that subsequently modify histone-DNA interactions [7]. Acetylation is a major type of histone modification involving the addition or removal of an acetyl group. This process is catalysed by proteins known as histone acetyltransferases (HATs) and histone

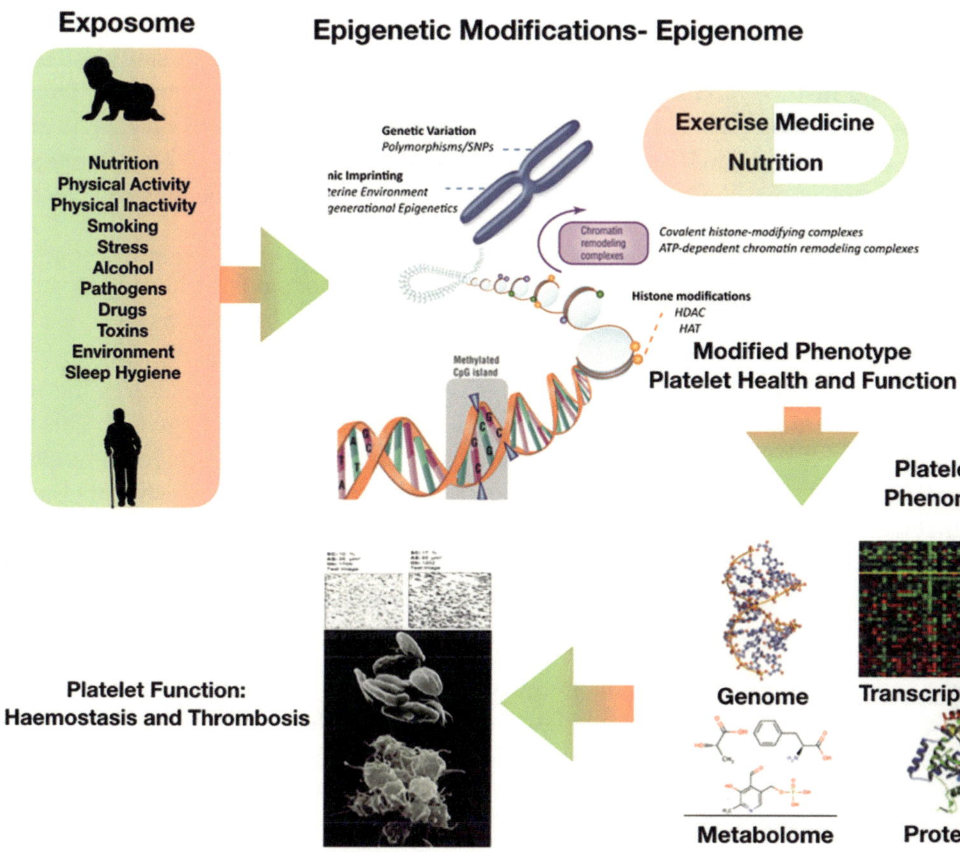

Figure 1. Modifiable risk factors covalently alter the static genome by processes involving epigenetic writers, erasers and readers, which fine-tune gene expression in an age and lifestyle dependent manner, a process known as *epigenetic drift*. The term epigenetics refers to stable patterns of gene expression and they are not dependent on dynamic changes in coding DNA. These gene expression states are encoded in the epigenome—a collection of chemical marks on DNA or on histones that are established during embryogenesis and are modified by age and environment over a person's lifetime. Studies have uncovered stochastic DNA methylation drift that reflects imperfect maintenance of epigenetic marks. We hypothesise that *drift* creates epigenetic mosaicism in ageing haemopoietic stem cells in the bone marrow. This in turn impacts on HSC differentiation, maturation and platelet production. This may highlight platelets as a functional diagnostic index of cardiovascular competence. The fact that the initiation and progression of CVD is characterised by a lag time between onset and clinical manifestation provides a window of opportunity for the implementation of intervention strategies to reduce the CVD burden. In addition, such studies will better inform primary, secondary and tertiary preventive strategies—promoting an ageing well paradigm and optimising a person's disease free years.

deacetyltransferases (HDACs). This mechanism alters chromatin structure (heterochromatin versus euchromatin) to influence gene expression [3].

RNA-based epigenetic processes involve non-coding RNA (ncRNA) and are factors in the chromatin-based regulation of gene expression [8]. ncRNAs can be classified as either long or short. Whilst long ncRNAs are a major form of RNA-based epigenetic regulation, some small ncRNAs also have a function in chromatin-based silencing. For example, microRNA (miRNA) is a subset of ncRNA that negatively regulates gene transcription by degrading or repressing target mRNA [9]. miRNA can control the expression of important epigenetic regulators such as histone deacetylases and DNA methyl-transferases and similarly, DNA methylation and histone modification can control the expression of some miRNA, thereby forming a feedback loop [10]. This complex crosstalk between miRNA and epigenetic pathways forms an epigenetic-miRNA regulatory circuit, arranging the whole gene expression profile. Disruption of this circuit interferes with normal physiological functions and can contribute to disease process.

Individuals age differently and lifestyle factors such as exercise or smoking have been shown to delay or accelerate the ageing process, respectively [11]. These observations have resulted in the search for molecular markers to predict and monitor age-associated disease. DNA methylation is associated with chronological age over time [12]. *Epigenetic drift* is the term given to epigenetic modifications as they occur as a direct consequence of age [13]. This was previously observed when DNA methylation marks in identical twins differed increasingly as a function of age [1]. Monozygous twins share a common genotype and while this study found that the twins were epigenetically synonymous during childhood, older twins showed significant differences in their total content and dispersal of histone acetylation and DNA methylation. Disparity in these epigenetic marks between twins may be as a result of lifestyle influences such as diet, physical activity levels, stress and smoking.

Epigenetic drift affects the majority of the genome over time leading to biological ageing (**Figure 2**). Ageing is a natural process associated with the de-regulation of histone tags, senescence-associated lncRNA, a gradual de-regulation of DNA methylation, in a potential linear fashion depicted by age-predictive linear models [14]. However, an individual exposed to either environmental or genetic risk factors may show signs of premature ageing as a result of either lifestyle or environmental risk factors. De-regulation of DNA methylation can increase the susceptibility to chronic diseases like CVD. Furthermore, it has been hypothesised that a healthy lifestyle may reserve a more intact epigenome, promoting longevity [15]. In a recent compelling study by Horvath and colleagues, a novel, sensitive and highly robust DNAm age estimator (based on 391 CpGs) was developed for human fibroblasts, keratinocytes, buccal cells, endothelial cells, lymphoblastoid cells, skin, blood and saliva samples [16]. This seminal research builds upon and overcomes the limitations of the two ground-breaking studies on the epigenetic clock (biological/molecular age) and its relationship with chronological age. These studies, one a blood-based age estimator [14] and the other a pan-tissue estimator [17], facilitated age estimates (DNAmAge) that are widely used in epidemiological studies. The novel 'skin and blood clock' overcomes the technical and sensitivity limitations of the previous DNAmAge biomarker panels.

DNA methylation undergoes extensive changes during differentiation of self-renewing stem cells [18, 19]. Indeed, DNA methylation is involved in the production of MKs and subsequent transcription. Lifestyle components such as physical inactivity and obesity may incur epigenetic changes in the production of platelets from megakaryocytes. Thus, platelets could signify a marker of megakaryocyte epigenetic drift, holding substantial predictive potential of disease. Epigenetic changes in the megakaryocyte genome such as hypomethylation of genes

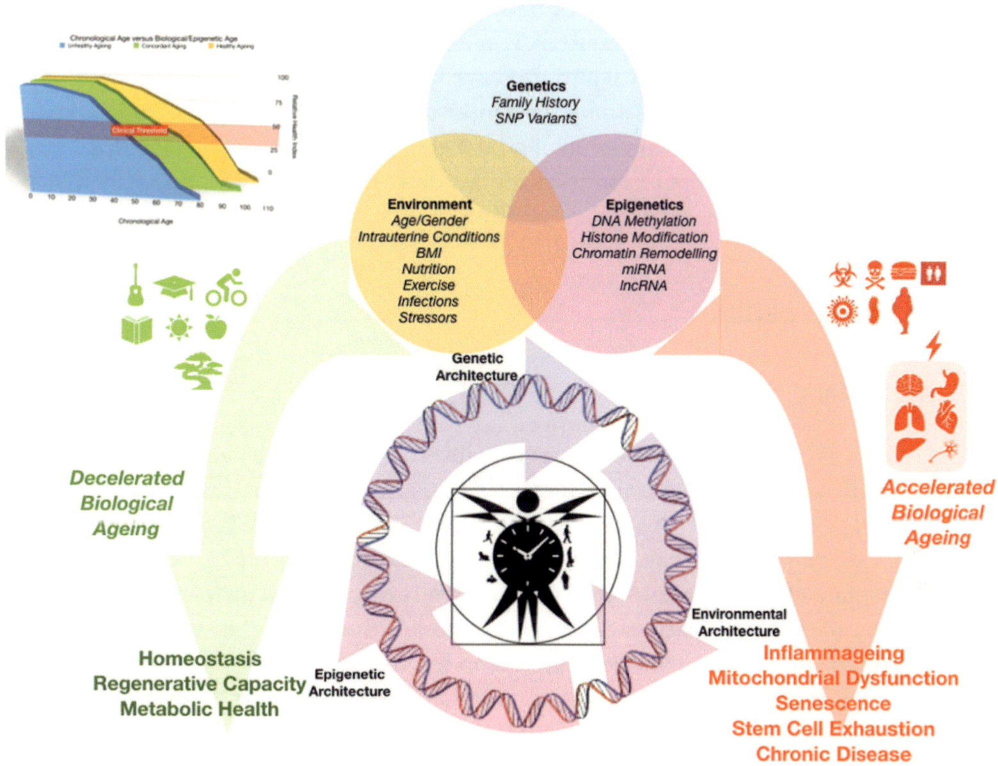

Figure 2. Schematic representation of ageing well versus accelerated ageing. Poor lifestyle choices such as inactivity and diet are rapidly becoming a global pandemic, accelerating many chronic illnesses such as cardiovascular disease, metabolic disorder, diabetes, Alzheimer's disease and cancer. Understanding their pathophysiology is important for the development of future therapeutic interventions, stratification of clinical trials and to challenge our current perception of engaging with cost effective measures such as lifestyle management. The aetiology of age-related chronic illnesses involves a complex interplay between many biological processes and is modulated by non-modifiable and modifiable risk factors.

determining PLT or changes in histone acetylation with aging have been suggested to play an important role in platelet function [20].

miRNAs are short [18–24] nucleotide long, non-coding RNAs that function in post-transcriptional regulation of gene expression. They inhibit translation by binding with the 3′-untranslated (UTR) regions of their target mRNA. Here, the miRNA promotes silencing of various genes [21, 22], hence now termed 'fine-tuners' of cellular phenotypes. They are thought to be involved in the regulation of ~60% of human genes [23, 24]. miRNA can be classed as intronic, exonic and intergenic miRNA, according to the location of their encoding genes [25]. Intronic miRNAs account for approximately 70% of all transcribed miRNA [26, 27]. Intergenic miRNAs are found between two protein-coding genes and employ their own promoters and regulatory molecules. As miRNAs target mRNA by imperfect binding, each miRNA has multiple targets, enabling miRNAs to regulate over half of the human genome [28]. The miRNA population within a cell can be highly concentrated, with tens of thousands of miRNA copies per cell.

They possess a long half-life (a half-life of between 28 and 220 h has been reported) and are very stable [29]. Turnover of mature miRNA is required for rapid changes in miRNA expression profiles. Regulation of miRNA maturation occurs during various steps throughout their biogenesis at both a transcriptional and post-transcriptional level [30]. Transcriptional regulation involves alterations to the expression of a host gene such as epigenetic regulation (where miRNA genes located near CpG islands in the genome are found to be hyper-methylated). Post-transcriptional mechanisms define modifications in miRNA processing and stability [31].

2. Platelet Epigenetics

2.1. Platelet miRNA

Platelet function is a highly regulated process. Despite their anucleate nature, platelets accommodate a small but competent transcriptome that is employed for translation of various proteins with significant physiological functions. Platelets have been shown to retain genetic material derived from their megakaryocyte precursor. Approximately 32% of all human genes are present in platelets at the mRNA level [23]. It is well accepted that platelets contain the necessary splicing machinery, rough ER and polyribosomes that allow the synthesis of proteins required for their functioning [32]. Perhaps due to the requirement of sustaining a proteome over an ~8-day life span, the fact that the average half-life of a cellular protein is 46 h, or the necessity to adapt to environmental stimuli, it is equitable to assume that the platelet must also retain its transcriptome, as well as processes of nucleated cells such as splicing, translating and post-transcriptional RNA mechanisms. The fact that platelets contain mRNA and are capable of protein synthesis has raised the issue of how these mRNAs are regulated. Notably, stored platelets in blood banks can synthesis integrin β3 [33]. The existence and functionality of a miRNA pathway in the anucleate human platelet was first described in a landmark study by Landry and co-workers [34], who showed by locked nucleic acid (LNA) microarray profiling, that platelets harboured an impressive number (219) of miRNA. Further analysis discovered the presence of functional processing miRNA machinery in platelets—Dicer and Ago2—suggesting that partial biogenesis of mature miRNA from pre-miRNA could occur within platelets themselves [34, 35] as pre-miRNAs have been identified at low levels (21 transcripts) in platelets [23]. Star or passenger strand miRNA have also been identified [36]. Accordingly, the detection of nuclear miRNA microprocessor Drosha and DGCR8 in platelets has not been observed, consistent with their anucleate nature. Moreover, miRNA-associated Ago2 complexes were identified, in addition to the presence of $P2Y_{12}$ in Ago2 precipitates, suggesting a regulation of $P2Y_{12}$ by miRNAs [34].

The next breakthrough study in platelet miRNA biology revealed that a protein involved in platelet granule release, platelet vesicle-associated membrane protein 8 (VAMP8), was associated with distinctly different platelet aggregation responses to epinephrine in healthy donors, and that VAMP8 was regulated by miR-86 [37]. Since then, the platelet miRNA field has grown exponentially, whereby a number of studies have suggested a physiological role for miRNA in the regulation of platelet function. Most notably, research by Nagalla et al. [38], who focused on the roles of miRNA as biomarkers of platelet reactivity and controllers of platelet mRNA disparity, demonstrated that miRNA profiles of healthy subjects (n = 19) were associated with the response of platelet aggregation to epinephrine [38]. They also employed a computational approach to produce possible miRNA-mRNA pairs (miR-200b: PRKAR2B, miR-495: KLHL5 and miR-107: CLOCK), pairings which were experimentally validated in cell lines. Networks of miRNA-mRNA pairs also associated with age, gender and race [39, 40]. Other reports on agonist-induced platelet activation by thrombin (and ADP) show differential expression of platelet miRNA compared to resting platelets [41].

Progression in miRNA detection techniques has led to the revelation of 40 new miRNA sequences, expanding the total amount of platelet expressed miRNAs to more than twice that (544) of the initial finding [23, 35, 42]. Transcriptomic approaches show that miRNA make up the majority (80%) of all small RNAs in platelets. Furthermore, comparison of RNA and miRNA by cell type showed that despite low RNA yields, platelets express high quantities of miRNA compared to their nucleated counterparts.

A number of highly expressed miRNA have been characterised in human platelets, some of which are involved in myeloid cell differentiation, megakaryocytopoiesis and thrombopoiesis. miR-223 has been identified as the most highly expressed platelet miRNA [38, 43, 44] and has roles in thrombopoiesis and megakaryocyte differentiation [24]. miR-223 regulates ADP $P2Y_{12}$, a target for existing anti-platelet drug therapy. The 3′-UTR of $P2Y_{12}$ mRNA has been identified as complementary to the miR-223 seed region. Platelet miR-223 has also been observed to be decreased in subjects who show high levels of platelet activation whilst on clopidogrel therapy. Furthermore, miR-223-deficient mice show reduced bleeding times, larger thrombi and elevated sensitivity to low doses of thrombin, suggesting an important role of miR-223 in modulating platelet function [45]. miR-126 plays central roles in vascular inflammation and is thought to be the second most highly expressed miRNA in platelets [46]. miR-126 was found to correlate with circulating P-Selectin levels in T2DM subjects and this level was sensitive to aspirin treatment, signifying a platelet origin. miR-126 is postulated to regulate ADAM9 and $P2Y_{12}$ receptor expression in platelets and inhibition of miR-126 in mice distinctly reduces platelet aggregation [47].

Existence of miRNA in platelets is multifaceted. Besides their obvious function as regulators of platelet protein expression, platelet miRNAs have been labelled as biomarkers of disease and platelet activation, markers of mature megakaryocyte miRNA and as a means of understanding megakaryocyte/platelet gene expression [48]. The majority of platelet miRNA are supposedly formed in the megakaryocyte and packaged into platelets upon formation. For example, miR-146b positively regulates megakaryopoieses by targeting and down regulating the megakaryopoieses inhibitor PDGFRA [49, 50]. miR-142 has also been reported to inhibit megakaryocyte production. In miR-142 knockout mice, platelet counts are decreased and MK differentiation is modified, including reduced proplatelet network establishment [51]. The total extent to which MK and platelet mature miRNA patterns correlate remains an area of active investigation. A significant correlation between the miRNA levels was found using three separate studies [34, 52, 53].

Perhaps, the most intriguing feature regarding platelet miRNA is their extracellular function. miRNA can be packaged and delivered to distant cells in the form of platelet microvesicles (PMV) and/or microparticles (PMP), fulfilling novel processes of gene regulation in target cells [54]. Initial studies by Laffont and Gidlöf demonstrated the functionality of platelet miRNA [24, 55]. Functional complexes of miR-223 and Argonaute 2 protein (Ago2) packaged in MVs from activated platelets were found to modulate the expression of targeted endothelial cell endogenous mRNA transcripts FBXW7 and EFNA1. This miR-223/Ago2 complex has also been shown to reduce expression levels of insulin-like growth factor 1 receptor in endothelial cells, and to promote human umbilical vein endothelial cell (HUVEC) apoptosis [56].

Gidlof et al. suggested that platelet miRNA could modulate vascular endothelial inflammatory responses [55]. They described a down regulation of intercellular adhesion molecule 1 (ICAM-1) gene expression in cultured human microvascular endothelial cells after exposure to miR-320b, which is secreted upon platelet activation and reduced in platelet thrombi aspirated from patients with ST-segment elevation myocardial infarction (STEMI). The relevance of this intercellular transfer was further reinforced when Liang et al., showed that platelet-released miR-223 through platelet MPs can encourage lung cancer cell invasion by targeting the tumour suppressor EPB41L3 [57].

Novel research has shown that platelet MPs containing miRNA can also be internalised by primary human macrophages and deliver functional miR-126-3p. miR-126-3p caused a down-regulation in the expression of four predicted mRNA targets of miR-126-3p and a reduction in macrophage cytokine release. This suggests that platelet miRNA-containing MPs can modify the macrophage transcriptome and potentially reprogram their function [58]. Finally, platelet-derived exosomes have recently been shown to carry miR-223, miR-339 and miR-21, which can be transferred to SMCs affecting PDGRFβ [13].

2.2. Platelets, lifestyle and miRNA in the aetiology of CVD

Efforts in coping with CVD require further understanding of its aetiology in order to develop effective management strategies. Epidemiological studies in adults have acknowledged a set of characteristic risk factors that predict the probability of a person developing clinical manifestations of disease [59, 60]. CVD risk factors are classed as modifiable or non-modifiable. Non-modifiable risk factors include age, ethnicity, gender and family history. Modifiable risk factors include hypertension, smoking, diabetes, unhealthy diet, cholesterol, physical inactivity (PI, sedentary lifestyle and low cardiorespiratory fitness) and overweight/obesity. Risk factors for CVD track from childhood into adulthood [61] and are strong predictors of subclinical atherosclerosis in early adulthood. The majority of CVD is caused by modifiable risk factors and up to 80% of CVD may be prevented if risk factors are avoided [62]. Physical inactivity and obesity are primary potent risk factors, both of which can severely impact platelet physiology.

Platelets have central roles in CVD [63] contributing to both early stages of endothelial dysfunction and advanced stages of the plaque rupture [64]. Platelets participate in early stage disease initiation through multiple mechanisms enabling adhesion to dysfunctional endothelium. Activated platelets express high levels of adhesion receptors (e.g., ICAM1, P-Selectin, CD40L) associated with oxidised-LDL (ox-LDL) that contributes to vascular inflammation

[65]. TLR signalling may also play a role in the progression of atherosclerosis by binding of lipopolysaccharides (LPS) to TLR4 on platelets and also mediating platelet-neutrophil interactions.

Direct cell-cell communication through platelet P-Selectin and CD40 ligand (CD40L) encourages inflammatory processes [66]. CD40L is thought to be at the heart of the atherosclerotic process, with 90% of circulating CD40L residing in platelets. CD40L is sent to the platelet surface upon activation, where it can initiate numerous inflammatory processes. The release of CD40L is intrinsically linked to αIIbβ3 as αIIbβ3 antagonists can block the release of sCD40L from activated platelets *in vitro*. Recently, platelet CD40 was shown to mediate the formation of platelet-leukocyte aggregates (stimulates leukocyte activation) and release inflammatory chemokines that activate endothelial cells, supporting atherosclerosis [66]. The significance of P-Selectin in atherosclerosis has been demonstrated in P-Selectin deficient animals that were protected from the disease. The role of platelet P-Selectin was clarified further by Huo et al., who illustrated that the introduction of P-Selectin expressing platelets into ApoE (−/−) mice accelerated atherosclerosis, whereas mice injected with platelets lacking P-Selectin formed smaller plaques [67].

Platelet-derived microparticles released upon activation may further amplify the progression of atherosclerosis through processes of adhesion, coagulation, inflammation and lipid metabolism [68]. Platelets also provide a huge repertoire of additional inflammatory mediators including a vast array of chemokines and cytokines that contribute to the crosstalk of platelets with other inflammatory cells—e.g., endothelial cells, monocytes, neutrophils, dendritic cells and T-cells [66]. The major function of platelets in atherosclerosis is the recruitment of leukocytes through direct receptor-ligand interactions or amplification of leukocyte recruitment through chemokine release. This bidirectional relationship is extremely important as platelets encourage leukocyte differentiation into a pro-adhesive and pro-migratory phenotype, and the leukocytes secrete mediators that reciprocally activate platelets.

Following atherosclerotic plaque rupture in severe CVD states, the exposure of thrombogenic substrates to circulating platelets instantly triggers platelet adhesion, activation and aggregation, forming a prothrombotic surface and subsequently encouraging thrombosis, vasoconstriction and vascular occlusion. Activated platelets expose phospholipids on their surface, which also promotes the coagulation cascade and subsequent fibrin production [64]. Given the critical roles of platelets in the pathogenesis of atherosclerosis and the development of acute thrombotic events, anti-platelet therapy has been widely employed in the primary and secondary prevention of CVD. Some of the current anti-platelet therapy drugs include Aspirin, which irreversibly inhibits cyclooxygenase to subsequently decrease TxA_2 production and limit platelet aggregation. Clopidogrel and Prasugrel are examples of $P2Y_{12}$ receptor antagonists that inhibit the soluble agonist ADP, whilst Tirofiban and Abciximab block αIIbβ3-ligand interactions. Other anti-platelet therapies include thrombin and phosphodiesterase inhibitors (block degradation of cyclic nucleotides) [69, 70].

Given the impact of miRNA gene regulation, it is unsurprising that the dysregulation of miRNA is implicated in CVD. miRNAs are central players in modulating gene expression of cells/platelets collectively involved in CVD, and mediate inflammation, lipid uptake and cell differentiation in atherosclerosis. Platelet miRNA signatures (miR-25-3p, miR-221-3p and miR-374b-5) alter between patients with ST-segment elevation myocardial infarction (STEMI) and those with non-STEMI [71] suggesting that levels of platelet miRNA could impact platelet thrombogenicity and type of infarction. Furthermore, circulating miRNAs associated with the risk of MI (miR-126, miR-150, miR-223 and miR-197) are abundantly expressed in platelets. Platelet miRNA are implicated in premature CAD as two miRNAs in platelets are up-regulated in patients compared to controls (miR-340* and miR-624*), although whether or not they are the cause or consequence is currently unknown [72]. Besides their roles as mediators and biomarkers of CVD, platelet miRNA act as novel surrogate measures of the responsiveness to anti-platelet therapies used in CVD [73]. miR-223 levels are significantly down regulated in low responders to anti-platelet therapy [45, 74]. Furthermore, expression of platelet miR-26a has been linked with clopidogrel resistance during coronary stenting [75]. This theory is strengthened by research demonstrating how the switch from dual anti-platelet treatment with clopidogrel to ticagrelor is linked with significant changes in the level of platelet-specific circulating miRNAs, namely miR-223, miR-126 and miR-150 and miR-96 [76]. Other research investigating the effects of anti-platelet therapy on platelet miRNA levels showed that *in vitro* platelet activation resulted in transfer of miR-126 from platelets to plasma, whereas in aspirin-treated platelets, this process was not observed. *In vivo*, aspirin intake resulted in platelet inhibition and lower circulating platelet-derived miR-126 levels than were seen in untreated subjects [77]. Greater understanding of the meaning of platelet miRNA in CVD patients could aid in the diagnosis and treatment of these diseases.

2.3. Effect of obesity on platelet function

Obesity is a multifactorial condition involving a plethora of interrelated processes such as alterations in lipid metabolism, insulin resistance, inflammation, endothelial dysfunction, adipokine imbalance and oxidative stress. These metabolic aberrations have been postulated to be involved in platelet hyper-aggregability. Indeed, platelet activation markers are described as elevated in obesity, contributing to the inflammatory and prothrombotic state [78]. Subjects with overweight and obesity display increased platelet activation markers urinary-11-dehydro-TXB2 [79], MPV [80] and PLT [81]. Greater platelet activation (P-Selectin and PMP) is also linked to central arterial stiffness and carotid wall thickness amongst other atherosclerotic risk factors in overweight and obese subjects [82, 83]. The major mechanisms behind platelet function in obesity include a reduced sensitivity to insulin and resistance to their main inhibitory mediators PGI_2 and NO, elevated oxidative stress and an altered intracellular environment with increased cytosolic Ca^{2+} [84]. Platelets have insulin receptors which impact platelet function by regulating platelet response and sensitisation of platelets to inhibitory mechanism of PGI_2 and NO. In obese subjects, the anti-aggregating effect of insulin is diminished [85, 86].

Elevated oxidative stress also plays important roles in obesity-related platelet dysfunction. Oxidative stress results from an imbalance between the generation of free radicals and antioxidant enzymes [87]. High reactive oxygen species (ROS) generation by excess adipose tissue reduces NO bioavailability, enhancing surface expression of adhesion molecules, and enabling platelet activation and adhesion. Increased ROS also converts arachidonic acid into F_2-isoprostanes such as 8-iso-$PGF_{2\alpha}$ that can modulate platelet adhesive function [88]. Activated platelets also produce ROS [89], amplifying their own aggregatory potential by increasing αIIbβ3 and CD40L expression [90, 91] and stimulating intraplatelet F_2-isoprostanes production. Both decreased NO synthesis and bioavailability from ECs and platelets contribute to the pathogenesis of obesity, likely promoting thrombosis. Research by Leite et al. describes a decrease of nitric oxide synthase (NOS) activity and cGMP levels with simultaneous platelet hyper-aggregability in obese subjects compared to healthy controls with impaired antioxidant responses as potential contributors [92]. Anfossi et al. showed that platelet sensitivity to anti-aggregatory effects of PGI_2 and NO is reduced in obesity [84]. Importantly, weight loss in obese subjects marks a reduction in platelet activation markers and can potentially reverse the platelet responsiveness to NO and prostacyclin [93, 94]. A 10% weight reduction in obese subjects resulted in significant reductions in BMI, endothelial dysfunction and platelet aggregation. The changes in platelet function were associated with improvement in insulin sensitivity, indicating a tight relationship between the two. Weight loss also resulted in reduction in lipid peroxidation markers [95] and P-Selectin expression in overweight CAD patients [96].

Although an association between obesity and platelet activation is evident, the molecular mechanisms responsible have only begun to surface [97]. Platelet RNA is reflective of pathological disease states where inflammatory transcript profiles (e.g., INFG, IL1R1, IL6 and TLR2) correlate significantly with increasing BMI [98], supporting the hypothesis that surplus fat could unfavourably alter the inflammatory potential of platelets. However, obesity can also cause dysregulation of other factors that control haemostasis such as microRNA (miRNA). There is increasing evidence to show that miRNA is involved in the pathogenesis of obesity [99], where plasma levels of miR-223 are reduced in obese compared to lean subjects, suggesting that the miR-223/$P2Y_{12}$ alliance could signify a contributing mechanism of platelet activation in obesity [36].

2.4. Role of physical activity on platelet function

Those who engage in regular physical activity or exercise have a reduced prevalence of CVD. PA has been extensively studied due to its beneficial effects on all-cause mortality. Evidence to support the inverse relationship between PA and either CVD, cancer or depression continues to accumulate. With regard to CVD, regular PA/exercise reduces blood pressure, serum triglycerides, total body fat and visceral fat and LDL cholesterol [100]. Differences in these known factors have been demonstrated to explain a large proportion of the inverse relationship between physical activity and CVD risk [101, 102]. However, over 40% of the inverse association remains unexplained. Although the beneficial effects of regular exercise on blood lipids and blood pressure have been well documented, research focusing on platelet function has only recently gained greater attention. Since platelets play a key role in the pathogenesis of CVD, the protective effect of exercise against CVD may be partially due to alterations of platelet function [103].

Aerobic fitness is measured by maximal oxygen uptake (VO_2 max) during incremental exercise and is globally acknowledged as the best assessment of cardiovascular fitness [104]. VO_2 max represents the maximal amount of oxygen that an individual can take in and use to produce energy. VO_2 max is a function of the ability of the cardiovascular system to deliver blood and oxygen to skeletal muscle, and the ability of skeletal muscle to extract this oxygen and use it to produce energy. Exercise effects on platelet function in both diseased and healthy populations have elicited profound interest in the last decade. The majority of research surrounding platelet function and physical activity/exercise has focused on acute (single bout) aerobic exercise. Potential effects of acute exercise on platelet function (mainly aggregation) have been investigated through various studies in adult subjects with varying intra- and inter-individual results, making interpretation problematic. Differences in population type (e.g., CVD versus healthy), methods employed to assess platelet function and techniques to examine reactivity are the main reasons for discrepancies and lack of consistency between research groups [105]. Different platelet adhesion experimental protocols have provided no definitive consensus on the platelet response to acute exercise in healthy adult subjects [106–110]. High levels of plasma fibrinogen after exercise result in elevated blood viscosity and this along with increased vWF binding, αIIbβ3 and P-Selectin expression all contribute to the increased platelet aggregation after acute exercise [111]. In general, it appears that acute vigorous exercise induces a hyper-reactive haemostatic state [112] and a transient increase in agonist-induced platelet adhesion and aggregation *in vitro* and *ex vivo*. However, there is no definitive consensus regarding the short-term effects of exercise on platelet function.

Cardiorespiratory fitness (CRF) is the ability to perform large muscle, moderate to high intensity exercise for prolonged periods and depends on the respiratory, cardiovascular and skeletal systems. CRF represents the adaptation to long-term exercise. High CRF levels are also linked with reduced CVD risk factors such as hypertension, obesity in the general population and CVD patients [113–116]. CRF was first postulated as a significant determinant for changes in platelet function in response to acute exercise after observations that acute strenuous exercise increased platelet activation in sedentary, but not physically active, subjects [106, 117]. The actual relationship between CRF and platelet function has been referred to in a recent breakthrough study by Heber et al., who investigated platelet function and CRF in 62 young women [118]. Platelet function was assessed by determination of P-Selectin and CD40L expression and quantification of platelet ROS generation in platelet-rich plasma (PRP). Basal platelet activation (reflected by CD62P expression) and agonist-induced platelet activation (ROS, CD62P and CD40L) were higher in the LF compared to the MF and HF. The group found no difference between basal CD40L expressions (non-agonist induced). Interestingly, basal platelet function in the MF and HF were almost equal, indicating a definite influence of CRF on platelet function. A high CRF level is a result of exercise training and habitual physical activity. Therefore, research on the effects of longitudinal exercise training on platelet function has mainly shown that habitual exercise has favourable effects on platelet function. Eight weeks of exercise training (60% VO_2 max 5×/week 30 min/day), reduced shear stress-induced platelet activation and ox-LDL-potentiated platelet function [109, 111]. Importantly, after 12 weeks of de-conditioning, the beneficial effects of exercise on platelets were non-existent and platelet function returned to its pre-training state.

De Meirelles et al. reported that chronic physical activity had favourable effects on platelet activation in hypertensive patients at rest [119]. Twelve weeks of regular exercise (75–85% VO_2 max 5×/week for 45–60 min) reduced platelet aggregation in response to collagen. Santilli et al. investigated the effects of regular high intensity (60–75%) aerobic exercise for 2 months in low and intermediate CVD risk sedentary subjects [120]. Exercise training was associated with reductions in TxA_2, plasma P-Selectin and platelet-derived CD40L, despite no reduction in CRP (representing systemic inflammation). Evidently, physical activity and exercise affects nearly all facets of platelet function [121]. Studies on the effects of acute exercise appear to heighten platelet reactivity. Regular exercise can improve this response, seems to have an anti-thrombotic effect on platelets and could represent a portion of the protective effects of exercise on CVD risk factors. Moreover, effects of exercise are not maintained with cessation of training. Of importance, all of these studies discussed were performed in adults and not adolescents when the CVD risk factors and atherosclerotic process has begun. Platelet function and exercise in children or adolescents is in its infancy, an area that requires urgent research [122].

2.5. Role of physical inactivity and sedentary lifestyle on platelet function

In contrast to physical activity, physical inactivity/sedentary behaviour is a universal leading cause of death and independent CVD risk factor [123, 124]. Sedentary behaviour refers to any waking activity characterised by an energy expenditure ≤1.5 metabolic equivalents in a sitting or reclining posture [125]. However, in contrast to the evidence supporting the benefits of acute and chronic exercise, relatively little is understood about the mechanisms underlying the physiological, cellular and molecular responses to physical inactivity. Incomplete understanding of this relationship is a huge barrier to combating the development of CVD and its ancillary risk factors. Our knowledge of physical inactivity is somewhat indirect and is mainly based on the positive effects of exercise training on the sedentary population. As a sedentary lifestyle is often associated with obesity [126], some mechanisms involved in the pathogenesis of physical inactivity are similar to that of obesity such as insulin resistance [127], hypertension and increased inflammation [128]. However, distinct factors associated with sedentary behaviour include reduced muscular activity of lower extremities, decreased blood flow and reduction of shear stress, which increases oxidative stress, endothelial dysfunction [129] and arterial remodelling [130, 131].

2.5.1. Physical activity/inactivity-specific miRNA

The plasticity of platelets and other blood cells is vital for responding to environmental changes in response to physical (in)activity patterns. However, the molecular factors influencing platelet function/response/adaptation to physical (in)activity remain poorly understood. Recently identified miRNAs have gained attention as modulators of platelet function [34]. Evidence for miRNA involvement in exercise-associated gene expression changes in a number of cell types including peripheral blood mononuclear cell, neutrophil and skeletal muscle in non-trained and trained subjects has been illustrated [132–134]. Work by Baggish et al. showed altered expression of circulating miRNA (c-miRNA) in response to acute and chronic exercise in athletes [135]. Eight c-miRNA involved in cellular processes related to exercise adaptation (muscle contractility, inflammation, and angiogenesis) were examined. They observed four distinctive signatures of c-miRNA; c-miRNA up-regulated by acute exhaustive exercise pre- and post-exercise intervention, c-miRNA responsive to acute exercise pre- but not post-intervention, c-miRNA only responsive to exercise intervention and non-responsive miRNA. Moreover, evidence of these physical activity-specific microRNA signatures [136–138] has ingrained concepts of physical inactivity-specific miRNA profiles. Epigenetic variation could therefore be a potential mechanism allowing for independent or synergistic effects of physical inactivity on platelet function. Hibler et al. recently described indications for epigenetic variation (by miRNA expression) as a link between physical activity and sedentary lifestyle [139]. An epigenetic adaptation to habitual exercise has been described [140, 141]. Similarly, an epigenetic adaptation to physical inactivity may exist.

2.6. Physical activity/inactivity and platelet epigenetic drift

It has been well recognised that regular exercise may reduce risk of major vascular thrombotic events and protect against CVD [123]. Differences in known factors explain a large percentage of the inverse relationship between physical activity and CVD risk [101, 102]. Nevertheless, over 40% of the inverse association remains unexplained. Although the beneficial effects of regular exercise on blood lipids and blood pressure have been well accepted, research focusing on platelet function has only recently gained greater attention. Whilst it is known that platelet function and platelet indices (markers of platelet activation) are altered in pathological states such as CVD, only a minority of studies have solely examined the relationship between overall physiological health and platelet function in healthy subjects [120, 142]. Therefore, it is imperative that future studies explore the feasibility of platelet indices and whole blood platelet function measurements, as useful, non-invasive initial biomarkers of early/subclinical CVD risk and lifestyle parameters.

Low cardiorespiratory fitness is associated with physical inactivity [143]. This has major health effects globally, with approximately 3.2 million deaths each year attributable to inadequate physical activity. Evidence has shown that physical inactivity and sedentary behaviour have direct effects on CVD risk factors [144]. Moreover, in contrast to the accumulating evidence supporting the benefits of regular exercise, relatively little is understood about the deleterious mechanisms underlying the physiological, cellular and molecular responses to PI, specifically with regard to platelet function.

3. Future research avenues

Our group, in collaboration with the European Space Agency (ESA), the Centre National d'Etudes Spatiales (*CNES*) and *MEDES* (Institute for Space Medicine and Physiology, Toulouse, France) have employed ground-based models of microgravity, i.e., dry water immersion (DI) to study the effects of spaceflight on human physiology in a precisely controlled environment. DI involves immersing a subject in a bath of thermoneutral water covered by a waterproof fabric [145]. Several factors act simultaneously on the human body during immersion, including hydrostatic compression, supportlessness and extensive physical inactivity. Hypokinesia and hypodynamia are the major characteristics of physical inactivity induced by dry immersion. Hypodynamia involves a reduction in postural muscle load, whereas hypokinesia is a decline in motor activity. For these reasons, DI has been well accepted as a valuable tool to study physical inactivity [146]. DI presents a unique opportunity to analyse the specific effects of physical inactivity on platelet physiology/function and related biomarkers.

Recent studies reflect the first comprehensive attempts to evaluate the relationship between platelet function and physical activity, physical inactivity and overweight. While exploratory in nature to date, several questions remain unanswered and so further studies are warranted. The search for simple biomarkers that allow for early identification of subclinical/CVD risk is ongoing. Platelets can reflect changes in unhealthy lifestyle patterns. The Impact-R test is a relatively inexpensive test that can reliably detect changes in platelet adhesion and could be employed for CVD risk evaluation amongst subjects who are asymptomatic. Platelet indices and function markers should be further tested in larger populations to determine their reliability as surrogate markers for evaluating physiological health and to test during either pharmacological and lifestyle interventions. A relatively low dose of exercise has been shown to be sufficient to normalise platelet function in low fit females [118]. Larger studies incorporating exercise interventions at low doses over a lengthy period of time and examining more extensive aspects of platelet function in low fit subjects would develop this knowledge. The prescription of anti-platelet therapy is frequently used to treat CVD patients. However, the other residual risks (oxidative stress, inflammation etc.), which occur due to associations between CVD risk factors, are not eliminated efficiently by these therapies. In this sense, physical activity has been emphasised as it promotes favourable physiological adaptations, which may attenuate the cardiovascular risk factors and residual risks. Regular exercise may also impact platelet function in CVD patients. Exercise interventions in these populations could be beneficial in terms of reducing anti-platelet therapy dosage or combining anti-platelet therapy with exercise [147], i.e., prescriptive exercise medicine as an adjuvant management strategy/therapy.

The investigation of epigenetic processes is almost a completely unexplored area in platelet biology as analyses of these mechanisms require DNA [2]. Platelets have functionally active mitochondria [148]. Like nuclear DNA, mitochondrial DNA (mtDNA) can also be methylated, moderating control of mitochondrial gene expression. Understanding epigenetic regulation of mitochondrial genes in platelets is proving crucial to understanding their implication in CVD development [6]. Furthermore, miRNA have recently been linked with platelet mitochondrial health in stored platelets [149]. Platelets contain the machinery to process pre-miRNA to mature miRNA [34]. Platelets contain higher levels of pre-miRNA than other blood cells [56], and the maturation of pre-miRNA could contributed to altered miRNA profiles due to physical activity and inactivity. This may represent a more focused and efficient method of monitoring platelet function. Targeting levels of other biogenesis molecules in the miRNA pathway would also be an interesting avenue of platelet miRNA biology. Recently, Elgheznawy et al. showed that Dicer was decreased in patients with TD2M compared to healthy controls, whereas interestingly, Argonaute 2 levels did not differ [150]. Experiments investigating levels of miRNA processing machinery such as Dicer and Argonaute 2 in physically active and sedentary populations would be of major interest.

Long-term lifestyle choices such as physical inactivity may incur epigenetic penalties in megakaryocytes, and in the biological processes of *megakaryocytopoiesis* and *thrombopoiesis*. Thus, platelet miRNA could reflect these epigenetic changes, holding substantial predictive potential of both health and disease. Epigenetic changes in the megakaryocyte genome such as methylation of genes determining platelet biogenesis or changes in histone acetylation with aging have been suggested to play an important role in platelet function [20]. Prescribed exercise could induce epigenetic changes in megakaryocytes to produce a healthier phenotype of platelets with a direct change in platelet reactivity.

4. Conclusion

It is evident that lifestyle factors such as physical activity, physical inactivity and overweight do impact platelet function. Platelets are indeed reflective of physiological and lifestyle changes, making them sensitive biomarkers of human health. Platelets represent a tangible link to physiological and pathological changes within the body. Future research in this area, will no doubt contribute to a greater mechanistic understanding of the relationship between epigenetics, cardiovascular health, lifestyle factors and platelet biology.

Acknowledgements

This work was supported through the Irish Higher Education Authority Programme for Research in Third Level Institutes (HEA-PRTLI Cycle 4: T3 Targeted Therapeutics & Theranostics), the Health Research Board of Ireland (HRB-RP/2005/184, HRA/2009/122, HRA/2009/122/R), Science Foundation Ireland (SFI-04/BR/B0577, 11/TIDA/B1927) to Dr. R.P. Murphy, and the International Visitor Award, Dublin City University, Dublin, Ireland, to Drs. B. Degryse and R.P. Murphy, ASU-DCU Transatlantic Research Fund to M.M and R.P.M. R.P. Murphy is a management committee member of the EU H2020 COST action-CM1406, EpiChemBio. The ground-based protocol and studies of microgravity (3-Day Dry Immersion and 5-Day Dry Immersion) were undertaken in MEDES, the Institute for Space Medicine and Physiology, *Toulouse* University Hospital, France, and was supported and funded by the Centre National d'Etudes Spatiales (*CNES*-National Centre for Space Studies) to M-A.C., R.P.M., N.N., and L.T.

Conflict of interest

We wish to confirm that there are no known conflicts of interest associated with this publication, and there has been no significant financial support for this work that could have influenced its outcome.

Author details

Laura Twomey[1], Robert G. Wallace[1], Marco Mangone[2], Bernard Degryse[1], Sinead Sheridan[3], Michael Harrison[4], Niall Moyna[1], Gerardene Meade-Murphy[5], Nastassia Navasiolava[6], Marc-Antoine Custaud[6] and Ronan P. Murphy[1]*

*Address all correspondence to: ronan.murphy@dcu.ie

1 School of Health & Human Performance, Dublin City University, Glasnevin, Dublin, Ireland

2 School of Life Sciences, Arizona State University, Tempe Campus, USA

3 School of Public Health, Li Ka Shing Faculty of Medicine, The University of Hong Kong, Hong Kong

4 School of Health Sciences—Health Sport and Exercise Science, Waterford Institute of Technology, Ireland

5 Department of Pharmacology and Therapeutics, University College Cork, Cork, Ireland

6 Centre de Recherche Clinique du CHU d'Angers, Angers, France

References

[1] Fraga MF, Ballestar E, Paz MF, Ropero S, Setien F, Ballestar ML, et al. Epigenetic differences arise during the lifetime of monozygotic twins. Proceedings of the National Academy of Sciences of the United States of America. 2005;**102**(30):10604-10609

[2] Freson K, Izzi B, Van Geet C. From genetics to epigenetics in platelet research. Thrombosis Research. 2012;**129**(3):325-329

[3] Weinhold B. Epigenetics: The science of change. Environmental Health Perspectives. 2006;**114**(3):A160-A167

[4] Lund G, Zaina S. Atherosclerosis: An epigenetic balancing act that goes wrong. Current Atherosclerosis Reports. 2011;**13**(3):208-214

[5] Zhong Z, Liang S, Sanchez-Lopez E, He F, Shalapour S, Lin XJ, et al. New mitochondrial DNA synthesis enables NLRP3 inflammasome activation. Nature. Aug 2018;**560**(7717): 198-203

[6] Baccarelli AA, Byun HM. Platelet mitochondrial DNA methylation: A potential new marker of cardiovascular disease. Clinical Epigenetics. 2015;**7**:44

[7] Webster AL, Yan MS, Marsden PA. Epigenetics and cardiovascular disease. The Canadian Journal of Cardiology. 2013;**29**(1):46-57

[8] Zaratiegui M, Irvine DV, Martienssen RA. Noncoding RNAs and gene silencing. Cell. 2007;**128**(4):763-776

[9] Calin GA, Liu CG, Ferracin M, Hyslop T, Spizzo R, Sevignani C, et al. Ultraconserved regions encoding ncRNAs are altered in human leukemias and carcinomas. Cancer Cell. 2007;**12**(3):215-229

[10] Sato F, Tsuchiya S, Meltzer SJ, Shimizu K. MicroRNAs and epigenetics. The FEBS Journal. 2011;**278**(10):1598-1609

[11] Gabbianelli R, Damiani E. Epigenetics and neurodegeneration: Role of early-life nutrition. The Journal of Nutritional Biochemistry. 2018;**57**:1-13

[12] Bollati V, Schwartz J, Wright R, Litonjua A, Tarantini L, Suh H, et al. Decline in genomic DNA methylation through aging in a cohort of elderly subjects. Mechanisms of Ageing and Development. 2009;**130**(4):234-239

[13] Tan H, Liu T, Zhang J, Zhou T. Random positioning of nucleosomes enhances heritable bistability. Molecular BioSystems. 2016;**13**(1):132-141

[14] Hannum G, Guinney J, Zhao L, Zhang L, Hughes G, Sadda S, et al. Genome-wide methylation profiles reveal quantitative views of human aging rates. Molecular Cell. 2013;**49**(2): 359-367

[15] Teschendorff AE. Epigenetic aging: Insights from network biology. Aging. 2013;**5**(10): 719-720

[16] Horvath S, Oshima J, Martin GM, Lu AT, Quach A, Cohen H, et al. Epigenetic clock for skin and blood cells applied to Hutchinson Gilford Progeria Syndrome and ex vivo studies. Aging. 2018;**10**(7):1758-1775

[17] Horvath S. DNA methylation age of human tissues and cell types. Genome Biology. 2013;**14**(10):R115

[18] Oshima M, Iwama A. Epigenetics of hematopoietic stem cell aging and disease. International Journal of Hematology. 2014;**100**(4):326-334

[19] Choudry FA, Frontini M. Epigenetic control of haematopoietic stem cell aging and its clinical implications. Stem Cells International. 2016;**2016**:5797521

[20] Daly ME. Determinants of platelet count in humans. Haematologica. 2011;**96**(1):10-13

[21] van Rooij E, Olson EN. MicroRNAs: Powerful new regulators of heart disease and provocative therapeutic targets. The Journal of Clinical Investigation. 2007;**117**(9):2369-2376

[22] van Rooij E, Olson EN. microRNAs put their signatures on the heart. Physiological Genomics. 2007;**31**(3):365-366

[23] Ple H, Landry P, Benham A, Coarfa C, Gunaratne PH, Provost P. The repertoire and features of human platelet microRNAs. PLoS One. 2012;**7**(12):e50746

[24] Laffont B, Corduan A, Ple H, Duchez AC, Cloutier N, Boilard E, et al. Activated platelets can deliver mRNA regulatory Ago2*microRNA complexes to endothelial cells via microparticles. Blood. 2013;**122**(2):253-261

[25] Liang R, Bates DJ, Wang E. Epigenetic control of MicroRNA expression and aging. Current Genomics. 2009;**10**(3):184-193

[26] Bartel DP, Chen CZ. Micromanagers of gene expression: The potentially widespread influence of metazoan microRNAs. Nature Reviews. Genetics. 2004;**5**(5):396-400

[27] Rodriguez A, Griffiths-Jones S, Ashurst JL, Bradley A. Identification of mammalian microRNA host genes and transcription units. Genome Research. 2004;**14**(10A):1902-1910

[28] Bartel DP. MicroRNAs: Target recognition and regulatory functions. Cell. 2009;**136**(2):215-233

[29] van Rooij E, Sutherland LB, Qi X, Richardson JA, Hill J, Olson EN. Control of stress-dependent cardiac growth and gene expression by a microRNA. Science. 2007;**316**(5824):575-579

[30] Finnegan EF, Pasquinelli AE. MicroRNA biogenesis: Regulating the regulators. Critical Reviews in Biochemistry and Molecular Biology. 2013;**48**(1):51-68

[31] Gulyaeva LF, Kushlinskiy NE. Regulatory mechanisms of microRNA expression. Journal of Translational Medicine. 2016;**14**(1):143

[32] Bijak M, Saluk J, Ponczek MB, Nowak P, Wachowicz B. The synthesis of proteins in unnucleated blood platelets. Postępy Higieny i Medycyny Doświadczalnej (Online). 2013;**67**:672-679

[33] Thon JN, Devine DV. Translation of glycoprotein IIIa in stored blood platelets. Transfusion. 2007;**47**(12):2260-2270

[34] Landry P, Plante I, Ouellet DL, Perron MP, Rousseau G, Provost P. Existence of a microRNA pathway in anucleate platelets. Nature Structural & Molecular Biology. 2009;**16**(9):961-966

[35] Dangwal S, Thum T. MicroRNAs in platelet physiology and pathology. Hämostaseologie. 2013;**33**(1):17-20

[36] Bray PF, McKenzie SE, Edelstein LC, Nagalla S, Delgrosso K, Ertel A, et al. The complex transcriptional landscape of the anucleate human platelet. BMC Genomics. 2013;**14**:1

[37] Kondkar AA, Bray MS, Leal SM, Nagalla S, Liu DJ, Jin Y, et al. VAMP8/endobrevin is overexpressed in hyperreactive human platelets: Suggested role for platelet microRNA. Journal of Thrombosis and Haemostasis. 2010;**8**(2):369-378

[38] Nagalla S, Shaw C, Kong X, Kondkar AA, Edelstein LC, Ma L, et al. Platelet microRNA-mRNA coexpression profiles correlate with platelet reactivity. Blood. 2011;**117**(19): 5189-5197

[39] Edelstein LC, McKenzie SE, Shaw C, Holinstat MA, Kunapuli SP, Bray PF. MicroRNAs in platelet production and activation. Journal of Thrombosis and Haemostasis. 2013; **11**(Suppl 1):340-350

[40] Simon LM, Edelstein LC, Nagalla S, Woodley AB, Chen ES, Kong X, et al. Human platelet microRNA-mRNA networks associated with age and gender revealed by integrated plateletomics. Blood. 2014;**123**(16):e37-e45

[41] Osman A, Falker K. Characterization of human platelet microRNA by quantitative PCR coupled with an annotation network for predicted target genes. Platelets. 2011;**22**(6): 433-441

[42] Teruel-Montoya R, Kong X, Abraham S, Ma L, Kunapuli SP, Holinstat M, et al. MicroRNA expression differences in human hematopoietic cell lineages enable regulated transgene expression. PLoS One. 2014;**9**(7):e102259

[43] Edelstein LC, Bray PF. MicroRNAs in platelet production and activation. Blood. 2011;**117**(20): 5289-5296

[44] Halkein J, De Windt LJ. miR-223: Sailing to terra incognita for microRNAs in platelets. Thrombosis and Haemostasis. 2013;**110**(6):1112-1113

[45] Shi R, Zhou X, Ji WJ, Zhang YY, Ma YQ, Zhang JQ, et al. The emerging role of miR-223 in platelet reactivity: Implications in antiplatelet therapy. BioMed Research International. 2015;**2015**:981841

[46] Fish JE, Santoro MM, Morton SU, Yu S, Yeh RF, Wythe JD, et al. miR-126 regulates angiogenic signaling and vascular integrity. Developmental Cell. 2008;**15**(2):272-284

[47] Kaudewitz D, Skroblin P, Bender LH, Barwari T, Willeit P, Pechlaner R, et al. Association of MicroRNAs and YRNAs with platelet function. Circulation Research. 2016;**118**(3): 420-432

[48] Emmrich S, Henke K, Hegermann J, Ochs M, Reinhardt D, Klusmann JH. miRNAs can increase the efficiency of ex vivo platelet generation. Annals of Hematology. 2012;**91**(11): 1673-1684

[49] Kamat V, Paluru P, Myint M, French DL, Gadue P, Diamond SL. MicroRNA screen of human embryonic stem cell differentiation reveals miR-105 as an enhancer of megakaryopoiesis from adult CD34+ cells. Stem Cells (Dayton, Ohio). 2014;**32**(5):1337-1346

[50] Zhai PF, Wang F, Su R, Lin HS, Jiang CL, Yang GH, et al. The regulatory roles of microRNA-146b-5p and its target platelet-derived growth factor receptor alpha (PDGFRA) in erythropoiesis and megakaryocytopoiesis. The Journal of Biological Chemistry. 2014;**289**(33): 22600-22613

[51] Chapnik E, Rivkin N, Mildner A, Beck G, Pasvolsky R, Metzl-Raz E, et al. miR-142 orchestrates a network of actin cytoskeleton regulators during megakaryopoiesis. eLife. 2014;**3**:e01964

[52] Hussein K, Theophile K, Buhr T, Beller A, Kreipe H, Bock O. Different lineage involvement in myelodysplastic/myeloproliferative disease with combined MPLW515L and JAK2V617F mutation. British Journal of Haematology. 2009;**145**(5):673-675

[53] Edelstein LC, Simon LM, Montoya RT, Holinstat M, Chen ES, Bergeron A, et al. Racial differences in human platelet PAR4 reactivity reflect expression of PCTP and miR-376c. Nature Medicine. 2013;**19**(12):1609-1616

[54] Diehl P, Fricke A, Sander L, Stamm J, Bassler N, Htun N, et al. Microparticles: Major transport vehicles for distinct microRNAs in circulation. Cardiovascular Research. 2012; **93**(4):633-644

[55] Gidlof O, van der Brug M, Ohman J, Gilje P, Olde B, Wahlestedt C, et al. Platelets activated during myocardial infarction release functional miRNA, which can be taken up by endothelial cells and regulate ICAM1 expression. Blood. 2013;**121**(19):3908-3917, S1-26

[56] Pan Y, Liang H, Liu H, Li D, Chen X, Li L, et al. Platelet-secreted microRNA-223 promotes endothelial cell apoptosis induced by advanced glycation end products via targeting the insulin-like growth factor 1 receptor. Journal of Immunology (Baltimore, Md. : 1950). 2014;**192**(1):437-446

[57] Liang H, Yan X, Pan Y, Wang Y, Wang N, Li L, et al. MicroRNA-223 delivered by platelet-derived microvesicles promotes lung cancer cell invasion via targeting tumor suppressor EPB41L3. Molecular Cancer. 2015;**14**:58

[58] Laffont B, Corduan A, Rousseau M, Duchez AC, Lee CH, Boilard E, et al. Platelet microparticles reprogram macrophage gene expression and function. Thrombosis and Haemostasis. 2016;**115**(2):311-323

[59] O'Donnell CJ, Nabel EG. Cardiovascular genomics, personalized medicine, and the National Heart, Lung, and Blood Institute: Part I: The beginning of an era. Circulation. Cardiovascular Genetics. 2008;**1**(1):51-57

[60] O'Donnell CJ, Nabel EG. Genomics of cardiovascular disease. The New England Journal of Medicine. 2011;**365**(22):2098-2109

[61] Juhola J, Magnussen CG, Viikari JS, Kahonen M, Hutri-Kahonen N, Jula A, et al. Tracking of serum lipid levels, blood pressure, and body mass index from childhood to adulthood:The Cardiovascular Risk in Young Finns Study. The Journal of Pediatrics. 2011; **159**(4):584-590

[62] McNeal CJ, Dajani T, Wilson D, Cassidy-Bushrow AE, Dickerson JB, Ory M. Hypercholesterolemia in youth: Opportunities and obstacles to prevent premature atherosclerotic cardiovascular disease. Current Atherosclerosis Reports. 2010;**12**(1):20-28

[63] Angiolillo DJ, Jakubowski JA, Ferreiro JL, Tello-Montoliu A, Rollini F, Franchi F, et al. Impaired responsiveness to the platelet P2Y12 receptor antagonist clopidogrel in patients with type 2 diabetes and coronary artery disease. Journal of the American College of Cardiology. 2014;**64**(10):1005-1014

[64] Badimon L, Padro T, Vilahur G. Atherosclerosis, platelets and thrombosis in acute ischaemic heart disease. European Heart Journal Acute Cardiovascular Care. 2012;**1**(1):60-74

[65] Daub K, Seizer P, Stellos K, Kramer BF, Bigalke B, Schaller M, et al. Oxidized LDL-activated platelets induce vascular inflammation. Seminars in Thrombosis and Hemostasis. 2010; **36**(2):146-156

[66] Lievens D, Zernecke A, Seijkens T, Soehnlein O, Beckers L, Munnix IC, et al. Platelet CD40L mediates thrombotic and inflammatory processes in atherosclerosis. Blood. 2010;**116**(20):4317-4327

[67] Andre P, Prasad KS, Denis CV, He M, Papalia JM, Hynes RO, et al. CD40L stabilizes arterial thrombi by a beta3 integrin-dependent mechanism. Nature Medicine. 2002;**8**(3): 247-252

[68] Xu XR, Zhang D, Oswald BE, Carrim N, Wang X, Hou Y, et al. Platelets are versatile cells: New discoveries in hemostasis, thrombosis, immune responses, tumor metastasis and beyond. Critical Reviews in Clinical Laboratory Sciences. 2016;**53**(6):409-430

[69] Papp J, Kenyeres P, Toth K. Clinical importance of antiplatelet drugs in cardiovascular diseases. Clinical Hemorheology and Microcirculation. 2013;**53**(1-2):81-96

[70] Metharom P, Berndt MC, Baker RI, Andrews RK. Current state and novel approaches of antiplatelet therapy. Arteriosclerosis, Thrombosis, and Vascular Biology. 2015;**35**(6): 1327-1338

[71] Ward JA, Esa N, Pidikiti R, Freedman JE, Keaney JF, Tanriverdi K, et al. Circulating cell and plasma microRNA profiles differ between non-ST-segment and ST-segment-elevation myocardial infarction. Family Medicine & Medical Science Research. 2013;**2**(2):108

[72] Sondermeijer BM, Bakker A, Halliani A, de Ronde MW, Marquart AA, Tijsen AJ, et al. Platelets in patients with premature coronary artery disease exhibit upregulation of miRNA340* and miRNA624*. PLoS One. 2011;**6**(10):e25946

[73] Zhang YY, Zhou X, Ji WJ, Shi R, Lu RY, Li JL, et al. Decreased circulating microRNA-223 level predicts high on-treatment platelet reactivity in patients with troponin-negative non-ST elevation acute coronary syndrome. Journal of Thrombosis and Thrombolysis. 2014;**38**(1):65-72

[74] Shi R, Ge L, Zhou X, Ji WJ, Lu RY, Zhang YY, et al. Decreased platelet miR-223 expression is associated with high on-clopidogrel platelet reactivity. Thrombosis Research. 2013;**131**(6):508-513

[75] Chen S, Qi X, Chen H, Li M, Gu J, Liu C, et al. Expression of miRNA-26a in platelets is associated with clopidogrel resistance following coronary stenting. Experimental and Therapeutic Medicine. 2016;**12**(1):518-524

[76] Carino A, De Rosa S, Sorrentino S, Polimeni A, Sabatino J, Caiazzo G, et al. Modulation of circulating microRNAs levels during the switch from clopidogrel to ticagrelor. BioMed Research International. 2016;**2016**:3968206

[77] de Boer HC, van Solingen C, Prins J, Duijs JM, Huisman MV, Rabelink TJ, et al. Aspirin treatment hampers the use of plasma microRNA-126 as a biomarker for the progression of vascular disease. European Heart Journal. 2013;**34**(44):3451-3457

[78] Bordeaux BC, Qayyum R, Yanek LR, Vaidya D, Becker LC, Faraday N, et al. Effect of obesity on platelet reactivity and response to low-dose aspirin. Preventive Cardiology. 2010;**13**(2):56-62

[79] Davi G, Guagnano MT, Ciabattoni G, Basili S, Falco A, Marinopiccoli M, et al. Platelet activation in obese women: Role of inflammation and oxidant stress. Journal of the American Medical Association. 2002;**288**(16):2008-2014

[80] Coban E, Ozdogan M, Yazicioglu G, Akcit F. The mean platelet volume in patients with obesity. International Journal of Clinical Practice. 2005;**59**(8):981-982

[81] Samocha-Bonet D, Justo D, Rogowski O, Saar N, Abu-Abeid S, Shenkerman G, et al. Platelet counts and platelet activation markers in obese subjects. Mediators of Inflammation. 2008;**2008**:834153

[82] Csongradi E, Nagy B Jr, Fulop T, Varga Z, Karanyi Z, Magyar MT, et al. Increased levels of platelet activation markers are positively associated with carotid wall thickness and other atherosclerotic risk factors in obese patients. Thrombosis and Haemostasis. 2011;**106**(4):683-692

[83] Cooper JN, Evans RW, Mori Brooks M, Fried L, Holmes C, Barinas-Mitchell E, et al. Associations between arterial stiffness and platelet activation in normotensive overweight and obese young adults. Clinical and Experimental Hypertension (New York, NY: 1993). 2014;**36**(3):115-122

[84] Anfossi G, Russo I, Massucco P, Mattiello L, Doronzo G, De Salve A, et al. Impaired synthesis and action of antiaggregating cyclic nucleotides in platelets from obese

subjects: Possible role in platelet hyperactivation in obesity. European Journal of Clinical Investigation. 2004;**34**(7):482-489

[85] Trovati M, Anfossi G, Massucco P, Mattiello L, Costamagna C, Piretto V, et al. Insulin stimulates nitric oxide synthesis in human platelets and, through nitric oxide, increases platelet concentrations of both guanosine-3′, 5′-cyclic monophosphate and adenosine-3′, 5′-cyclic monophosphate. Diabetes. 1997;**46**(5):742-749

[86] Anfossi G, Russo I, Trovati M. Platelet resistance to the anti-aggregating agents in the insulin resistant states. Current Diabetes Reviews. 2006;**2**(4):409-430

[87] Otani H. Oxidative stress as pathogenesis of cardiovascular risk associated with metabolic syndrome. Antioxidants & Redox Signaling. 2011;**15**(7):1911-1926

[88] Patrono C, Falco A, Davi G. Isoprostane formation and inhibition in atherothrombosis. Current Opinion in Pharmacology. 2005;**5**(2):198-203

[89] Victor VM, Rocha M, Sola E, Banuls C, Garcia-Malpartida K, Hernandez-Mijares A. Oxidative stress, endothelial dysfunction and atherosclerosis. Current Pharmaceutical Design. 2009;**15**(26):2988-3002

[90] Pignatelli P, Sanguigni V, Lenti L, Ferro D, Finocchi A, Rossi P, et al. gp91phox-dependent expression of platelet CD40 ligand. Circulation. 2004;**110**(10):1326-1329

[91] Begonja AJ, Gambaryan S, Geiger J, Aktas B, Pozgajova M, Nieswandt B, et al. Platelet NAD(P)H-oxidase-generated ROS production regulates alphaIIbbeta3-integrin activation independent of the NO/cGMP pathway. Blood. 2005;**106**(8):2757-2760

[92] Leite NR, Siqueira de Medeiros M, Mury WV, Matsuura C, Perszel MB, Noronha Filho G, et al. Platelet hyperaggregability in obesity: Is there a role for nitric oxide impairment and oxidative stress? Clinical and Experimental Pharmacology & Physiology. 2016;**43**(8):738-744

[93] Basili S, Pacini G, Guagnano MT, Manigrasso MR, Santilli F, Pettinella C, et al. Insulin resistance as a determinant of platelet activation in obese women. Journal of the American College of Cardiology. 2006;**48**(12):2531-2538

[94] Russo I, Traversa M, Bonomo K, De Salve A, Mattiello L, Del Mese P, et al. In central obesity, weight loss restores platelet sensitivity to nitric oxide and prostacyclin. Obesity (Silver Spring, Md). 2010;**18**(4):788-797

[95] Vazzana N, Guagnano MT, Cuccurullo C, Ferrante E, Lattanzio S, Liani R, et al. Endogenous secretory RAGE in obese women: Association with platelet activation and oxidative stress. The Journal of Clinical Endocrinology and Metabolism. 2012;**97**(9):E1726-E1730

[96] Keating FK, Schneider DJ, Savage PD, Bunn JY, Harvey-Berino J, Ludlow M, et al. Effect of exercise training and weight loss on platelet reactivity in overweight patients with coronary artery disease. Journal of Cardiopulmonary Rehabilitation and Prevention. 2013;**33**(6):371-377

[97] Blokhin IO, Lentz SR. Mechanisms of thrombosis in obesity. Current Opinion in Hematology. 2013;**20**(5):437-444

[98] Freedman JE, Larson MG, Tanriverdi K, O'Donnell CJ, Morin K, Hakanson AS, et al. Relation of platelet and leukocyte inflammatory transcripts to body mass index in the Framingham heart study. Circulation. 2010;**122**(2):119-129

[99] Zampetaki A, Kiechl S, Drozdov I, Willeit P, Mayr U, Prokopi M, et al. Plasma microRNA profiling reveals loss of endothelial miR-126 and other microRNAs in type 2 diabetes. Circulation Research. 2010;**107**(6):810-817

[100] Li J, Siegrist J. Physical activity and risk of cardiovascular disease—A meta-analysis of prospective cohort studies. International Journal of Environmental Research and Public Health. 2012;**9**(2):391-407

[101] Mora S, Cook N, Buring JE, Ridker PM, Lee IM. Physical activity and reduced risk of cardiovascular events: Potential mediating mechanisms. Circulation. 2007;**116**(19): 2110-2118

[102] Kwasniewska M, Kostka T, Jegier A, Dziankowska-Zaborszczyk E, Leszczynska J, Rebowska E, et al. Regular physical activity and cardiovascular biomarkers in prevention of atherosclerosis in men: A 25-year prospective cohort study. BMC Cardiovascular Disorders. 2016;**16**:65

[103] Kumar R, Bouskill V, Schneiderman JE, Pluthero FG, Kahr WH, Craik A, et al. Impact of aerobic exercise on haemostatic indices in paediatric patients with haemophilia. Thrombosis and Haemostasis. 2016;**115**(6):1120-1128

[104] Shepard RJ. Evaluation of the Canadian home fitness test in middle-aged men. Canadian Medical Association Journal. 1977;**117**(10):1136-1138

[105] Aurigemma C, Fattorossi A, Sestito A, Sgueglia GA, Farnetti S, Buzzonetti A, et al. Relationship between changes in platelet reactivity and changes in platelet receptor expression induced by physical exercise. Thrombosis Research. 2007;**120**(6):901-909

[106] Wang JS, Jen CJ, Kung HC, Lin LJ, Hsiue TR, Chen HI. Different effects of strenuous exercise and moderate exercise on platelet function in men. Circulation. 1994;**90**(6):2877-2885

[107] Wang JS, Jen CJ, Chen HI. Effects of chronic exercise and deconditioning on platelet function in women. Journal of Applied Physiology (Bethesda, Md: 1985). 1997;**83**(6): 2080-2085

[108] Wang JS, Yang CF, Wong MK, Chow SE, Chen JK. Effect of strenuous arm exercise on oxidized-LDL-potentiated platelet activation in individuals with spinal cord injury. Thrombosis and Haemostasis. 2000;**84**(1):118-123

[109] Wang JS, Li YS, Chen JC, Chen YW. Effects of exercise training and deconditioning on platelet aggregation induced by alternating shear stress in men. Arteriosclerosis, Thrombosis, and Vascular Biology. 2005;**25**(2):454-460

[110] Wang JS. Exercise prescription and thrombogenesis. Journal of Biomedical Science. 2006;**13**(6):753-761

[111] Wang JS, Chow SE. Effects of exercise training and detraining on oxidized low-density lipoprotein-potentiated platelet function in men. Archives of Physical Medicine and Rehabilitation. 2004;**85**(9):1531-1537

[112] Huskens D, Roest M, Remijn JA, Konings J, Kremers RM, Bloemen S, et al. Strenuous exercise induces a hyperreactive rebalanced haemostatic state that is more pronounced in men. Thrombosis and Haemostasis. 2016;**115**(6):1109-1119

[113] Franklin BA, Lavie CJ, Squires RW, Milani RV. Exercise-based cardiac rehabilitation and improvements in cardiorespiratory fitness: Implications regarding patient benefit. Mayo Clinic Proceedings. 2013;**88**(5):431-437

[114] Swift DL, Lavie CJ, Johannsen NM, Arena R, Earnest CP, O'Keefe JH, et al. Physical activity, cardiorespiratory fitness, and exercise training in primary and secondary coronary prevention. Circulation Journal: Official Journal of the Japanese Circulation Society. 2013;**77**(2):281-292

[115] DeFina LF, Haskell WL, Willis BL, Barlow CE, Finley CE, Levine BD, et al. Physical activity versus cardiorespiratory fitness: Two (partly) distinct components of cardiovascular health? Progress in Cardiovascular Diseases. 2015;**57**(4):324-329

[116] Myers J, McAuley P, Lavie CJ, Despres JP, Arena R, Kokkinos P. Physical activity and cardiorespiratory fitness as major markers of cardiovascular risk: Their independent and interwoven importance to health status. Progress in Cardiovascular Diseases. 2015;**57**(4):306-314

[117] Kestin AS, Ellis PA, Barnard MR, Errichetti A, Rosner BA, Michelson AD. Effect of strenuous exercise on platelet activation state and reactivity. Circulation. 1993;**88**(4 Pt 1): 1502-1511

[118] Heber S, Assinger A, Pokan R, Volf I. Correlation between cardiorespiratory fitness and platelet function in healthy women. Medicine and Science in Sports and Exercise. 2016;**48**(6):1101-1110

[119] de Meirelles LR, Mendes-Ribeiro AC, Mendes MA, da Silva MN, Ellory JC, Mann GE, et al. Chronic exercise reduces platelet activation in hypertension: upregulation of the L-arginine-nitric oxide pathway. Scandinavian Journal of Medicine & Science in Sports. 2009;**19**(1):67-74

[120] Santilli F, Vazzana N, Liani R, Guagnano MT, Davi G. Platelet activation in obesity and metabolic syndrome. Obesity Reviews: An Official Journal of the International Association for the Study of Obesity. 2012;**13**(1):27-42

[121] Heber S, Volf I. Effects of physical (in)activity on platelet function. BioMed Research International. 2015;**2015**:165078

[122] Ribeiro J, Almeida-Dias A, Ascensao A, Magalhaes J, Oliveira AR, Carlson J, et al. Hemostatic response to acute physical exercise in healthy adolescents. Journal of Science and Medicine in Sport. 2007;**10**(3):164-169

[123] Blair SN, Morris JN. Healthy hearts—And the universal benefits of being physically active: Physical activity and health. Annals of Epidemiology. 2009;**19**(4):253-256

[124] Kohl HW 3rd, Craig CL, Lambert EV, Inoue S, Alkandari JR, Leetongin G, et al. The pandemic of physical inactivity: Global action for public health. Lancet (London, England). 2012;**380**(9838):294-305

[125] Chevance G, Foucaut AM, Bernard P. State of knowledge on sedentary behaviors. Presse Medicale (Paris, France: 1983). 2016;**45**(3):313-318

[126] Van Gaal LF, Mertens IL, De Block CE. Mechanisms linking obesity with cardiovascular disease. Nature. 2006;**444**(7121):875-880

[127] Hamburg NM, McMackin CJ, Huang AL, Shenouda SM, Widlansky ME, Schulz E, et al. Physical inactivity rapidly induces insulin resistance and microvascular dysfunction in healthy volunteers. Arteriosclerosis, Thrombosis, and Vascular Biology. 2007;**27**(12): 2650-2656

[128] Nosova EV, Yen P, Chong KC, Alley HF, Stock EO, Quinn A, et al. Short-term physical inactivity impairs vascular function. The Journal of Surgical Research. 2014;**190**(2): 672-682

[129] Thosar SS, Johnson BD, Johnston JD, Wallace JP. Sitting and endothelial dysfunction: The role of shear stress. Medical Science Monitor: International Medical Journal of Experimental and Clinical Research. 2012;**18**(12):RA173-RA180

[130] Thijssen DH, Maiorana AJ, O'Driscoll G, Cable NT, Hopman MT, Green DJ. Impact of inactivity and exercise on the vasculature in humans. European Journal of Applied Physiology. 2010;**108**(5):845-875

[131] Thijssen DH, Dawson EA, van den Munckhof IC, Birk GK, Timothy Cable N, Green DJ. Local and systemic effects of leg cycling training on arterial wall thickness in healthy humans. Atherosclerosis. 2013;**229**(2):282-286

[132] Radom-Aizik S, Zaldivar F Jr, Oliver S, Galassetti P, Cooper DM. Evidence for microRNA involvement in exercise-associated neutrophil gene expression changes. Journal of Applied Physiology (Bethesda, Md: 1985). 2010;**109**(1):252-261

[133] Radom-Aizik S, Zaldivar F Jr, Leu SY, Adams GR, Oliver S, Cooper DM. Effects of exercise on microRNA expression in young males peripheral blood mononuclear cells. Clinical and Translational Science. 2012;**5**(1):32-38

[134] Nielsen S, Akerstrom T, Rinnov A, Yfanti C, Scheele C, Pedersen BK, et al. The miRNA plasma signature in response to acute aerobic exercise and endurance training. PLoS One. 2014;**9**(2):e87308

[135] Baggish AL, Hale A, Weiner RB, Lewis GD, Systrom D, Wang F, et al. Dynamic regulation of circulating microRNA during acute exhaustive exercise and sustained aerobic exercise training. The Journal of Physiology. 2011;**589**(Pt 16):3983-3994

[136] Bye A, Rosjo H, Aspenes ST, Condorelli G, Omland T, Wisloff U. Circulating microRNAs and aerobic fitness—The HUNT-Study. PLoS One. 2013;**8**(2):e57496

[137] Kangas R, Pollanen E, Rippo MR, Lanzarini C, Prattichizzo F, Niskala P, et al. Circulating miR-21, miR-146a and Fas ligand respond to postmenopausal estrogen-based hormone replacement therapy—A study with monozygotic twin pairs. Mechanisms of Ageing and Development. 2014;**143-144**:1-8

[138] Altana V, Geretto M, Pulliero A. MicroRNAs and physical activity. MicroRNA (Shariqah, United Arab Emirates). 2015;**4**(2):74-85

[139] Hibler E. Epigenetics and colorectal neoplasia: The evidence for physical activity and sedentary behavior. Current Colorectal Cancer Reports. 2015;**11**(6):388-396

[140] Ling C, Ronn T. Epigenetic adaptation to regular exercise in humans. Drug Discovery Today. 2014;**19**(7):1015-1018

[141] Pareja-Galeano H, Sanchis-Gomar F, Garcia-Gimenez JL. Physical exercise and epigenetic modulation: Elucidating intricate mechanisms. Sports Medicine (Auckland, NZ). 2014;**44**(4):429-436

[142] Santilli F, Vazzana N, Iodice P, Lattanzio S, Liani R, Bellomo RG, et al. Effects of high-amount-high-intensity exercise on in vivo platelet activation: Modulation by lipid peroxidation and AGE/RAGE axis. Thrombosis and Haemostasis. 2013;**110**(6):1232-1240

[143] Wei M, Gibbons LW, Kampert JB, Nichaman MZ, Blair SN. Low cardiorespiratory fitness and physical inactivity as predictors of mortality in men with type 2 diabetes. Annals of Internal Medicine. 2000;**132**(8):605-611

[144] Prentice A, Jebb S. TV and inactivity are separate contributors to metabolic risk factors in children. PLoS Medicine. 2006;**3**(12):e481

[145] Coupe M, Fortrat JO, Larina I, Gauquelin-Koch G, Gharib C, Custaud MA. Cardiovascular deconditioning: From autonomic nervous system to microvascular dysfunctions. Respiratory Physiology & Neurobiology. 2009;**169**(Suppl 1):S10-S12

[146] Widlansky ME. The danger of sedenterism: Endothelium at risk. American Journal of Physiology. Heart and Circulatory Physiology. 2010;**299**(2):H243-H244

[147] Martins CC, Bagatini MD, Cardoso AM, Zanini D, Abdalla FH, Baldissarelli J, et al. Regular exercise training reverses ectonucleotidase alterations and reduces hyperaggregation of platelets in metabolic syndrome patients. Clinica Chimica Acta; International Journal of Clinical Chemistry. 2016;**454**:66-71

[148] Antony PM, Boyd O, Trefois C, Ammerlaan W, Ostaszewski M, Baumuratov AS, et al. Platelet mitochondrial membrane potential in Parkinson's disease. Annals of Clinical Translational Neurology. 2015;**2**(1):67-73

[149] Dahiya N, Sarachana T, Kulkarni S, Wood WH III, Zhang Y, Becker KG, et al. miR-570 interacts with mitochondrial ATPase subunit g (ATP5L) encoding mRNA in stored platelets. Platelets. 2017;**28**(1):74-81

[150] Elgheznawy A, Shi L, Hu J, Wittig I, Laban H, Pircher J, et al. Dicer cleavage by calpain determines platelet microRNA levels and function in diabetes. Circulation Research. 2015;**117**(2):157-165

Section 4

Homeostasis and Disease

Chapter 7

Reactive Oxygen Species, Cellular Redox Homeostasis and Cancer

Rabiatul Basria S.M.N. Mydin and Simon I. Okekpa

Additional information is available at the end of the chapter

http://dx.doi.org/10.5772/intechopen.76096

Abstract

Redox homeostasis is attained by the cautious regulation of both reactive oxygen species (ROS) formation and removal from the body system. A shift in ROS balance promotes oxidative injury and tumour development by inflicting damage to DNA and inducing inconsistencies in the genome. The sources of endogenous ROS in a cell include mETC, NOX, LOX, cytochrome P450 and XO. The exogenous risk factors of ROS are pollutants, chemicals/drugs, radiation and heavy metals. Oxidative phosphorylation in the mitochondria produces ROS with unpaired electrons. Superoxide anion is the major ROS produced in the human mitochondria. Bulk of the ROS generation in the mitochondria occurs at the electron transport chain as derivatives of respiration. Cancer cells sustain ROS production by suppressing the antioxidant-generation system. Balance between ROS production and subsequent detoxification is regulated by scavenging enzymes and antioxidant agents. Failure in sirtuin-3 (SIRT3), ATM and p53 activities elevates the intracellular levels of ROS. PKCα induces the expression of NOX (DUOX) during cancer development and the consequent increase in ROS production. The PI3K/AKT signalling pathway activates NOX with consequent ROS production and subsequent induction of instability in the genome, leading to cancer. In conclusion, the interruption of the redox pathways that regulate ROS and its redox signalling activities affects cell physiology and can ultimately result in abnormal signalling, uncontrolled oxidative impairment and tumorigenesis.

Keywords: homeostasis, cancer, reactive oxygen species, mitochondrial electron transport chain (ETC), NOX, GSH, glutathione oxidase (GPX), superoxide dismutase (SOD), thioredoxin (TRX), sirtuins

1. Introduction

Reactive oxygen species (ROS) are known as oxygen free radicals, which greatly contribute in complex cellular pathways, such as metabolism, immune system regulation, proliferation, differentiation and vascular transformations [1, 2]. ROS have a short life span and possess unpaired electrons [3]. Oxidative stress, DNA damage and cancer occur as a result of ROS imbalance due to dysregulated generation of free radicals (ROS) from oxygen and inability to neutralise and detoxify the harmful effects caused by the free radicals in the body through counteracting their oxidative effect by antioxidants [4–6]. Under normal and healthy circumstances, ROS production and removal are strictly regulated and controlled by very effective defensive machinery that blocks excessive ROS production. Some ROS, such as superoxide anion radical (O_2^-) and hydrogen peroxide (H_2O_2), are necessary life functions because they play a vital role in the regulation of cell defence mechanisms necessary for signalling, steroid synthesis, G-protein-coupled-receptor activation, gene expression and transcription factor regulation [7, 8]. Therefore, ROS can act both as good and bad molecules because of their dual nature and can either induce regulation of cellular physiology or promote the induction of cytotoxicity depending on generation levels, site of generation and magnitude of generation [9]. However, high ROS levels make cells vulnerable to damage. The derivatives from oxygen contain free radicals, such as superoxide anion (O_2^-) and hydroxyl radical (OH^-) plus non-radical molecules, such as hypochlorous acid and H_2O_2 [3], which have been linked to oxidative injury due to their high reactivity potential against proteins, lipids and DNA [10].

The generation of ROS can be activated by either various endogenous or exogenous factors. The major source of endogenous ROS in cells of mammals is the mitochondrial electron transport chain (ETC). Other endogenous ROS sources are from the activities of NADPH oxidases (NOX), lipoxygenases (LOX), cytochrome P450 and xanthine oxidase (XO) [11]. Exogenous factors that contribute to ROS production are pollutants, chemicals/drugs, radiation and heavy metals [12]. Redox homeostasis is attained by cautious regulation of both ROS formation and removal from the body system [10]. Maintenance of homeostasis and signalling event of redox require the significant regulation of synthesis and detoxification. An interruption to the redox route that regulates ROS and its redox signalling activities affects cell physiology and can ultimately result in abnormal signalling, uncontrolled toxic by-product accumulation, oxidative impairment and cytotoxicity [2]. High oxidative stress levels are normally linked to abnormalities, which characterise tumour-specific modification that exposes the cancer cells to additional raise of ROS depending on the strength of their antioxidant defence system [13].

Cancer is one of the leading causes of death globally. Recent evidence suggests altered redox stability and dysregulated redox signalling as the two frequent hallmarks of cancers, which are implicated in the progression of malignancy and treatment resistance [13]. Cancer cells have been postulated to persistently exhibit high levels of reactive oxygen species (ROS) as a result of alterations in microenvironment, genetic mutations and dysregulation of metabolic processes [13]. This shift in pro-oxidant balance promotes tumour development by inflicting damage to DNA and causing inconsistency in the genome [1]. The DNA damage and instability induced to the genome activate an inflammatory reaction leading to stability of hypoxia inducible factor-1 and subsequent metabolic reprogramming [13, 14]. The ROS detoxification mechanism has provided selective advantage for its survival during pro-oxidation situations. Balance between ROS production and its quick detoxification are regulated by scavenging enzymes and antioxidant agents that limit the accumulation of ROS in the body.

2. Roles of mitochondria in ROS formation

Mitochondria generates 90% of the energy required for cells and tissues to function effectively and serves as the core site for energy metabolism in the cells, since it is involved in the generation of ATP via oxidative phosphorylation (OXPHOS) [15]. This process liberates electron from reducing substrates and delivers the electron to O_2 leading to the establishment of electrochemical gradient which triggers the ATP synthesis. Oxidative phosphorylation in the mitochondria produces ROS with unpaired electrons due to electron reduction from the oxygen [16–18]. Superoxide anion ($O_2^{-}\bullet$) is the major ROS produced in the human mitochondria which is formed due to monoelectronic O_2 reduction. Most ROS originate from the superoxide anion which also mediates oxidative chain reactions.

In vivo production of $O_2^{-}\bullet$ could either be synthesised enzymatically by cyP450-dependent oxygenases, NADPH oxidase and xanthine oxidase or non-enzymatically by transferring an electron directly to O_2 [18]. $O_2^{-}\bullet$ is capable of reacting with free radicals such as nitric oxide (NO•) to produce reactive nitrogen species (RNS) [19]. $O_2^{-}\bullet$ dismutation can occur spontaneously or through superoxide dismutases (SODs) catalysed reaction to generate hydrogen peroxide (H_2O_2) [20–22]. The mitochondrial generated H_2O_2 has numerous probable fates. H_2O_2 is fairly stable and permeable to the membrane, and therefore, it can diffuse inside the cell and get eliminated by mitochondrial or cytosolic antioxidant systems, which are catalase, thioredoxin-peroxidase and glutathione-peroxidase [23]. Mitochondrially produced H_2O_2, also function as a cytosolic signalling molecule, thereby, affecting the networks that control energy metabolism, stress response, redox balance and cell cycle [24–26]. None metabolised H_2O_2 in the mitochondria undergoes Fenton reaction and then transformed subsequently into hydroxyl radical (•OH) which is naturally a very strong oxidant with high damaging impact on molecules due to its high reactive nature [27]. The above reason has made researchers to believe that mitochondria have developed competent systems for H_2O_2 removal and also mechanisms for metal chelating (chaperone proteins) which prevents the formation of radical. Bulk of the ROS generation in the mitochondria occurs at the electron transport chain (ETC) as derivatives of respiration [17, 18, 28]. The ETC terminal component known as cytochrome c oxidase (Complex IV) acquires four (4) electrons from cytochrome c and then reduces one molecule of O_2 to form two H_2O. All the intermediates that is partially reduced are retained until reduction is fully achieved [16].

2.1. ROS and mitochondrial activation of apoptosis

High exposure of the mitochondria to ROS results to injurious consequences such as inflicting oxidative mitochondrial DNA damage. It has also been suggested that ROS is deeply involved in the extrinsic pathway of apoptosis. Extrinsic receptor-mediated pathway for cell death requires active engagement of the death receptors on the cell membrane surface with their corresponding ligands [29]. Receptor-mediated apoptotic pathway comprises of death receptors such as CD95 (Fas), TNF-related apoptosis-inducing ligand (TRAIL) receptors and TNF. Activation of Fas as well as TNFR1 generate ROS due to superoxide ($O_2\bullet-$) production and formation of NADPH oxidase daises derived from lipid raft. Induction of apoptosis or necrosis is linked with lipid raft-mediated downstream ROS generation [30, 31]. Downregulation of FLIP (FLICE inhibitory protein), a strong inhibitor of apoptosis is mediated by ROS via ubiquitination and consequent proteasome degradation or by scavenging of nitric oxide (NO) to prevent FLIP S-nitrosation and cytoprotection [32]. ROS sensitises cancer cells to apoptosis induced by TRAIL [33]. CD95 and TRAIL death receptors have been observed to be highly upregulated in reaction in the presence of hydrogen peroxide via NF-kappa B activation [34]. ROS promote apoptosis via JNK activation, inducing either intrinsic or extrinsic apoptotic signalling [35]. TNFα induced ROS perpetrates oxidation of JNK, thereby, inactivating-phosphatases via catalytic transformation of their cysteine into sulfenic acid resulting to prolonged activation of JNK which is necessary for the release of cytochrome c and cleavage of caspase 3 as well as cell death [36]. TNFα activates MAPK cascade. ASK1, a redox-sensitive MAPK kinase, is located at the JNK upstream. Reduced thioredoxin1 (Trx1) binds to ASK1 during non-oxidising circumstances to form a complex known as ASK1 signalosome (Trx1/ASK1 complex) which perform redox switch functions. Persistent cellular ROS causes detachment of the oxidised Trx1 from the Trx1/ASK1 complex leading to full ASK1 activation through TRAF2/6 recruitment [37]. ASK2 a member of ASK family attaches to ASK1 and stabilises it in mitochondria, nucleus and cytosol. Saxena et al. have revealed that redox protein known as thioredoxin interacting protein (TXNIP) with apoptosis promoting potential under oxidative stress, shuttles from the nucleus to the mitochondria leading to the removal of TXNIP from ASK1 and formation of a compound with mitochondrial Trx2. This suppression of the inhibition is mediated by Trx2 results in ASK1phosphorylation and induction of the mitochondrial pathway for apoptosis with caspase-3 cleavage and cytochrome c release [38]. The major target for ROS inside the mitochondria is the permeability transition pore (mPTP) in which the oxidative modification of its proteins has significant influence on the anion fluxes within the mitochondria [39]. This could cause overload of Ca^{2+} and ROS in reaction to pro-apoptotic stimuli causing mPTP to assume a very high state of conductance allowing unrestrained entry of solutes along the electrochemical gradient into the matrix of the mitochondria. The above phenomenon is termed mitochondrial permeability transition (MPT), which results in mitochondrial membrane potential dissipation and consequent osmotic swelling of the matrix of the mitochondria due to fluid influx [40]. The early phase of the mitochondrial swelling involves water movement from inter-cristae spaces into the mitochondrial matrix. Persistent movement of this water exacts pressure on the outer membrane due to increased volume of the matrix leading to mPTP opening and/or rupturing of the outer membrane of the mitochondria allowing the matrix to expand further [41]. This causes cytochrome c to be released with consequent activation of the downstream effector caspases by Apaf-1-pro-caspase 9-apoptosome complex.

3. Antioxidant system responsible for the redox homeostasis

The antioxidant systems are either enzymatic or non-enzymatic. The enzymatic antioxidant system consists of peroxiredoxin (Prx) system, catalase, SOD and the glutathione peroxidase (GPx) system, while the non-enzymatic antioxidant systems consist of α-tocopherol, lipoic acid and ascorbic acid [42–45].

3.1. Superoxide dismutases (SOD)

Intracellular ROS levels are regulated by the balance between ROS generating enzymes and antioxidant enzymes, which include superoxide dismutases (SOD), catalase, thioredoxin and glutathione peroxidase (GPX) [42]. SOD functions to convert O_2^- into H_2O_2, which is later converted into water by glutathione peroxidase or catalase. Human cells express three types of SOD: MnSOD (manganese SOD) expressed by the mitochondria, CuZnSOD (copper-zinc SOD) expressed by the cytoplasm and third is the extracellular SOD. A study has demonstrated that lack of MnSOD in mice generated excessive oxidative stress causing their mortality [46]. Another study also revealed that mice with a deficiency of CuZnSOD developed hepatocellular carcinoma due to sustained oxidative damage [47]. Lack of MnSOD has also been linked to elevated risk of lung cancer, prostate cancer, non-Hodgkin's lymphoma and ovarian cancer [48–51].

3.2. Glutathione oxidase (GPX)

GPX is a selenium-dependent antioxidant enzyme, which regulates hydrogen and lipid peroxide levels. Lack of GPX in the body increases tissue damage by ROS [43] and low GPX levels, which results in increased LDL oxidation [44]. GPX catalyses the reduction of hydrogen peroxide to form glutathione disulphide (GSSG) with glutathione (GSH) functioning as the substrate. An increased risk of bladder cancer, lung cancer and breast cancer has been associated with the substitution of proline-leucine at codon 198 in human GPX [52–55].

3.3. Thioredoxins (Trx)

The protective function of thioredoxins (Trxs) in cells against oxidative stress is via the reaction between their active site known as 2-cysteine and ROS resulting in reduction of oxidised proteins. Trxs also function as hydrogen donors to thioredoxin-dependent peroxide reductases. Trx possesses a Cys-Gly-Pro-Cys active site, which is essential for redox regulatory functions of Trx. Trx, when combined with Trx reductase and NADPH, forms a redox-sensitive machinery, which controls the levels of oxidised cysteine on proteins. The antioxidant properties of Trx can be attributed to the reduction of the oxidised form of Trx peroxidase by Trx, while the reduced peroxidase scavenges H_2O_2 [45]. The two isoforms of Trx are Trx1 (expressed in the cytoplasm and the nucleus) and Trx2 (expressed in the mitochondria), which are very crucial for cell survival [56]. Trx1 is a redox-sensitive binding protein that controls the activity of NF-κB through the reduction of cys62 on the p50 of NF-κB [57]. Trx1 has been linked to breast tumours, colon cancer, cervical cancer, gastric cancer, lung cancer, liver cancer and melanoma and carcinomas of the pancreas [58–62].

3.4. GSH

GSH is a thiol protein consisting of cysteine, glutamine and glycine, which functions as an important antioxidant in detoxification of metabolic processes [63]. Elevated levels of GSH in the cancerous patient tissue are reinforced by improved accessibility to the biosynthetic

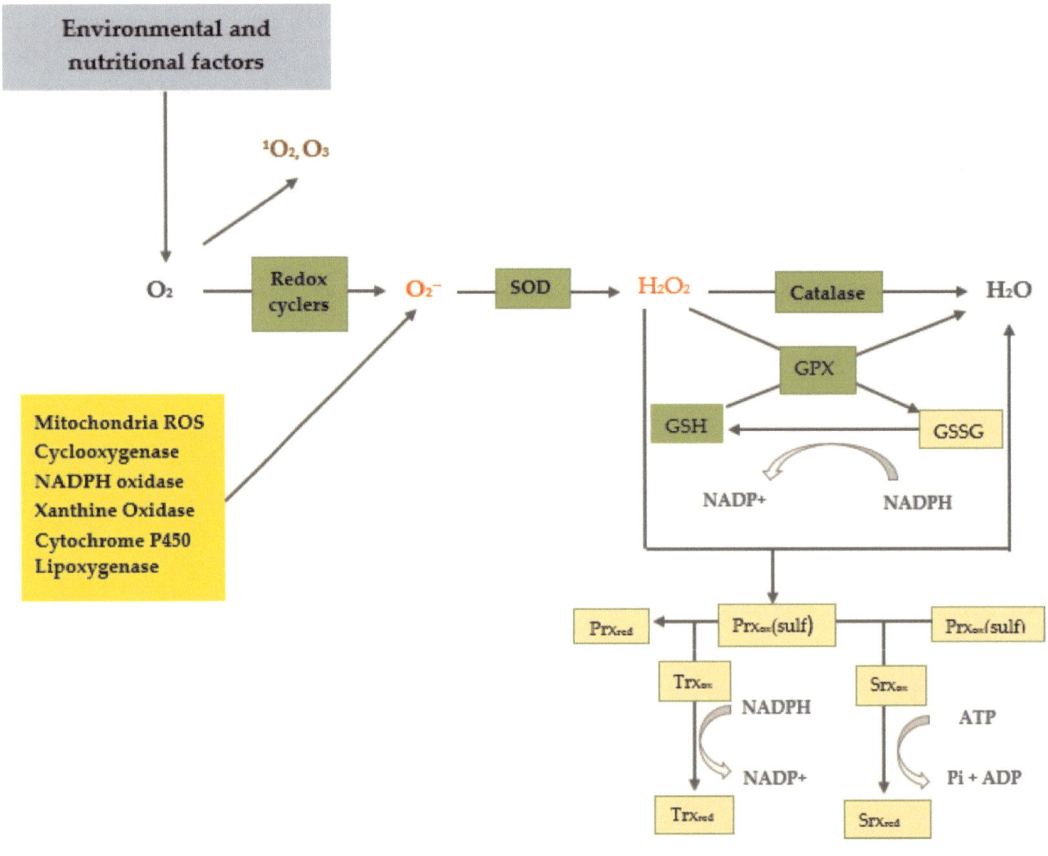

Figure 1. Mechanism of ROS formation and disposal. Enzymatically or non-enzymatically generated superoxide reacts with other radicals. The generated superoxide and H_2O_2 form hydroxyl radicals and singlet oxygen. SODs catalyse the dismutation of the superoxide, while H_2O_2 decomposition to water is catalysed by catalases in conjunction with glutathione peroxidase/GPx and Prxs. Thiol, a component of cysteine found in Prx, is oxidatively converted into Cys-sulfenic acid by (Prx_{ox}) and subsequently reduced by thioredoxin (Trx_{red}). H_2O_2 further oxidises Cys-sulfenic acid (Prx_{ox}) into Cys-sulfinic acid and reduced subsequently back by ATP and reduced sulfiredoxin (Srx_{red}) to Cys-sulfenic acid (Cys-SOH). Cys-SOH formed functions to regulate protein activities by absorbing the oxidative insults leading to the deflection of injurious oxidative impairment [24]. Cys-SOH can be converted into oxidative post-translational modification (Ox-PTM) in redox environment. This modification reacts with tripeptide glutathione (Glu-Cys-Gly) to form S-glutathionylated Cys (GSSG) or react with thiol to form disulphide bond. Formation of GSSG protects the Cys of the host from further oxidative reactions.

elements of GSH, such as glutamate, cysteine and glycine [64, 65]. The negative regulator of cysteine/glutamate known as SLC7A11 is always upregulated in human tumours [66]. Glutamate cysteine ligase modifier subunit (GCLM) is also upregulated in many types of human cancer but also requisite for effective GSH synthesis [67]. The cellular levels of GSH and its regeneration are modulated by NADPH and GR catalysing the reduction of oxidised GSSG back to GSH, in a process facilitated by the upregulation of NADPH production by cancer cells (**Figure 1**). Maintenance and elevation of GSH levels in cells are critical for the initiation and proliferation of tumours [67, 68]. Loss of GSH, or decrease in the ratio of glutathione to glutathione disulphide (GSH:GSSG), results in increased oxidative stress susceptibility and cancer development. Also, elevated levels of GSH increase antioxidant activities against numerous cancer cells, thereby enhancing the resistance of the cancer cells against oxidative stress [69].

3.5. Peroxiredoxins (Prxs)

Prxs are made up of six isoenzyme families capable of reducing H_2O_2 and alkyl hydroperoxides to their resultant H_2O or alcohol. Prxs are essential antioxidants that mediate the balancing mechanism of cellular H_2O_2 production, which is necessary for signalling and cell metabolism [70]. Nrf2 upregulates Prxs in oxidative stress circumstances [71]. PRDX1 plays the role of tumour suppressor in the development of breast cancer by interacting with oncogene (c-Myc) suppressing its transcriptional action [72, 73]. Contrarily, PRDX1 has promotional activities on pancreatic carcinoma, hepatocellular cancer, oesophageal cancer, oral cancer and lung cancer via upregulation of heme-oxygenase 1 and NF-κB pathway activation [74–77]. PRDX2 stimulates colorectal carcinoma by upregulating Wnt/β catenin levels, while it stimulates prostate cancer by upregulating the receptive activities of androgen [78, 79].

4. Enzymes responsible for the redox homeostasis

Some of the enzymes associated with redox homeostasis are NADPH oxidase, ATM kinase and sirtuin-3 among others.

4.1. NADPH oxidase

NADPH oxidase is hetero-proteins, which consist of seven isoforms. ROS production by NADPH oxidase is via the NOX protein. NADPH oxidases are referred collectively as the NOX family. The NOX family is comprised of NOX (NOX1, NOX2, NOX3, NOX4 andNOX5) and dual oxidases (DUOX1 and DUOX2) [80–82]. The isoforms DUOX1 and DUOX2 contain additional peroxidase domains, which exert the catalytic dismutation of superoxide anion to yield H_2O_2 [74]. Cytosolic electron transfer from NADPH across the cell membrane is catalysed by NADPH oxidase, thereby oxidising the molecular oxygen, which is later reduced to generate ROS species called superoxide anion radical (O_2^-). Generation of NADPH in the mitochondria plays a critical role in metastasis. In many tumour cells, reductive carboxylation in the mitochondria to generate NADPH is powered by mitochondrial citrate transporter (mCTP), cytosolic isocitrate dehydrogenase (IDH1) and mitochondrial isocitrate dehydrogenase (IDH2). This development assists cells to retain redox balance in the mitochondria averting the oxidative trauma received as a result of the detachment from the extracellular matrix [75].

4.2. ATM kinase

ROS production can be inhibited by ATM kinase and the same ATM kinase also serves as the function of redox-regulated DNA damage sensing protein [76]. ATM-regulated tumour suppressor works by interfering with KEAP1-facilitated NRF2 ubiquitination, thereby activating and stabilising the major regulators of antioxidants [77]. ATM facilitates the upregulation of glucose-6-phosphate dehydrogenase in order to promote NADPH production, thereby suppressing ROS levels [83].

4.3. Sirtuin-3

ROS levels are inherently elevated in cancer cells owing to mitochondrial defective oxidative metabolism [84]. Upregulated oxidative signals are implicated in the development and advancement of different cancer types [85]. Raised levels of ROS contribute to the initiation of cancer, transformation to malignancy and therapy resistance. ROS inflicted damage is more frequently seen in mitochondrial DNA than nuclear DNA because of its closeness to the main ROS source of generation and inadequate restoring capacities. For instance, silent information regulators of gene transcription-3 (sirtuin-3) are crucial in ROS regulation and effective flow of electrons via ETC. Failure in sirtuin-3 activities elevates intracellular levels of ROS, thereby inducing instability to the DNA of the mitochondria [86]. Sirtuins are an enzyme family, which is dependent on NAD-class III histone deacetylase. Seven homologues of Sirtuins (SIRT1–7) exist in mammals [87]. SIRT1 deacetylate gene regulates proteins like p53, forkhead proteins and NF-κB, which modulate resistance of cells to stress [36]. The deacetylase function of sirtuin proteins depends on the intracellular or endogenous content of NAD^+ [87]. Sirtuins are involved in the catalysis of exclusive reactions that lead to the formation of deacetylated substrate, acetyl ADP-ribose (AADPR) and nicotinamide [87]. Also, SIRT1 interrupts apoptosis, rescuing vulnerable cells after repetitive oxidative stress exposure [60]. SIRT2 deacetylate cytoskeletal proteins, such as forkhead proteins, histones, etc. SIRT3 reacts to redox status changes in the mitochondria by influencing the enzyme activities of manganese superoxide dismutase (MnSOD), which in turn scavenges ROS in the mitochondria and thus modulating the levels of ROS and metabolic homeostatic reliability [87].

5. Pathways implicated in redox homeostasis

Dysregulations associated with various tumour proliferations, autophagy and apoptosis depend on the activation of targets sensitive to redox reactions, such as PKC, Akt, PTEN, p53, etc. [88].

5.1. PKC pathways

PKC has isoenzymes, such as PKCα, PKCβ and PKCδ, with conflicting actions in different cancers [89]. PKCβ is the isoenzyme of PKC responsible for the stimulation and phosphorylation of p66/shc, which binds to cytochrome c in order to activate ROS generation [90]. Recent studies demonstrated that PKCα induces the expression of DUOX (a member of NOX family) during cancer development and subsequent ROS production [91, 92]. PKCδ has been demonstrated to be involved in the activation of NOX through the alteration of redox balance, thereby influencing the differentiation of tumour cells [90].

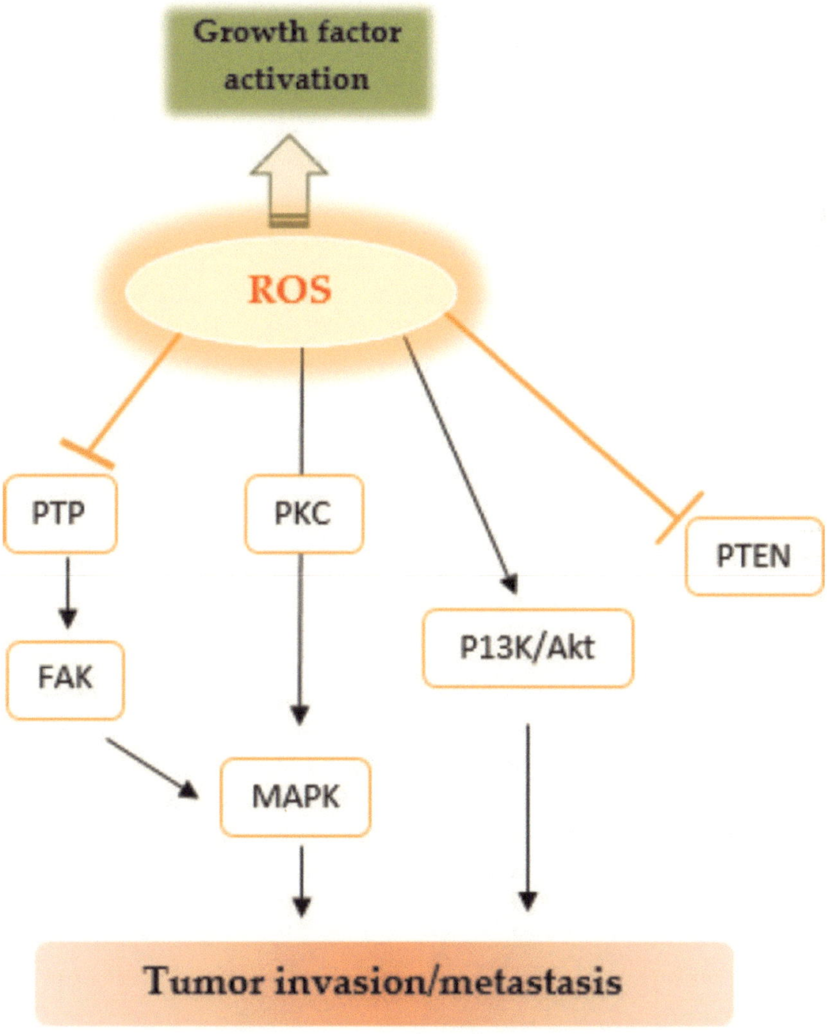

Figure 2. Effect of ROS imbalance in some pathways. Imbalance in the levels of ROS in the cell causes inhibition of PTEN and PTP dependent phosphorylation and consequent inactivation of FAK. The PI3K/Akt and PKC signals are activated in the process leading to invasion/metastasis.

5.2. PI3K/AKT pathway

PI3K/AKT signalling pathway activates NOX with consequent ROS production and subsequent induction of instability to the genome of cancer cells [91]. Upregulation of PTEN suppresses ROS synthesis, thus regulating PI3K/AKT pathway [92]. ROS-mediated PTEN inactivity alters kinase-phosphatase stability favouring the signalling of tumorigenic-tyrosine kinase receptor via Akt (**Figure 2**) leading to the inhibition of apoptosis due to phosphorylation and inactivation of Bad and caspase-9 [93]. Akt improves cell survival by negatively regulating the activities of Bcl-2 homology domain 3 (BH3)-only proteins via binding and inactivation of pro-survival Bcl-2 family members. Akt survival effects on cells depend on the S136 phosphorylation on BAD [94]. Akt-mediated BAD phosphorylation is stimulated by survival factors on S136 leading to the creation of 14-3-3 protein binding site causing BAD to miss its protein target [94]. Akt phosphorylates FOXO proteins (FOXO1, FOXO3a and FOXO4) on T24 and S256 attaching onto 14-3-3 proteins in the cell nucleus causing displacement of transcription factors of FOXO from their gene target and consequent export out of the nucleus. This results in the blocking FOXO facilitated transcription of genes that can stimulate apoptotic processes and cell-cycle arrest, thereby encouraging cell survival. Akt also promotes survival by targeting HDM2 causing inhibition of BH3-only proteins by triggering degradation of p53. Akt induces phosphorylation of HDM2 on S166 and S186, causing HDM2 translocation to the nucleus to regulate p53 function negatively [94]. Deficiency of p53 in cancer results in higher cytokine transcription and consequent accumulation of ROS [95]. p53 is another tumour suppressor that has the potential to activate NRF2 and elevate antioxidant enzyme (SOD, GPX1 and NADPH) expression, thereby reactivating the antioxidant system (**Figure 1**) [10, 96]. Previous study has shown that 53 plays a pro oxidant role through the reduction of SLC7A11 expression, which is responsible for cysteine uptake during GSH synthesis [87]. Thus, the antioxidant activity of p53 is necessary because of its ability to avert cancer, thus implicating loss of tumour suppressors in upregulated intracellular ROS expression.

6. Conclusion

Redox homeostasis is achieved by the regulation of both ROS formation and removal. Shifts in ROS balance induce oxidative injury and tumour development. The balance between ROS production and subsequent detoxification is regulated by scavenging enzymes and antioxidant agents. Targeting the ROS generation pathway with anticancer medications can aid patient recuperation. Therefore, the modulation of ROS levels is a modern anticancer therapy. Further studies are needed to determine when ROS inhibition and activation can be applied in clinical cancer treatment.

Acknowledgements

The authors would like to thank Universiti Sains Malaysia Bridging Grant: 304/CIPPT/6316184 and Dr. Norehan Mokhtar, The Director of Advanced Medical and Dental Institute, Universiti Sains Malaysia for supporting this work.

Conflict of interest

All authors declared that there is no conflict of interests.

Abbreviations

ROS	reactive oxygen species
SOD	superoxide dismutases
TRX	thioredoxin
GPX	glutathione oxidase
GSH	glutathione
mETC	mitochondrial electron transport chain
NOX	NADPH oxidases
LOX	lipoxygenases
XO	xanthine oxidase
Prx	peroxiredoxin
MnSOD	manganese SOD
CuZnSOD	copper-zinc SOD
GCLM	glutamate cysteine ligase modifier subunit
GSSG	S-glutathionylated Cys
PRDXs	peroxiredoxins
GR	glutathione reductase
GCLM	glutamate cysteine ligase modifier subunit
Ox-PTM	oxidative post-translational modification
DUOX	dual oxidases
mCTP	mitochondrial citrate transporter
IDH	isocitrate dehydrogenase
Sirtuin-3	silent information regulators of gene transcription-3
ATM	ataxia telangiectasia mutated
FAK	focal adhesion kinase
BAD	Bcl-2-associated death

FOXO	forkhead box
PI3K	phosphoinositide-3 kinase
PTP	protein tyrosine phosphatase
PTEN	phosphatase and tensin homologue
PKC	protein kinase C
Nrf2	nuclear factor erythroid 2-related factor 2

Author details

Rabiatul Basria S.M.N. Mydin[1]* and Simon I. Okekpa[1,2]

*Address all correspondence to: rabiatulbasria@usm.my

1 Oncological and Radiological Sciences Cluster, Advanced Medical and Dental Institute, Universiti Sains Malaysia, Kepala Batas, Pulau Pinang Malaysia

2 Department of Medical Laboratory Science, Faculty of Health Sciences, Ebonyi State University, Abakaliki, Ebonyi State, Nigeria

References

[1] Chouchani ET, Pell VR, Gaude E, Aksentijević D, Sundier SY, Robb EL, et al. Ischaemic accumulation of succinate controls reperfusion injury through mitochondrial ROS. Nature. 2014;**515**(7527):431

[2] Holmström KM, Finkel T. Cellular mechanisms and physiological consequences of redox-dependent signalling. Nature Reviews. Molecular Cell Biology. 2014;**15**(6):411

[3] Manda G, Isvoranu G, Comanescu MV, Manea A, Butuner BD, Korkmaz KS. The redox biology network in cancer pathophysiology and therapeutics. Redox Biology. 2015;**5**: 347-357

[4] Schramm A, Matusik P, Osmenda G, Guzik TJ. Targeting NADPH oxidases in vascular pharmacology. Vascular Pharmacology. 2012;**56**(5–6):216-231

[5] Cretu E, Trifan A, Aprotosoaie AC, Miron A. 15-Lipoxygenase inhibition, superoxide and hydroxyl radicals scavenging activities of *Cedrus brevifolia* bark extracts. Revista Medico-Chirurgicală a Societăţii de Medici şi Naturalişti din Iaşi. 2013;**117**:250-256

[6] Ciocoiu M, Miron A, Bădescu M. New polyphenolic extracts for oxidative stress treatment in experimental diabetes. Revista Medico-Chirurgicală a Societăţii de Medici şi Naturalişti din Iaşi. 2008;**112**(3):757-763

[7] Amanso AM, Griendling KK. Differential roles of NADPH oxidases in vascular physiology and pathophysiology. Frontiers in Bioscience (Scholar Edition). 2012;**4**:1044

[8] Sedeek M, Nasrallah R, Touyz RM, Hébert RL. NADPH oxidases, reactive oxygen species, and the kidney: Friend and foe. Journal of the American Society of Nephrology. 2013; **24**(10):1512-1518

[9] Bar-Peled L, Kemper EK, Suciu RM, Vinogradova EV, Backus KM, Horning BD, et al. Chemical proteomics identifies druggable vulnerabilities in a genetically defined cancer. Cell. 2017;**171**(3):696-709

[10] Kong H, Chandel NS. Regulation of redox balance in cancer and T cells. Journal of Biological Chemistry. 2018;**293**(20):7499-7507

[11] Li X, Fang P, Mai J, Choi ET, Wang H, Yang XF. Targeting mitochondrial reactive oxygen species as novel therapy for inflammatory diseases and cancers. Journal of Hematology & Oncology. 2013;**6**(1):19

[12] Galadari S, Rahman A, Pallichankandy S, Thayyullathil F. Reactive oxygen species and cancer paradox: To promote or to suppress? Free Radical Biology and Medicine. 2017;**104**: 144-164

[13] Panieri E, Santoro MM. ROS homeostasis and metabolism: A dangerous liason in cancer cells. Cell Death & Disease. 2016;**7**(6):e2253

[14] Gao P, Tchernyshyov I, Chang TC, Lee YS, Kita K, Ochi T, et al. c-Myc suppression of miR-23a/b enhances mitochondrial glutaminase expression and glutamine metabolism. Nature. 2009;**458**(7239):762

[15] Brown GC. Control of respiration and ATP synthesis in mammalian mitochondria and cells. The Biochemical Journal. 1992;**284**(1):1-3

[16] Kowaltowski AJ, de Souza-Pinto NC, Castilho RF, Vercesi AE. Mitochondria and reactive oxygen species. Free Radical Biology & Medicine. 2009;**47**(4):333-343

[17] Liu Y, Fiskum G, Schubert D. Generation of reactive oxygen species by the mitochondrial electron transport chain. Journal of Neurochemistry. 2002;**80**(5):780-787

[18] Turrens JF. Mitochondrial formation of reactive oxygen species. The Journal of Physiology. 2003;**52**(2):335-344

[19] Radi R, Cassina A, Hodara R. Nitric oxide and peroxynitrite interactions with mitochondria. Biological Chemistry. 2002;**383**(3–4):401-409

[20] Loschen G, Azzi A, Richter C, Flohé L. Superoxide radicals as precursors of mitochondrial hydrogen peroxide. FEBS Letters. 1974;**42**(1):68-72

[21] Weisiger RA, Fridovich I. Superoxide dismutase organelle specificity. The Journal of Biological Chemistry. 1973;**248**(10):3582-3592

[22] Boveris A, Chance B. The mitochondrial generation of hydrogen peroxide. General properties and effect of hyperbaric oxygen. The Biochemical Journal. 1973;**134**(3):707-716

[23] Nordberg J, Arner ES. Reactive oxygen species, antioxidants, and the mammalian thioredoxin system1. Free Radical Biology & Medicine. 2001;**31**(11):1287-1312

[24] Droge W. Free radicals in the physiological control of cell function. Physiological Reviews. 2002;**82**(1):47-95

[25] Hamanaka RB, Chandel NS. Mitochondrial reactive oxygen species regulate cellular signaling and dictate biological outcomes. Trends in Biochemical Sciences. 2010;**35**(9):505-513

[26] Starkov AA. The role of mitochondria in reactive oxygen species metabolism and signaling. Annals of the New York Academy of Sciences. 2008;**1147**(1):37-52

[27] Aruoma OI. Free radicals, oxidative stress, and antioxidants in human health and disease. Journal of the American Oil Chemists' Society. 1998;**75**(2):199-212

[28] Murphy MP. How mitochondria produce reactive oxygen species. The Biochemical Journal. 2009;**417**(1):1-3

[29] Scaffidi C, Fulda S, Srinivasan A, Friesen C, Li F, Tomaselli KJ, et al. Two CD95 (APO-1/Fas) signaling pathways. The EMBO Journal. 1998;**17**(6):1675-1687

[30] Morgan MJ, Liu ZG. Reactive oxygen species in TNFα-induced signaling and cell death. Molecules and Cells. 2010;**30**(1):1-2

[31] Li PL, Gulbins E. Lipid rafts and redox signaling. Antioxidants & Redox Signaling. 2007;**9**(9):1411-1416

[32] Wang L, Azad N, Kongkaneramit L, Chen F, Lu Y, Jiang BH, et al. The Fas death signaling pathway connecting reactive oxygen species generation and FLICE inhibitory protein down-regulation. The Journal of Immunology. 2008;**180**(5):3072-3080

[33] Izeradjene K, Douglas L, Tillman DM, Delaney AB, Houghton JA. Reactive oxygen species regulate caspase activation in tumor necrosis factor–related apoptosis-inducing ligand–resistant human colon carcinoma cell lines. Cancer Research. 2005;**65**(16):7436-7445

[34] Woo SH, Park IC, Park MJ, An S, Lee HC, Jin HO, et al. Arsenic trioxide sensitizes CD95/Fas-induced apoptosis through ROS-mediated upregulation of CD95/Fas by NF-κB activation. International Journal of Cancer. 2004;**112**(4):596-606

[35] Pantano C, Shrivastava P, McElhinney B, Janssen-Heininger Y. Hydrogen peroxide signaling through tumor necrosis factor receptor 1 leads to selective activation of c-Jun N-terminal kinase. The Journal of Biological Chemistry. 2003;**278**(45):44091-44096

[36] Kamata H, Honda SI, Maeda S, Chang L, Hirata H, Karin M. Reactive oxygen species promote TNFα-induced death and sustained JNK activation by inhibiting MAP kinase phosphatases. Cell. 2005;**120**(5):649-661

[37] Fujino G, Noguchi T, Matsuzawa A, Yamauchi S, Saitoh M, Takeda K, et al. Thioredoxin and TRAF family proteins regulate reactive oxygen species-dependent activation of ASK1 through reciprocal modulation of the N-terminal homophilic interaction of ASK1. Molecular and Cellular Biology. 2007;**27**(23):8152-8163

[38] Saxena G, Chen J, Shalev A. Intracellular shuttling and mitochondrial function of thioredoxin-interacting protein. Journal of Biological Chemistry. 2010;**285**(6):3997-4005

[39] Zoratti M, Szabò I. The mitochondrial permeability transition. Biochimica et Biophysica Acta (BBA)–Reviews on Biomembranes. 1995;**1241**(2):139-176

[40] Fulda S, Galluzzi L, Kroemer G. Targeting mitochondria for cancer therapy. Nature Reviews. Drug Discovery. 2010;**9**(6):447

[41] Kaasik A, Safiulina D, Zharkovsky A, Veksler V. Regulation of mitochondrial matrix volume. American Journal of Physiology. Cell Physiology. 2007;**292**(1):C157-C163

[42] Kim HW, Lin A, Guldberg RE, Ushio-Fukai M, Fukai T. Essential role of extracellular SOD in reparative neovascularization induced by hindlimb ischemia. Circulation Research. 2007;**101**(4):409-419

[43] Lewis P, Stefanovic N, Pete J, Calkin AC, Giunti S, Thallas-Bonke V, et al. Lack of the antioxidant enzyme glutathione peroxidase-1 accelerates atherosclerosis in diabetic apolipoprotein e–deficient mice. Circulation. 2007;**115**(16):2178-2187

[44] Hamanishi T, Furuta H, Kato H, Doi A, Tamai M, Shimomura H, et al. Functional variants in the glutathione peroxidase-1 (GPx-1) gene are associated with increased intima-media thickness of carotid arteries and risk of macrovascular diseases in Japanese type 2 diabetic patients. Diabetes. 2004;**53**(9):2455-2460

[45] Kang SW, Chae HZ, Seo MS, Kim K, Baines IC, Rhee SG. Mammalian peroxiredoxin isoforms can reduce hydrogen peroxide generated in response to growth factors and tumor necrosis factor-α. The Journal of Biological Chemistry. 1998;**273**(11):6297-6302

[46] Li Y, Huang TT, Carlson EJ, Melov S, Ursell PC, Olson JL, et al. Dilated cardiomyopathy and neonatal lethality in mutant mice lacking manganese superoxide dismutase. Nature Genetics. 1995;**11**(4):376

[47] Pham CG, Bubici C, Zazzeroni F, Papa S, Jones J, Alvarez K, et al. Ferritin heavy chain upregulation by NF-κB inhibits TNFα-induced apoptosis by suppressing reactive oxygen species. Cell. 2004;**119**(4):529-542

[48] Liu G, Zhou W, Wang LI, Park S, Miller DP, Xu LL, et al. MPO and SOD2 polymorphisms, gender, and the risk of non-small cell lung carcinoma. Cancer Letters. 2004;**214**(1):69-79

[49] Kang D, Lee KM, Park SK, Berndt SI, Peters U, Reding D, et al. Functional variant of manganese superoxide dismutase (SOD2 V16A) polymorphism is associated with prostate cancer risk in the prostate, lung, colorectal, and ovarian cancer study. Cancer Epidemiology and Prevention Biomarkers. 2007;**16**(8):1581-1586

[50] Wang SS, Davis S, Cerhan JR, Hartge P, Severson RK, Cozen W, et al. Polymorphisms in oxidative stress genes and risk for non-Hodgkin lymphoma. Carcinogenesis. 2006;**27**(9): 1828-1834

[51] Olson SH, Carlson MD, Ostrer H, Harlap S, Stone A, Winters M, et al. Genetic variants in SOD2, MPO, and NQO1, and risk of ovarian cancer. Gynecologic Oncology. 2004;**93**(3): 615-620

[52] Ichimura Y, Habuchi T, Tsuchiya N, Wang L, Oyama C, Sato K, et al. Increased risk of bladder cancer associated with a glutathione peroxidase 1 codon 198 variant. The Journal of Urology. 2004;**172**(2):728-732

[53] Raaschou-Nielsen O, Sørensen M, Hansen RD, Frederiksen K, Tjønneland A, Overvad K, et al. GPX1 Pro198Leu polymorphism, interactions with smoking and alcohol consumption, and risk for lung cancer. Cancer Letters. 2007;**247**(2):293-300

[54] Ravn-Haren G, Olsen A, Tjønneland A, Dragsted LO, Nexø BA, Wallin H, et al. Associations between GPX1 Pro198Leu polymorphism, erythrocyte GPX activity, alcohol consumption and breast cancer risk in a prospective cohort study. Carcinogenesis. 2005;**27**(4):820-825

[55] Arsova-Sarafinovska Z, Matevska N, Eken A, Petrovski D, Banev S, Dzikova S. Glutathione peroxidase 1 (GPX1) genetic polymorphism, erythrocyte GPX activity, and prostate cancer risk. International Urology and Nephrology. 2009;**41**(1):63

[56] Tanaka T, Hosoi F, Yamaguchi-Iwai Y, Nakamura H, Masutani H, Ueda S, et al. Thioredoxin-2 (TRX-2) is an essential gene regulating mitochondria-dependent apoptosis. The EMBO Journal. 2002;**21**(7):1695-1703

[57] Matthews JR, Wakasugi N, Virelizier JL, Yodoi J, Hay RT. Thiordoxin regulates the DNA binding activity of NF-χB by reduction of a disulphid bond involving cysteine 62. Nucleic Acids Research. 1992;**20**(15):3821-3830

[58] Turunen N, Karihtala P, MÄntyniemi A, Sormunen R, Holmgren A, Kinnula VL, et al. Thioredoxin is associated with proliferation, p53 expression and negative estrogen and progesterone receptor status in breast carcinoma. APMIS. 2004;**112**(2):123-132

[59] Fujii S, Nanbu Y, Nonogaki H, Konishi I, Mori T, Masutani H, et al. Coexpression of adult T-cell leukemia-derived factor, a human thioredoxin homologue, and human papillomavirus DNA in neoplastic cervical squamous epithelium. Cancer. 1991;**68**(7):1583-1591

[60] Nakamura H, Masutani H, Tagaya Y, Yamauchi A, Inamoto T, Nanbu Y, et al. Expression and growth-promoting effect of adult t-cell leukemia-derived factor a human thioredoxin homologue in hepatocellular carcinoma. Cancer. 1992;**69**(8):2091-2097

[61] Soini Y, Kahlos K, Näpänkangas U, Kaarteenaho-Wiik R, Säily M, Koistinen P, et al. Widespread expression of thioredoxin and thioredoxin reductase in non-small cell lung carcinoma. Clinical Cancer Research. 2001;**7**(6):1750-1757

[62] Lincoln DT, Ali EE, Tonissen KF, Clarke FM. The thioredoxin-thioredoxin reductase system: Over-expression in human cancer. Anticancer Research. 2003;**23**(3B):2425-2433

[63] Sies H. Glutathione and its role in cellular functions. Free Radical Biology & Medicine. 1999;**7**(9–10):916-921

[64] Quinlan CL, Orr AL, Perevoshchikova IV, Treberg JR, Ackrell BA, Brand MD. Mitochondrial complex II can generate reactive oxygen species at high rates in both the forward and reverse reactions. The Journal of Biological Chemistry. 2012;**287**(32):27255-27264

[65] Gamcsik MP, Kasibhatla MS, Teeter SD, Colvin OM. Glutathione levels in human tumors. Biomarkers. 2012;**17**(8):671-691

[66] Jiang L, Kon N, Li T, Wang SJ, Su T, Hibshoosh H, et al. Ferroptosis as a p53-mediated activity during tumour suppression. Nature. 2015;**520**(7545):57

[67] Harris IS, Treloar AE, Inoue S, Sasaki M, Gorrini C, Lee KC, et al. Glutathione and thioredoxin antioxidant pathways synergize to drive cancer initiation and progression. Cancer Cell. 2015;**27**(2):211-222

[68] Cramer SL, Saha A, Liu J, Tadi S, Tiziani S, Yan W, et al. Systemic depletion of L-cyst (e) ine with cyst (e) inase increases reactive oxygen species and suppresses tumor growth. Nature Medicine. 2017;**23**(1):120

[69] Traverso N, Ricciarelli R, Nitti M, Marengo B, Furfaro AL, Pronzato MA, et al. Role of glutathione in cancer progression and chemoresistance. Oxidative Medicine and Cellular Longevity. 2013;**2013**:972913

[70] Perkins A, Nelson KJ, Parsonage D, Poole LB, Karplus PA. Peroxiredoxins: Guardians against oxidative stress and modulators of peroxide signaling. Trends in Biochemical Sciences. 2015;**40**(8):435-445

[71] Park MH, Jo M, Kim YR, Lee CK, Hong JT. Roles of peroxiredoxins in cancer, neurodegenerative diseases and inflammatory diseases. Pharmacology & Therapeutics. 2016;**163**:1-23

[72] Egler RA, Fernandes E, Rothermund K, Sereika S, de Souza-Pinto N, Jaruga P, et al. Regulation of reactive oxygen species, DNA damage, and c-Myc function by peroxiredoxin 1. Oncogene. 2005;**24**(54):8038

[73] Cao J, Schulte J, Knight A, Leslie NR, Zagozdzon A, Bronson R, et al. Prdx1 inhibits tumorigenesis via regulating PTEN/AKT activity. The EMBO Journal. 2009;**28**(10):1505-1517

[74] Sumimoto H. Structure, regulation and evolution of Nox-family NADPH oxidases that produce reactive oxygen species. The FEBS Journal. 2008;**275**(13):3249-3277

[75] Segal AW. The function of the NADPH oxidase of phagocytes and its relationship to other NOXs in plants, invertebrates, and mammals. The International Journal of Biochemistry & Cell Biology. 2008;**40**(4):604-618

[76] Jiang L, Shestov AA, Swain P, Yang C, Parker SJ, Wang QA, et al. Reductive carboxylation supports redox homeostasis during anchorage-independent growth. Nature. 2016; **532**(7598):255

[77] D D'Souza A, Parish IA, Krause DS, Kaech SM, Shadel GS. Reducing mitochondrial ROS improves disease-related pathology in a mouse model of ataxia-telangiectasia. Molecular Therapy. 2013;**21**(1):42-48

[78] Lu W, Fu Z, Wang H, Feng J, Wei J, Guo J. Peroxiredoxin 2 is upregulated in colorectal cancer and contributes to colorectal cancer cells' survival by protecting cells from oxidative stress. Molecular and Cellular Biochemistry. 2014;**387**(1–2):261-270

[79] Shiota M, Yokomizo A, Kashiwagi E, Takeuchi A, Fujimoto N, Uchiumi T, et al. Peroxiredoxin 2 in the nucleus and cytoplasm distinctly regulates androgen receptor activity in prostate cancer cells. Free Radical Biology & Medicine. 2011;**51**(1):78-87

[80] Cramer SL, Saha A, Liu J, Tadi S, Tiziani S, Yan W, et al. Systemic depletion of serum l-Cyst (e) ine with an engineered human enzyme induces production of reactive oxygen species and suppresses tumor growth in mice. Nature Medicine. 2017;**23**(1):120

[81] Segal BH, Veys P, Malech H, Cowan MJ. Chronic granulomatous disease: Lessons from a rare disorder. Biology of Blood and Marrow Transplantation. 2011;**17**(1):S123-S131

[82] Lassègue B, San Martín A, Griendling KK. Biochemistry, physiology, and pathophysiology of NADPH oxidases in the cardiovascular system. Circulation Research. 2012;**110**(10): 1364-1390

[83] Gorrini C, Baniasadi PS, Harris IS, Silvester J, Inoue S, Snow B, et al. BRCA1 interacts with Nrf2 to regulate antioxidant signaling and cell survival. The Journal of Experimental Medicine. 2013;**210**(8):1529-1544

[84] Cosentino C, Grieco D, Costanzo V. ATM activates the pentose phosphate pathway promoting anti-oxidant defence and DNA repair. The EMBO Journal. 2011;**30**(3):546-555

[85] Tafani M, Sansone L, Limana F, Arcangeli T, De Santis E, Polese M, et al. The interplay of reactive oxygen species, hypoxia, inflammation, and sirtuins in cancer initiation and progression. Oxidative Medicine and Cellular Longevity. 2016;**2016**:3907147

[86] Prasad S, Gupta SC, Tyagi AK. Reactive oxygen species (ROS) and cancer: Role of antioxidative nutraceuticals. Cancer Letters. 2017;**387**:95-105

[87] Haigis MC, Deng CX, Finley LW, Kim HS, Gius D. SIRT3 is a mitochondrial tumor suppressor: A scientific tale that connects aberrant cellular ROS, the Warburg effect, and carcinogenesis. Cancer Research. 2012;**72**(10):2468-2472

[88] Braidy N, Poljak A, Grant R, Jayasena T, Mansour H, Chan-Ling T, et al. Differential expression of sirtuins in the aging rat brain. Frontiers in Cellular Neuroscience. 2015;**9**:167

[89] Marengo B, Nitti M, Furfaro AL, Colla R, Ciucis CD, Marinari UM, Pronzato MA, Traverso N, Domenicotti C. Redox homeostasis and cellular antioxidant systems: Crucial players in cancer growth and therapy. Oxidative Medicine and Cellular Longevity. 2016; **2016**:6235641

[90] Antal CE, Hudson AM, Kang E, Zanca C, Wirth C, Stephenson NL, et al. Cancer-associated protein kinase C mutations reveal kinase's role as tumor suppressor. Cell. 2015;**160**(3):489-502

[91] Wang J, Shao M, Liu M, Peng P, Li L, Wu W, et al. PKCα promotes generation of reactive oxygen species via DUOX2 in hepatocellular carcinoma. Biochemical and Biophysical Research Communications. 2015;**463**(4):839-845

[92] Nitti M, Furfaro AL, Cevasco C, Traverso N, Marinari UM, Pronzato MA, et al. PKC delta and NADPH oxidase in retinoic acid-induced neuroblastoma cell differentiation. Cellular Signalling. 2010;**22**(5):828-835

[93] Leonarduzzi G, Sottero B, Gabriella Testa G, Biasi F, Poli G. New insights into redox-modulated cell signaling. Current Pharmaceutical Design. 2011;**17**(36):3994-4006

[94] Xu J, Tian W, Ma X, Guo J, Shi Q, Jin Y, et al. The molecular mechanism underlying morphine-induced Akt activation: Roles of protein phosphatases and reactive oxygen species. Cell Biochemistry and Biophysics. 2011;**61**(2):303-311

[95] Manning BD, Cantley LC. AKT/PKB signaling: Navigating downstream. Cell. 2007;**129**(7): 1261-1274

[96] Chen W, Sun Z, Wang XJ, Jiang T, Huang Z, Fang D, et al. Direct interaction between Nrf2 and p21 Cip1/WAF1 upregulates the Nrf2-mediated antioxidant response. Molecular Cell. 2009;**34**(6):663-673

Chapter 8

S6 Kinase: A Compelling Prospect for Therapeutic Interventions

Sheikh Tahir Majeed, Rabiya Majeed,
Ghazia Shah and Khurshid I Andrabi

Additional information is available at the end of the chapter

http://dx.doi.org/10.5772/intechopen.75209

Abstract

S6 kinase, a member of AGC family of protein kinases and a downstream effector of mTORC1 pathway has over the years found much relevance in maintaining a normal cellular state by virtue of its established role in regulation of cell growth and proliferation. S6 kinase activity has been linked to different cellular processes like glucose homeostasis, translational and transcriptional regulation. Hence any dysregulation in S6K1 leads to the emergence of various pathological conditions like diabetes, cancer and obesity. It is as such S6 kinase has emerged as a potential target for therapeutic interventions employed in curing such diseases. The Present Chapter reviews the regulation of S6K1, its structural organization and functions, besides highlighting its potential to act as an alternative therapeutic target for various cancerous situations exhibiting deranged mTOR signaling so as to overcome the possibility of relapses observed otherwise while using conventional drugs

Keywords: S6 kinase, mTOR, rapamycin, cancer, cell growth and proliferation

1. Introduction

Cell signaling pathways function as cohorts to contribute in maintaining cellular homeostasis-a phenomenon that remains pivotal in regulating cell growth and survival. The unique but immensely coordinated response of these diverse signals is primarily regulated by the availability of nutrients, growth factors and energy status of the cell. The signals destined to activate certain cellular functions may simultaneously be antagonistic for other functions; to strike a much required balance for proper functioning of the cell. Relentless efforts over the last several decades have led to recognize the ability of these signaling networks to promote cell growth and proliferation by way of their potential to phosphorylate one another to distinguish themselves as kinases. The kinases specifically transfer the phosphoryl group from ATP to their substrates such as to bring about a change in their function that helps transduce the upstream signals for the accomplishment of functions. Intrigued by the properties of these kinases, the scientific community over the years has shown great interest in identifying more kinases that lead to the identification of a family of protein kinases known as ribosomal S6 kinases (RSKs) that respond to nutrients and growth factors. The RSKs came to fore after studying the inducible phosphorylations on ribosomal protein S6 (rps6) and accordingly p90 ribosomal S6 kinase was identified from Xenopus oocytes. The homologous counterparts of p90 S6 kinase in humans were soon identified and grouped into family of four enzymes termed as RSK1-4. It was later on observed that the RSKs may play a minor role in rps6 phosphorylation [1] to give way for the emergence of a major rps6 kinase identified as 70 kDa ribosomal S6 kinase [1–4]. The 70 kDa ribosomal S6 kinase was purified from mitogen stimulated Swiss mouse 3T3 cells [5]. Cloning of 70 kDa ribosomal S6 kinase was soon accomplished after protein sequencing of a rat and rabbit cDNA [6–8]. Subsequent identification and cloning of human orthologs of S6K genes RPS6KB1 and RPS6KB2 encoding S6K1 and S6K2 thus, respectively, led to the emergence of a second family of S6 kinases, a principal kinase of rps6 in somatic cells [2, 9–14].

S6 kinases figure among important signaling molecules required for propagation of cell growth and proliferation. Adding to its importance is the fact that S6 kinase (S6K1) is a downstream effector of mTORC1 (mechanistic target of rapamycin), a master regulator of cell growth and proliferation. Evidences are strong enough to suggest that the mTOR-S6K1 axis controls fundamental cellular processes including transcription, translation and lipid synthesis. Further, this nexus partly regulates glucose homoeostasis, controls insulin sensitivity, adipocyte metabolism, tissue and organ size, learning, memory and aging [15]. S6K1 has therefore been considered to play a pivotal role in regulating cellular physiology. Any dysregulation of this signaling axis has deleterious implications ranging from metabolic disorders to various cancers. mTOR/S6K1 signaling network has hence remained central target for various medical interventions employed over the years for treating the pathological conditions. However, so far the therapeutic interventions involving mTOR as a target molecule for treating cancers, particularly renal carcinomas have reported a fair share of relapses. The resistance shown to these drugs post chronic exposures remain one of the major concerns the scientific community is faced upto. One of the reasons for any such observation would be the incomplete understanding vis-a-vis the regulation of mTOR/S6K1 axis. It is as such conceivable that improved understanding of S6K1 regulation would help develop novel therapeutics that should take care of these challenges. The present chapter reviews the regulation of S6K, its structure and functions.

2. S6K protein family; domain architecture and cellular localization

Based on the sequence alignment of their catalytic domains, S6Ks have been placed in AGC family of protein kinases that mainly comprise of protein kinase A (PKA) protein kinase G (PKG) and protein kinase C (PKC) [16]. The structural resemblance in AGC kinases allows them to exhibit more or less a similar mode of regulation. These kinases fundamentally exhibit bi-lobal structural organization around the kinase domain. This structural arrangement enables their amino and carboxyl termini to coordinate ATP binding [17, 18]. Further these kinases comprise of an activation loop (commonly known as T-loop) that precedes their C-terminus lobe. The phosphorylation of T-loop brings in conformational changes important for phosphoryl transfer. Two other important phosphorylation sites, the turn motif (TM) (so-named due to its location at the cusp of a structural turn in the PKA tail) and hydrophobic motif (HM) sequentially follow the kinase domain. The phosphorylated HM site engages a hydrophobic pocket within the N-lobe. The phosphorylated TM site stabilizes phospho-HM binding to the N-lobe hydrophobic pocket. Together, these three phosphorylations stand critical in stabilizing the catalytically active state [17, 18].

RPS6KB1, human ortholog of S6K1 genes, encodes two isoforms, p70S6K1 and p85S6K1 (formerly known as p70S6KII and p70S6KI, respectively) through alternative translational start sites [10]. p70S6K1, a 502 amino acid protein and its larger isoform p85S6K1, having nuclear localization sequence (NLS) within the 23 amino acid N-terminal extension, were earlier believed to remain localized in the cytoplasm and nucleus, respectively. However, the data generated of late, contests the nuclear localization of p85S6K1 while showing its sub-cellular localization to be cytoplasmic and at the same time claims p70S6K1 localization to be both nuclear and cytoplasmic [19]. S6K2, encoded by RPS6KB2 gene, constitutes

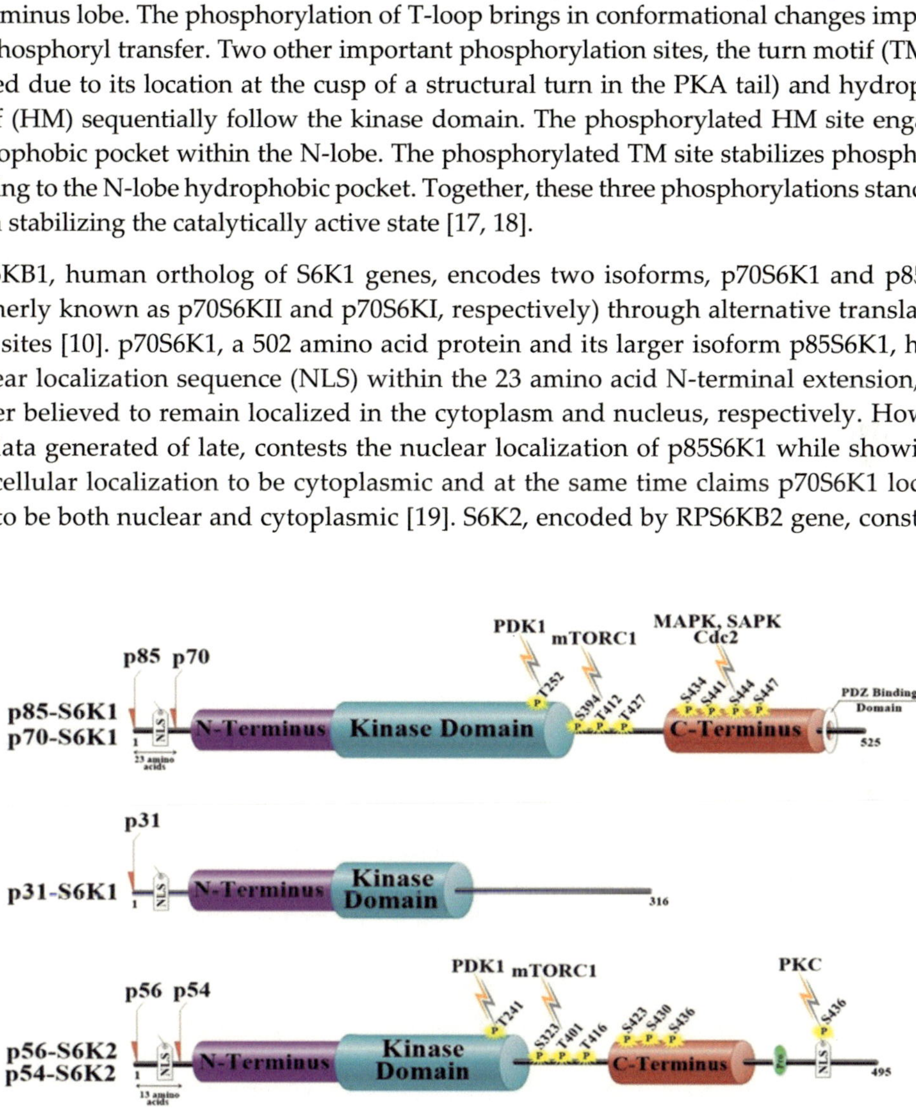

Figure 1. S6K isoforms, structure and domain organization with potential phosphorylating kinases along with their sites of action. S6K1 isoforms include p70-, p85- and p31-S6K1; alternative start site usage lengthens the p85- and p31-S6K1 N-termini by 23 amino acids (note that p31-S6K1 lacks most of the kinase domain). S6K2 isoforms include p54- and p56-S6K2; alternative start site usage lengthens the p56-S6K2 N-terminus by 13 amino acids. NLSs lie within the N-terminal extensions of p85-S6K1 and p56-S6K2, whereas S6K2 additionally contains an NLS within the C-terminus as well as a proline-rich domain (Pro). S6Ks contain an acidic N-terminal domain (NTD), kinase domain (KD), linker region and acidic C-terminal domain (CTD). The N-terminal domain contains the TOS motif, whereas the CTD contains the auto-inhibitory pseudosubstrate domain and RSPRR motif. mTORC1 phosphorylates the HM (hydrophobic motif) site (Thr412) in the linker region and PDK1 phosphorylates the T-loop site (Thr252) within the kinase domain. Other regulatory phosphorylation (P) sites, including the TM (turn motif) site (Ser394), are shown.

another member of S6 kinase family that, via alternative start sites, produces two nuclear isoforms, owing to the presence of nuclear localization sequence at the C-terminus [11]. The longer isoform, p56S6K2 comprises a 13 amino acid extension at N-terminus than its shorter isoform, p54S6K2 [9]. Moreover S6K2 isoforms have been characterized by the presence of proline-rich domain at the C-terminal [20]. S6K1 and S6K2 have been structurally dissected into several regulatory domains, namely an acidic N-terminus that contains the TOR signaling (TOS) motif; the kinase domain that contains the activation/T-loop; a linker region that contains the TM and HM sites; and a basic C-terminus containing an auto-inhibitory pseudosubstrate domain (**Figure 1**). These proteins are homologous to each other within appreciable limits; their catalytic domains share about 83% identity at the amino acid level [9]. However, the differences observed in the extreme N- and C-terminal regions, direct these kinases to distinct compartments or to different molecular targets. Thus, for example, the C-terminal PDZ binding domain in S6K1 allows recruitment to the actin cytoskeleton via binding to neurabin [21] and S6K2 containing a proline-rich region in its C-terminus facilitates interaction with SH3 domain or WW repeat containing protein [9]. Besides full length forms of S6K1 and S6K2, a novel kinase domain truncated splice variant, p31S6K (**Figure 1**) with potential oncogenic properties, has also been reported [22].

3. S6 kinase regulation; a historical perspective for prevalent mode of activation

Research carried out over the past 20 years identified diverse growth factors and mitogens with the ability to activate the S6 kinases [23, 24]. Seemingly among these multiple inputs, insulin/IGF pathway, which signals via phosphoinositide 3-kinase (PI3K) and mTORC1 appears to be the best studied activator of S6K1 [25, 26]. Besides PI3K, other independent signaling pathways like Ras/MAPK (mitogen activated protein kinase) have also been implicated in S6K1 activation [27, 28]. S6K1's contribution in controlling cell growth and proliferation being paramount made the basis for studying its structure and function. The data accumulated over a period of time, hence revealed the molecular insights that govern activation of S6K1 by mitogens and identified, to a large extent, the complex interactions between its specific domains and phosphorylation sites. These observations eventually became the reason for putting forth the models that explain the activation of S6K1 by stepwise multisite phosphorylations [29–33]. These models propose that during inactive state, the basic C-terminal pseudosubstrate domain of S6 K1 interacts with its acidic N-terminus. This interaction obstructs the phosphorylation at kinase domain and renders it inactive [6, 34]. Stimulation by growth factors besides other mitogens result in the phosphorylation of four proline-directed sites at C-terminal auto-inhibitory pseudosubstrate domain (Ser434, Ser441, Ser444 and Ser447) [30, 32]. These phosphorylations induce a conformational change thereby releasing the interaction between the two termini and enables access to the HM and T-loop sites. Phosphorylation at these two critical sites ultimately leads to the full activation of the kinase [6, 34]. Several proline-directed kinases, including ERK1/2, JNK1/2 and CDK1 have been implicated in phosphorylation of these sites; it however remains unclear as to which kinase(s) play a dominant role in-vivo [32]. Scientific observations during early 1990's propounded the theory that mTOR controls the in-vivo activation of S6K1 [35]. These observations were further augmented during 1998–1999 when mTORC1 was shown to directly phosphorylate the HM site (Thr412) of S6K1 in vitro [36, 37]. The indispensible nature of HM site phosphorylation is further supported by the evidence that mutation of Thr 412 to alanine (T412A) abolishes the S6K1 activity, whereas the phosphor mimicked variant (T412E) enhances basal activity even in absence of mitogens [20, 33, 38]. The findings put forward hence strengthened the notion that phosphorylation of Thr 412 at HM site remains pivotal for S6K1 activation. Co-ordinate phosphorylation of Thr252 in the activation/T-loop and Thr412 at HM site leads to maximal activation of S6K1 [29, 33, 39]. Set out to identify the potential kinases responsible for Thr 252 phosphorylation, in vitro followed by in-vivo studies show PDK1, also corroborated as a kinase for AKT, to directly phosphorylate Thr252 at T-loop site [29, 39, 40]. This was supported by the findings that PDK1 null embryonic stem cells, $PDK1^{-/-}$ or T252A mutation rendered S6K1 enzymatically dead [33, 41]. Further it is observed that PDK1 mediated phosphorylation of C-Terminus truncated variant of S6K1, S6K1-ΔCT remains significantly higher than full length S6K1 while mutation in c-terminal phosphorylation sites to alanine results in poor S6K1 activation [29]. Moreover, PDK1 poorly activates S6K1 T412A-ΔCT or T394A-ΔCT in vitro. All these findings reveal that unphosphorylated c-terminus puts restraint toward full activation of S6K1 by blocking the access of PDK1 to activation loop and the importance of HM and TM sites toward bringing PDK1 mediated S6K1 activation [29]. Phosphorylation of Ser394 (or Ser371 in P70S6K isoform) at turn motif site also remains of much significance for complete activation of S6K1, as S394A substitution of S6K1 renders it completely inactive [42]. Notwithstanding the importance of turn motif site, the data accumulated on regulation and function of this phosphorylation event has progressed, albeit rather slowly. Whatever little the quantum of data may be available for regulation around TM site phosphorylation till date, it concordantly reflects that this site does not represent an autophosphorylation event [42]. Some more interesting observations reveal that even the addition of T412E substitution fails to restore the activity of S394A mutant, thus supporting the notion that this site plays an important yet independent role in regulating the activity of S6K1 [42]. Evidences, though scant, have also emerged which corroborate mTOR in promoting Ser394 phosphorylation in vitro [43]. However, since data accumulated in certain other cases does not correlate phospho-Ser394 well with mTORC1 activity, intermediation of some other kinase besides mTORC1 is also speculated. Reports based on analogy to co-translational phosphorylation of TM site (Thr450) in AKT are tempting to speculate that S6K1 TM site phosphorylation also represents an early event that occurs co-translationally prior to T-loop and HM site phosphorylation [44]. Indeed, a recent report supports such an idea, as Ser394 phosphorylation occurs simultaneously with the production of S6K1 protein from a transfected plasmid [31]. Since the phosphorylations at the HM and T-loop sites exhibit strong cooperativity, the temporal order of these two phosphorylations has not yet been convincingly deciphered. However, the data available till date has put forth two models for S6K1 activation wherein the conventional and widely accepted model suggests mTORC1 mediated phosphorylation at Thr 412 to precede PDK1 mediated phosphorylation at Thr 252 [29, 38, 39]. An alternate model however, suggests that the phosphorylation at T-loop (Thr 252) precedes HM phosphorylation (Thr 412) [31, 33] (**Figure 2**).

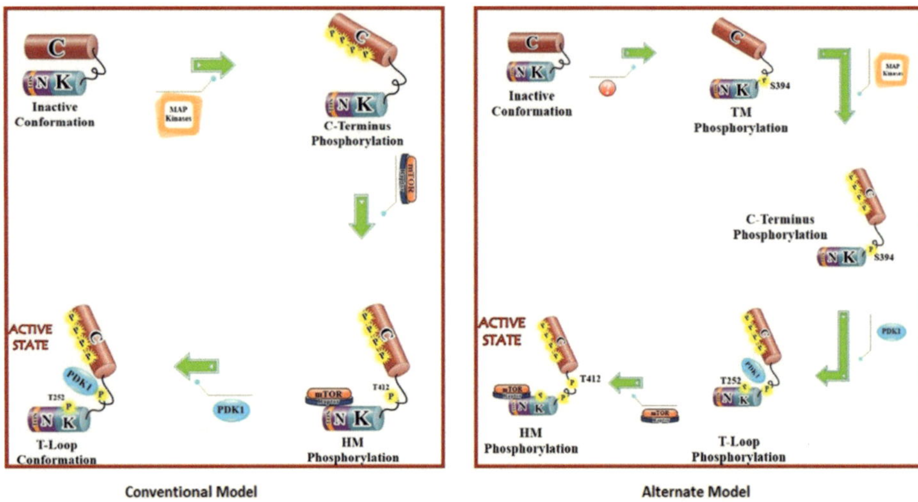

Figure 2. Stepwise activation of S6K1 via multi-site phosphorylation. Conventional model: the interaction of the C- and N-terminal domains results in auto-inhibition of S6K1. Step 1: mitogens promote C-terminal domain (C) phosphorylation on multiple sites to induce a more relaxed conformation. Step 2: the release of the auto-inhibitory C-terminal domain (CTD) enables mTORC1 access to the HM and thus phosphorylation of Thr^{412}. Step 3: the release of the auto-inhibitory CTD and phosphorylation on Thr^{412} enables PDK1-mediated phosphorylation of the T-loop on Thr^{252}, resulting in full activation of S6K1. Phospho-Thr^{412} serves as docking site for PDK1. Owing to insufficient data, the temporal order of TM site phosphorylation (Ser^{394}) is not depicted. Alternative model: Step 1: an unknown kinase phosphorylates the inactive form of S6K on the TM site Ser^{394}. Step 2: mitogens promote C-terminal domain (C) phosphorylation on multiple sites to induce a more relaxed conformation. Step 3: The release of the auto-inhibitory C-terminal domain enables PDK1 access to the T-loop. Step 4: PDK1-mediated phosphorylation of Thr^{252} promotes mTORC1-mediated phosphorylation on the HM site, Thr^{412}. KD, kinase domain; N,N-terminal domain.

S6K1 amino terminus serves a regulatory role in promoting phosphorylation at HM site and thereby at T-loop site that ultimately brings about its complete activation. The regulation by amino terminus is documented to be two pronged. Firstly, it acts as a receptor of an activating input critical for Thr 412 and Thr 252 phosphorylation; and, secondly, it suppresses an inhibitory function mediated by C-terminus. These observations have strong scientific support as the data generated over a period of time shows that amino terminus truncation of S6K1 (ΔNT) abolishes the phosphorylation of rapamycin-sensitive sites Thr 412, Thr 252, Ser 427 and renders it inactive [38, 45] whereas additional deletion of C-terminus (ΔNT/ΔCT) restores all the rapamycin-sensitive phosphorylations as well as the kinase activity of S6K1. The regulatory function of the amino terminus was mapped to a short stretch of amino acids (comprising of 5–9 amino acids and represented by a signature sequence FDIDL) present at extreme end known as TOS motif [20]. It has been shown that S6K1 truncated of TOS motif or its F5A variant (F5A mutation within the FDIDL sequence) abolishes S6K1 kinase activity as well as Thr412 and Thr252 phosphorylation. However, deletion of C-terminus from the F5A mutant (F5A-ΔCT) partially restores kinase activity and Thr 412 phosphorylation [20]. The TOS motif function has been related toward a launch pad required for mTORC1 to engage substrates and to mediate phosphorylation of rapamycin-sensitive sites [46, 47]. The data generated supports the view that mTOR interaction with TOS motif is indirect and mediated by another member of mTORC1 complex, a scaffold protein called Raptor [46, 47]. For phosphorylation of S6K1 and 4EBP-1, the other mTORC1 effector molecule, binding with raptor has been unequivocally stated to be critical. However, the precise mechanism by which raptor mediates efficient phosphorylation of S6K1 and 4EBP-1 remains largely debatable. Two models have however been proposed to explain this mechanism. The first model suggests that raptor and mTOR associate in two states with varying affinities governed by the nutrient availability. During nutrient starved state Raptor binds tightly to mTOR and renders it inactive. While as the loose-binding complex, formed during nutrient sufficiency, activates mTOR and promotes efficient phosphorylation of mTOR targets [48]. Furthermore, overexpression of raptor increases the amount of mTOR found in the tight-binding complex, thereby explaining the observation that overexpression of raptor inhibits mTOR activity. However, it is interesting to note that rapamycin is able to disrupt the raptor-mTOR interaction regardless of nutrient status [49], but it is phosphate dependent. The second model supports the existence of Raptor as a scaffolding protein for mTORC-1complex, wherein Raptor has been shown to preferentially bind unphosphorylated forms of mTOR targets and recruit the substrates to the mTOR complex for phosphorylation. Role of S6K1 amino terminus in suppressing the inhibitory function of C-terminus remained unclear till identification of an RSPRR motif (a short stretch of 5 amino acids, 433–437) in C-terminus way back in 2005 [47]. This motif has been suggested to negatively regulate S6K1 activation. Evidences, though scant, have propounded that RSPRR motif functions as a docking site for a negative regulator, such as a phosphatase, that is suppressed by mTORC1 [47].R3Amutation of RSPRR motif within the dead ΔNT or TOS motif-mutant (F5A) backbone (NT-R3A or F5A-R3A) has been shown to rescue insulin-stimulated Thr412 phosphorylation and S6K1 activation [15]. These findings though place TOS motif in exhibiting negative control over RSPPR motif inhibition, the exact mechanism behind this regulation still remains largely a mystery.

3.1. Rapamycin, S6K1 inhibition; a chronology of events

In 1970s, a bacterial strain, Streptomyces hygroscopicus was isolated from a soil sample from Easter Island ("Rapa Nui" in the native language) [50]. This strain was found to produce an anti-fungal metabolite [51]. Post purifications, the metabolite was characterized as a macrocyclic lactone and named "Rapamycin" after its birthplace (**Figure 3**) [52].

The strong ability of rapamycin to modulate cellular proliferation besides its promising role in immune suppression for treating various cancers lead to a desperate search in decoding its mode of action and simultaneously initiated hunt for nailing down its targets. The search continued for about two decades and finally culminated in early 1990s with identification of a target protein in yeast, whose mutant conferred resistance to the growth inhibitory effects of rapamycin. The protein was subsequently named as TOR (target of rapamycin) [53–56]. Shortly after, the mammalian counterpart of TOR (mTOR) was identified as physical target of rapamycin by three groups separately in 1994, 1995, respectively [57–59]. However effective mechanism of rapamycin action remained oblivious till intracellular cofactor, the peptidyl-prolyl cis/trans isomerase, immunophilin FK506-binding protein 12 (FKBP12) was shown to bind rapamycin as a gain-of-function component to mediate its inhibitory effect on TOR [56].

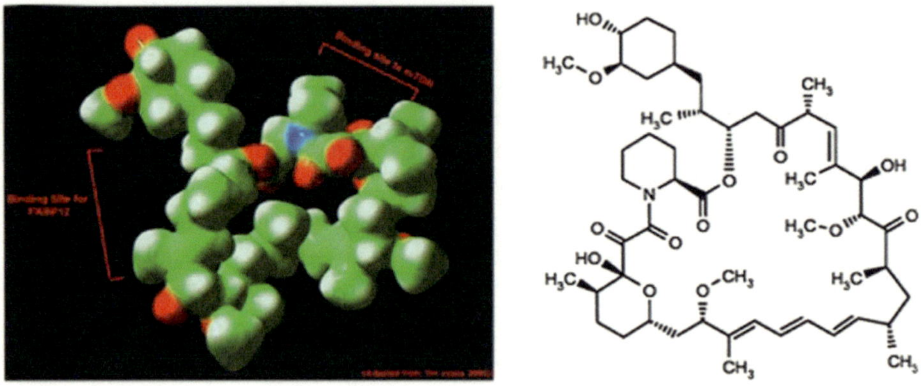

Figure 3. Electron-density model of a molecule of the immunosuppressant drug rapamycin: It is mainly used to prevent rejection in organ transplantation, and is used in kidney transplants. It also has anti-fungal and anti-cancer properties, and was originally derived from soil bacteria found on Easter Island. The chemical formula is C51.H79.N.O13. The atoms are represented here as color-coded blobs: carbon (green), hydrogen (white), nitrogen (blue) and oxygen (red) (adapted from Dr. Tim Evans Science photo library).

Thus, TOR is also referred to as FKBP12 rapamycin associated protein (FRAP) [60].Rapamycin-FKBP12 complex binds to the FKBP12-rapamycin binding (FRB) domain of TOR (**Figure 4**) [55, 60–62] to inhibit its intrinsic kinase activity, including autophosphorylation, thereby inhibiting access of TOR to its substrates [56]. This finding is however in contradiction with the earlier reports wherein it was shown that rapamycin has little effect on mTOR kinase activity [63]. In mammals, rapamycin in complex with FKBP12 acts as an allosteric inhibitor of mTOR complex1 (mTORC1). Only mTORC1 is acutely sensitive to inhibition of rapamycin. However, long term exposure to rapamycin has been shown to inhibit mTORC2 in certain cell types [64]. It is believed that the rapamycin-FKBP12 complex prevents the association between mTOR and raptor; therefore, downstream targets that depend on raptor binding are specifically inhibited [49].

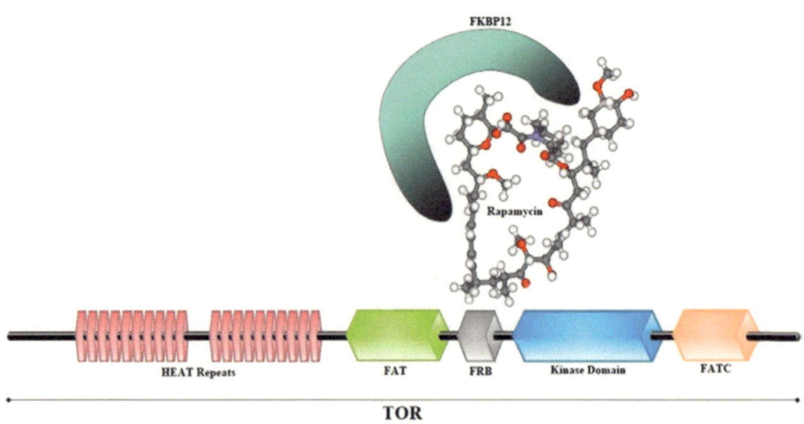

Figure 4. Association of rapamycin with TOR: schematic representation of structural organization of mTOR at domain level and its association with rapamycin–FKBP12 complex.

S6K1 being a downstream effector of mTOR, shows inhibitory response to rapamycin as is evidenced by loss of its ability to phosphorylate its substrate, ribosomal protein S6.

Early 1990s witnessed a rigorous search for identifying the rapamycin-sensitive regulatory phosphorylation sites in S6K1 and during 1995 three sites, T252 in the activation/T-loop, T412 in hydrophobic motif and S427 in the linker domain, which connects the auto-inhibitory domain to the catalytic domain were identified as principal rapamycin-sensitive sites [65]. These sites were shown to be dephosphorylated by rapamycin in hierarchical fashion T412 > S427 > T252 with T412 dephosphorylation most closely paralleling loss of kinase activity [65]. Besides, these sites were shown to be responsive to mitogenic stimulation as well [66]. Though all these three sites show rapamycin responsiveness, their conversions to either acidic or neutral amino acids reveal that T252 and T412 were critical regulatory sites where as T427 appeared to play a modulatory function [65]. Further T412 was shown to be principal site of rapamycin-induced S6K1 inactivation as T412E showed increased basal activity and was largely rapamycin resistant [65]. In 1995, Weng et al. reported the cooperativity among the two termini for exhibiting their regulatory effects on S6K1 activation by showing that amino- terminus truncated mutant, S6K1-ΔNT, was inactive. However, additional deletion of carboxy terminus, generating S6K1-ΔNT/ΔCT, rescued its phosphorylation and activation state. Surprisingly it was observed that the mutant was rapamycin resistant [45]. However, its responsiveness toward Wortmannin inhibition did not get compromised. These observations combined with the findings that the carboxy-terminal truncated S6K1 (S6K1-ΔCT) retains rapamycin sensitivity raised the speculations during latter half of 1990's that the inhibitory effect of the rapamycin–FKBP12 complex on S6K1 was either due to blockade of an upstream activator or activation of some phosphatase which mediate their influence through the two termini. The presence of Thr412 phosphorylation on S6K1-ΔNT/ΔCT isolated from rapamycin-treated cells questioned the idea that mTORC1 represents the sole S6K1 Thr412 kinase. In ΔNT/ΔCT, serum and insulin promote Thr412 phosphorylation and kinase activation in a completely rapamycin-resistant manner, suggesting that a rapamycin-insensitive kinase mediates Thr412 phosphorylation [38, 45]. This conundrum was soon resolved in 2005 by observing that rapamycin-insensitive mTORC2 mediates non-physiological S6K1 Thr412 phosphorylation in S6K1 mutants lacking a C-terminus [67]. It was also observed that unlike other AGC kinases S6Ks have atypically an extended C-terminus that imposes additional inhibitory influence on their state of activity [15]. Hence in the dead ΔNT allele, mTORC1 (mTOR/raptor) cannot dock to S6K1 and as such is incapacitated to phosphorylate Thr412, and mTORC2 (mTOR/rictor) cannot phosphorylate Thr412 due to steric hindrance imposed by the extended C-terminus [67]. In the partially rapamycin-resistant ΔCT mutant, both rapamycin-sensitive mTORC1 and rapamycin-insensitive mTORC2 cooperatively mediate Thr412 phosphorylation. In ΔNT/ΔCT (and F5A-ΔCT), only rapamycin-insensitive mTORC2 mediates Thr412 phosphorylation.

4. Other modes of S6K1 regulation

Studies of late propose the involvement of various post translational modifications (PTMs) in regulation of S6K1. Prominent among them are phosphatase-mediated dephosphorylation, acetylation and ubiquitination.

4.1. Dephosphorylation

S6K1 has been proposed to be regulated by co-ordinated action of phosphatases. This was observed while examining the action of rapamycin on phosphorylation state of S6K1.Whereas as rapamycin could dephosphorylate S6K1 through its inhibitory action on mTORC1 complex; the involvement of certain phosphatases to regulate this event was not ruled out. Indeed, S6Ks have been suspected to represent targets of PP2A (protein phosphatase 2A)-like phosphatases. This is supported by the evidence that regulation of several TOR substrates in *S. cerevisiae* occurs via suppression of PP2A-like phosphatases [68]. Further it has also been shown that PP2A interacts with wild type S6K1 and not ΔNT/ΔCT S6K1, a variant truncated of both the termini [69]. It is of late been observed that genetic ablation of PP2A regulatory subunit B (PP2A-B′) in *Drosophila melanogaster* leads to dS6K (Drosophila S6K) deregulation and a variety of metabolic defects [70]. Some further studies report that S6K1 phosphorylation (412) is enhanced in human cells upon knock down of PPP2R5C, the human PP2A-B′ orthologue [70]. All these studies support the role of phosphatases in regulating S6K1 activity. However, it is to be ascertained how mTOR influences these PP2A like phosphatase to modulate S6K1 Thr412 phosphorylation.

4.2. Acetylation and ubiquitination

Acetylation of S6K1 occurs reportedly by the action of two acetyltransferase enzymes, p300/CBP (cAMP-response-element-binding protein binding protein) and PCAF (p300/CBP-associated factor) [71].While as this PTM has been shown to occur on extreme C-terminus of S6K1 in response to the stimulation by mitogens, its dependence on phosphorylation has not been confirmed [72]. Acetylation has been primarily linked to the stabilization of S6K1 [71]. However, the concrete evidence to support this notion has remained largely elusive till date.

Polyubiquitination is another PTM, S6K1 is responsive to. S6K1 appears to experience the ubiquitination in response to mitogen stimulation [73, 74] through direct action of ubiquitin ligase, ROC1 [75]. RNA interference (RNAi) mediated knock down of ROC1 results in stabilization of S6K. These results are hence suggestive of the notion that polyubiquitination causes destabilization of S6K1 and a resultant attenuation of S6K1 downstream signals. Although regulation of S6K1 through ubiquitination still remains in its infancy; it however has opened a new window toward understanding detailed mechanistic inputs of proteasome mediated S6K1 regulation and function.

5. Functions associated with S6K1—a debatable discourse

S6K1, a downstream effector of mTOR is considered to be a multifaceted effector that regulates cell growth and proliferation by phosphorylating multiple ways. To add to its wide range of functions, S6K1 has much recently been shown to play a pivotal role in cellular senescence through its newly found substrate ZRF1 [76]. S6K1 gets activated in response to various signaling pathways including mTOR, PI3-kinase and MAPK, in a coordinated manner through sequential phosphorylation events directed at multiple sites [62]. While S6K1 extends its influence on various cellular functions, some of its associated functions still appear untenable, reason being the inadequate support data. Some of the S6K1 associated functions are briefly described below:

5.1. Cell growth and translational control—what is the connection?

S6K1 has been shown to control cell growth primarily by its ability to phosphorylate S6 protein of 40S ribosomal subunit (rpS6). These phosphorylations are shown to be localized at the C-terminus and happen in sequential order, namely S236 > S235 > S240 > S244 > S247. The dependence of S6 phosphorylation on S6K1 can traced by the observation that rpS6 deficient mice, rps6−/− show phenotypic resemblance to S6K1 deficient mice, S6K1−/−. Although these mice are viable and fertile, however both exhibit cell growth defects. It however is confusing to observe that $S6K1^{-/-}$ mice display minimal defects in rpS6 phosphorylation, while S6 K2, homolog of S6K1, deficient mice, that is, $S6K2^{-/-}$ grow to normal size despite exhibiting a significant reduction in rpS6 phosphorylation [77]. The conundrum can partly be resolved by the observation that claims selective recruitment of S6K1 and S6K2 to different cellular compartments and as such enables them to differentially phosphorylate rpS6 or alternatively by the preposition that rpS6 phosphorylation may ensue at a specific developmental stage where S6K1 but not S6K2 is active. Although a discordance seems to appear vis-a-vis S6K1 activation, rpS6 phosphorylation and cell growth, their complete understanding as far as therapeutics is concerned remains but pivotal. Although S6 phosphorylation was initially thought to be required for selectively translating mRNAs characterized by 5′-oligo Pyrimidine tract (5′-TOP). Later studies however confirmed that 5′ TOP mRNA translation remains unaffected in $S6K1^{-/-}/S6K2^{-/-}$ mice, which display minimal rpS6 phosphorylation [77], hence suggesting these two events to be independent [78].

S6K1 maintains its influence on cellular translation not only by the ability to phosphorylate ribosomal protein S6 (rpS6) but some studies provide evidences to establish S6K1 as a transcription factor kinase as well. One of the study shows S6K1 to phosphorylate eukaryotic initiation factor 4B (eIF4B), a component of cap-binding complex to control cap dependent translation. While as inactive S6K1 associates with eIF3 (a component of pre-initiation complex) and upon activation by mTORC1, via serum stimulation, dissociates to phosphorylate eIF4B, another component of pre-initiation complex [79, 80]. Besides it also controls translational initiation by phosphorylating PDCD4, a negative regulator of eIF4A (**Figure 5**). Phosphorylated PDCD4 gets ultimately degraded by ubiquitin ligase βTRCP [81]. The importance of S6K1 vis-a-vis translational control can further be demonstrated by the observation which implicates S6K1 in controlling translation elongation as well. The study carried shows S6K1 to phosphorylate and inactivate eEF2k (eukaryotic elongation factor-2 kinase, a negative regulator of translation [82].

5.2. Cell cycle progression-a reality check

S6K1 has been extensively discussed as a kinase that drives cells from G1 to S phase during cell cycle progress. However, the data in support does not seem to match this notion completely

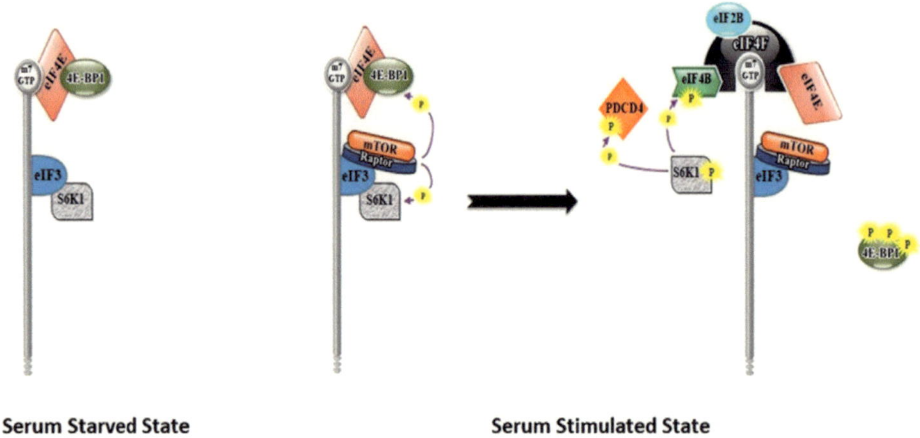

Figure 5. S6K1, a component of translation initiation complex: S6K1 is bound to eIF3 in basal (resting) state cells. Upon serum stimulation, mTORC1 is recruited to the complex, wherein it activates S6K1, which dissociates from eIF3 and phosphorylates several of its targets involved in the initiation of translation.

and in fact some of the data available even contests the veracity of these claims. Whereas Embryonic Stem (ES) cells devoid of S6K1 (S6K1−/−) and mouse embryonic fibroblasts (MEFs) obtained from S6K1 deficient mice do not show significant defects in cell proliferation [77], other studies implicate S6K1 in facilitating cell cycle progression through phosphorylation of estrogen receptor leading to the activation of its target genes that ultimately promotes cellular proliferation [83]. Additionally S6K1 exerts its control over cellular transcription through its ability to phosphorylate cAMP response element binding protein (CREB) isoform CREMr and transcription factor UBF-1 [84]. Phosphorylated UBF-1 in turn activates RNA Polymerase 1 driven transcription of genes encoding ribosomal RNAs and as such aid in ribosomal biogenesis. However, contrary to this belief McMullen et al. proposed no absolute requirement of S6K at either transcriptional or translational levels during ribosomal biogenesis. They showed that $S6K^{-}/^{-}$ mice do not exhibit impaired cardiac hypertrophy, a response dependent on ribosomal biogenesis [85]. These observations put a question mark on the actual involvement of S6K1 in controlling cell cycle progression. Howsoever discordant the views with respect to direct involvement of S6K1 may be but the role of S6K1 in accelerating the cell cycle from G1 to S remains undisputed.

5.3. Cellular metastasis—handy option for therapeutic interventions

S6 kinase, being a downstream effector of mTOR is upregulated in various cancerous situations and as such regulates cellular metastasis. Its involvement in brain tumor pathogenesis was established by the fact that mTOR/S6K axis remains constitutively activated in glioma cells and upon S6K knockdown, the transformed phenotype of these cells is partially rescued [86]. Additionally constitutive activation of S6K1 has been reported in esophageal squamous cell carcinomas (ESCC) [87]. These and several other examples highlight the importance of S6K1 in various cancerous state and as such makes it a potential target for therapeutic interventions to treat cancer. A number of pharmaceutical companies have already ventured in developing S6K specific inhibitors like ATP-competitive compounds that exhibit selective inhibition of S6K. This remains important in the backdrop of cancer relapses/resistance observed while treating them with mTOR inhibitor rapamycin or its rapalogues [88]. However, until recently it was not clear as to what contributes to the rapamycin resistance in these cells. This conundrum was to a greater extent resolved when MAPK interacting kinase (MNK) was found to be an active player in exerting rapamycin resistance in cancer cells. The study deliberated that cancer cell resistance to the mTORC1 inhibitor rapamycin involves MNK activation via a feedback signaling loop elicited by rapamycin [89, 90]. Much recently a study by Brown et al. reveals that post rapamycin treatment mTORC1 activity is sustained by MNK in cancer cells by way of promoting the association between MNK and mTORC1 to form a sub-complex. This way MNK facilitates binding of mTORC1 with its substrates and with phosphatidyl inositol 3'kinase-related kinase (PIKK) stabilizer, TELO2, while discouraging DEPTOR (endogenous mTOR inhibitor) binding [91]. In addition MNKs are also known for phosphorylating eukaryotic initiation factor 4E, eIF4e [92] and as such may play a role in cellular translation. Further the data generated in our lab hints out at the possible role of eIF4e in regulating S6K1 (unpublished data). We believe that prolonged exposure of cells to rapamycin results in activation of MNKs with the resultant activation of eIF4e. As a result S6K1 gets activated to counter the anti-proliferative/anti-cancerous effects of rapamycin (**Figure 6**). Thus therapeutic interventions envisaging S6K1 as a direct target would help overcome, to a larger extent, the resistance experienced toward mTORC1 inhibitors and as such emerge as more appropriate target for treating cancers.

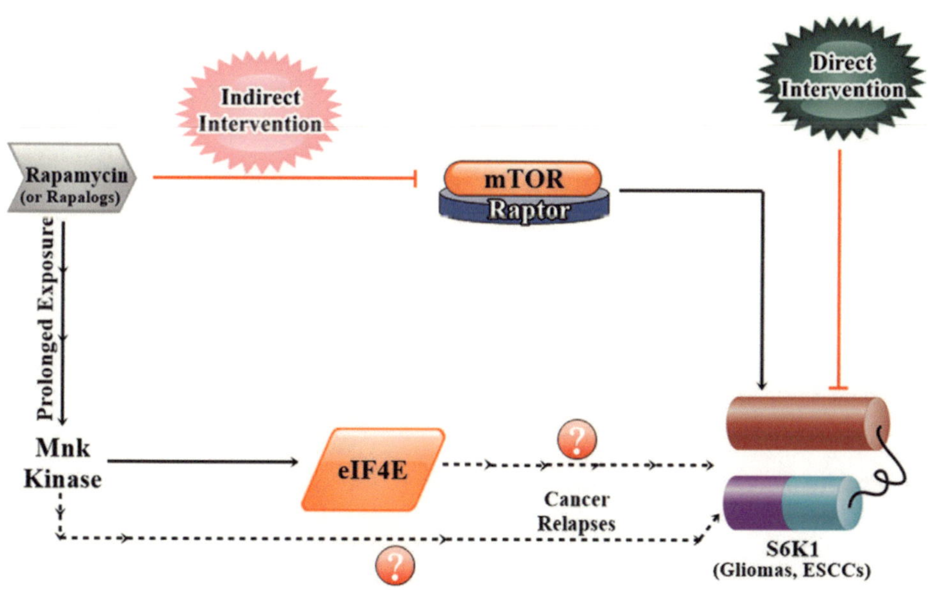

Figure 6. S6K1, a prudent target for treating cancers: an illustration representing S6K1 as a potential target for the specific inhibitors which could be used to overcome the resistance observed while using mTORC1 inhibitors during various cancerous situations.

6. Conclusion

Understanding S6K1 regulation and its associated functions over the last two decades has seen an upward trend. It has now been possible to establish S6K1 an important regulator of cell growth and proliferation. Further our understanding vis-à-vis S6K1 functions have also improved considerably to dissociate many of its functions from the ones related to phosphorylating rpS6 only. Instead its control over various cellular factors makes it more important candidate for regulating cellular physiology. Of late, S6K1 appears as a prudent target for treating various cancers and as such various trials have been initiated to use specific S6K1 inhibitors with the aim to treat different cancers where the routine practices of using generalized drugs appear less effective. However, achieving a complete success in this endeavor still seems difficult. A major challenge in this regard is to ensure that S6K1 s role is tumor specific regardless of its type or origin. Resolving such issues will of course be a task that would attract more research in the years to come.

Conflict of interest

The authors declare that there is no conflict of interest.

Author details

Sheikh Tahir Majeed[1], Rabiya Majeed[1,2], Ghazia Shah[1] and Khurshid I Andrabi[1]*

*Address all correspondence to: andrabik@uok.edu.in

1 Department of Biotechnology, Science block, University of Kashmir, Jammu and Kashmir, India

2 Department of Biochemistry, University of Kashmir, Jammu and Kashmir, India

References

[1] Anjum R, Blenis J. The RSK family of kinases: Emerging roles in cellular signalling. Nature Reviews. Molecular Cell Biology. 2008;**9**(10):747

[2] Chung J, Kuo CJ, Crabtree GR, Blenis J. Rapamycin-FKBP specifically blocks growth-dependent activation of and signaling by the 70 kd S6 protein kinases. Cell. 1992;**69**(7): 1227-1236

[3] Price DJ, Calvo V, Avruch J, Bierer BE. Rapamycin-induced inhibition of the 70-kilodalton S6 protein kinase. Science. 1992;**257**(5072):973-977

[4] Stewart MJ, Berry CO, Zilberman F, Thomas G, Kozma SC. The Drosophila p70s6k homolog exhibits conserved regulatory elements and rapamycin sensitivity. Proceedings of the National Academy of Sciences. 1996;**93**(20):10791-10796

[5] Jenö P, Ballou LM, Novak-Hofer I, Thomas G. Identification and characterization of a mitogen-activated S6 kinase. Proceedings of the National Academy of Sciences. 1988;**85**(2):406-410

[6] Banerjee P, Ahmad MF, Grove JR, Kozlosky C, Price DJ, Avruch J. Molecular structure of a major insulin/mitogen-activated 70-kDa S6 protein kinase. Proceedings of the National Academy of Sciences. 1990;**87**(21):8550-8554

[7] Harmann B, Kilimann MW. cDNA encoding a 59 kDa homolog of ribosomal protein S6 kinase from rabbit liver. FEBS Letters. 1990;**273**(1-2):248-252

[8] Kozma SC, Ferrari S, Bassand P, Siegmann M, Totty N, Thomas G. Cloning of the mitogen-activated S6 kinase from rat liver reveals an enzyme of the second messenger subfamily. Proceedings of the National Academy of Sciences. 1990;**87**(19):7365-7369

[9] Gout I, Minami T, Hara K, Tsujishita Y, Filonenko V, Waterfield MD, et al. Molecular cloning and characterization of a novel p70 S6 kinase, p70 S6 kinase β containing a proline-rich region. The Journal of Biological Chemistry. 1998;**273**(46):30061-30064

[10] Grove JR, Banerjee P, Balasubramanyam A, Coffer PJ, Price DJ, Avruch J, et al. Cloning and expression of two human p70 S6 kinase polypeptides differing only at their amino termini. Molecular and Cellular Biology. 1991;**11**(11):5541-5550

[11] Koh H, Jee K, Lee B, Kim J, Kim D, Yun Y-H, et al. Cloning and characterization of a nuclear S6 kinase, S6 kinase-related kinase (SRK); a novel nuclear target of Akt. Oncogene. 1999;**18**(36):5115

[12] Lee-Fruman KK, Kuo CJ, Lippincott J, Terada N, Blenis J. Characterization of S6K2, a novel kinase homologous to S6K1. Oncogene. 1999 Sep;**18**(36):5108-5114

[13] Saitoh M, Ten Dijke P, Miyazono K, Ichijo H. Cloning and characterization of p70S6KβDefines a novel family of p70 S6 kinases. Biochemical and Biophysical Research Communications. 1998;**253**(2):470-476

[14] Shima H, Pende M, Chen Y, Fumagalli S, Thomas G, Kozma SC. Disruption of the p70s6k/p85s6k gene reveals a small mouse phenotype and a new functional S6 kinase. The EMBO Journal. 1998;**17**(22):6649-6659

[15] Magnuson B, Ekim B, Fingar DC. Regulation and function of ribosomal protein S6 kinase (S6K) within mTOR signalling networks. The Biochemical Journal. 2012;**441**(1):1-21

[16] Hanks SK, Hunter T. Protein kinases 6. The eukaryotic protein kinase superfamily: Kinase (catalytic) domain structure and classification. The FASEB Journal. 1995;**9**(8):576-596

[17] Jacinto E, Lorberg A. TOR regulation of AGC kinases in yeast and mammals. The Biochemical Journal. 2008;**410**(1):19-37

[18] Pearce LR, Komander D, Alessi DR. The nuts and bolts of AGC protein kinases. Nature Reviews. Molecular Cell Biology. 2010;**11**(1):9

[19] Rosner M, Hengstschläger M. Nucleocytoplasmic localization of p70 S6K1, but not of its isoforms p85 and p31, is regulated by TSC2/mTOR. Oncogene. 2011;**30**(44):4509

[20] Schalm SS, Blenis J. Identification of a conserved motif required for mTOR signaling. Current Biology. 2002;**12**(8):632-639

[21] Burnett PE, Blackshaw S, Lai MM, Qureshi IA, Burnett AF, Sabatini DM, et al. Neurabin is a synaptic protein linking p70 S6 kinase and the neuronal cytoskeleton. Proceedings of the National Academy of Sciences. 1998;**95**(14):8351-8356

[22] Karni R, de Stanchina E, Lowe SW, Sinha R, Mu D, Krainer AR. The gene encoding the splicing factor SF2/ASF is a proto-oncogene. Nature Structural & Molecular Biology. 2007;**14**(3):185

[23] Fenton TR, Gout IT. Functions and regulation of the 70 kDa ribosomal S6 kinases. The International Journal of Biochemistry & Cell Biology. 2011;**43**(1):47-59

[24] Meyuhas O, Dreazen A. Ribosomal protein S6 kinase: From TOP mRNAs to cell size. Progress in Molecular Biology and Translational Science. 2009;**90**:109-153

[25] Foster KG, Fingar DC. Mammalian target of rapamycin (mTOR): Conducting the cellular signaling symphony. The Journal of Biological Chemistry. 2010;**285**(19):14071-14077

[26] Zoncu R, Efeyan A, Sabatini DM. mTOR: From growth signal integration to cancer, diabetes and ageing. Nature Reviews. Molecular Cell Biology. 2011;**12**(1):21

[27] Chou MM, Blenis J. The 70 kDa S6 kinase complexes with and is activated by the rho family G proteins Cdc42 and Rac1. Cell. 1996;**85**(4):573-583

[28] Romanelli A, Martin KA, Toker A, Blenis J. p70 S6 kinase is regulated by protein kinase Cζ and participates in a phosphoinositide 3-kinase-regulated signalling complex. Molecular and Cellular Biology. 1999;**19**(4):2921-2928

[29] Alessi DR, Kozlowski MT, Weng Q-P, Morrice N, Avruch J. 3-Phosphoinositide-dependent protein kinase 1 (PDK1) phosphorylates and activates the p70 S6 kinase in vivo and in vitro. Current Biology. 1998 Jan 15;**8**(2):69-81

[30] Ferrari S, Bannwarth W, Morley SJ, Totty NF, Thomas G. Activation of p70s6k is associated with phosphorylation of four clustered sites displaying Ser/Thr-Pro motifs. Proceedings of the National Academy of Sciences. 1992;**89**(15):7282-7286

[31] Keshwani MM, Von Daake S, Newton AC, Harris TK, Taylor SS. Hydrophobic motif phosphorylation is not required for activation loop phosphorylation of p70 ribosomal protein S6 kinase 1 (S6K1). The Journal of Biological Chemistry. 2011;**286**(26):23552-23558

[32] Mukhopadhyay NK, Price DJ, Kyriakis JM, Pelech S, Sanghera J, Avruch J. An array of insulin-activated, proline-directed serine/threonine protein kinases phosphorylate the p70 S6 kinase. The Journal of Biological Chemistry. 1992;**267**(5):3325-3335

[33] Weng Q-P, Kozlowski M, Belham C, Zhang A, Comb MJ, Avruch J. Regulation of the p70 S6 kinase by phosphorylation in vivo analysis using site-specific anti-phosphopeptide antibodies. The Journal of Biological Chemistry. 1998;**273**(26):16621-16629

[34] Price DJ, Mukhopadhyay NK, Avruch J. Insulin-activated protein kinases phosphorylate a pseudosubstrate synthetic peptide inhibitor of the p70 S6 kinase. The Journal of Biological Chemistry. 1991;**266**(25):16281-16284

[35] Brown EJ, Beal PA, Keith CT, Chen J, Shin TB, Schreiber SL. Control of p70 s6 kinase by kinase activity of FRAP in vivo. Nature. 1995;**377**(6548):441

[36] Burnett PE, Barrow RK, Cohen NA, Snyder SH, Sabatini DM. RAFT1 phosphorylation of the translational regulators p70 S6 kinase and 4E-BP1. Proceedings of the National Academy of Sciences. 1998;**95**(4):1432-1437

[37] Isotani S, Hara K, Tokunaga C, Inoue H, Avruch J, Yonezawa K. Immunopurified mammalian target of rapamycin phosphorylates and activates p70 S6 kinase α in vitro. The Journal of Biological Chemistry. 1999;**274**(48):34493-34498

[38] Dennis PB, Pullen N, Pearson RB, Kozma SC, Thomas G. Phosphorylation sites in the autoinhibitory domain participate in p70s6k activation loop phosphorylation. The Journal of Biological Chemistry. 1998;**273**(24):14845-14852

[39] Pullen N, Dennis PB, Andjelkovic M, Dufner A, Kozma SC, Hemmings BA, et al. Phosphorylation and activation of p70s6k by PDK1. Science. 1998;**279**(5351):707-710

[40] Alessi DR, James SR, Downes CP, Holmes AB, Gaffney PRJ, Reese CB, et al. Characterization of a 3-phosphoinositide-dependent protein kinase which phosphorylates and activates protein kinase Bα. Current Biology. 1997 Apr 1;**7**(4):261-269

[41] Williams MR, Arthur JSC, Balendran A, van der Kaay J, Poli V, Cohen P, et al. The role of 3-phosphoinositide-dependent protein kinase 1 in activating AGC kinases defined in embryonic stem cells. Current Biology. 2000;**10**(8):439-448

[42] Moser BA, Dennis PB, Pullen N, Pearson RB, Williamson NA, Wettenhall RE, et al. Dual requirement for a newly identified phosphorylation site in p70s6k. Molecular and Cellular Biology. 1997;**17**(9):5648-5655

[43] Saitoh M, Pullen N, Brennan P, Cantrell D, Dennis PB, Thomas G. Regulation of an activated S6 kinase 1 variant reveals a novel mammalian target of rapamycin phosphorylation site. The Journal of Biological Chemistry. 2002;**277**(22):20104-20112

[44] Oh WJ, Wu C, Kim SJ, Facchinetti V, Julien L-A, Finlan M, et al. mTORC2 can associate with ribosomes to promote cotranslational phosphorylation and stability of nascent Akt polypeptide. The EMBO Journal. 2010;**29**(23):3939-3951

[45] Cheatham L, Monfar M, Chou MM, Blenis J. Structural and functional analysis of pp70S6k. Proceedings of the National Academy of Sciences. 1995;**92**(25):11696-11700

[46] Nojima H, Tokunaga C, Eguchi S, Oshiro N, Hidayat S, Yoshino K, et al. The mammalian target of rapamycin (mTOR) partner, raptor, binds the mTOR substrates p70 S6 kinase and 4E-BP1 through their TOR signaling (TOS) motif. The Journal of Biological Chemistry. 2003;**278**(18):15461-15464

[47] Schalm SS, Tee AR, Blenis J. Characterization of a conserved C-terminal motif (RSPRR) in ribosomal protein S6 kinase 1 required for its mammalian target of rapamycin-dependent regulation. The Journal of Biological Chemistry. 2005;**280**(12):11101-11106

[48] Lee C-H, Inoki K, Guan K-L. mTOR pathway as a target in tissue hypertrophy. Annual Review of Pharmacology and Toxicology. 2007;**47**:443-467

[49] Kim D-H, Sarbassov DD, Ali SM, King JE, Latek RR, Erdjument-Bromage H, et al. mTOR interacts with raptor to form a nutrient-sensitive complex that signals to the cell growth machinery. Cell. 2002;**110**(2):163-175

[50] Vezina C, Kudelski A, Sehgal SN. Rapamycin (AY-22, 989), a new antifungal antibiotic. Journal of Antibiotics (Tokyo). 1975;**28**(10):721-726

[51] Sehgal SN, Baker H, Vézina C. Rapamycin (AY-22, 989), a new antifungal antibiotic. Journal of Antibiotics (Tokyo). 1975;**28**(10):727-732

[52] Vignot S, Faivre S, Aguirre D, Raymond E. mTOR-targeted therapy of cancer with rapamycin derivatives. Annals of Oncology. 2005;**16**(4):525-537

[53] Heitman J, Movva NR, Hall MN. Targets for cell cycle arrest by the immunosuppressant rapamycin in yeast. Science. 1991;**253**(5022):905-909

[54] Kunz J, Henriquez R, Schneider U, Deuter-Reinhard M, Movva NR, Hall MN. Target of rapamycin in yeast, TOR2, is an essential phosphatidylinositol kinase homolog required for G1 progression. Cell. 1993;**73**(3):585-596

[55] Petroulakis E, Mamane Y, Le Bacquer O, Shahbazian D, Sonenberg N. mTOR signaling: Implications for cancer and anticancer therapy. British Journal of Cancer. 2006;**94**(2):195

[56] Wullschleger S, Loewith R, Hall MN. TOR signaling in growth and metabolism. Cell. 2006;**124**(3):471-484

[57] Brown EJ, Albers MW, Shin TB, Keith CT, Lane WS, Schreiber SL. A mammalian protein targeted by G1-arresting rapamycin–receptor complex. Nature. 1994;**369**(6483):756

[58] Sabatini DM, Erdjument-Bromage H, Lui M, Tempst P, Snyder SH. RAFT1: A mammalian protein that binds to FKBP12 in a rapamycin-dependent fashion and is homologous to yeast TORs. Cell. 1994;**78**(1):35-43

[59] Sabers CJ, Martin MM, Brunn GJ, Williams JM, Dumont FJ, Wiederrecht G, et al. Isolation of a protein target of the FKBP12-rapamycin complex in mammalian cells. The Journal of Biological Chemistry. 1995;**270**(2):815-822

[60] Fingar DC, Blenis J. Target of rapamycin (TOR): An integrator of nutrient and growth factor signals and coordinator of cell growth and cell cycle progression. Oncogene. 2004;**23**(18):3151

[61] Gingras A-C, Raught B, Sonenberg N. Regulation of translation initiation by FRAP/mTOR. Genes & Development. 2001;**15**(7):807-826

[62] Schmelzle T, Hall MN. TOR, a central controller of cell growth. Cell. 2000;**103**(2):253-262

[63] Peterson RT, Beal PA, Comb MJ, Schreiber SL. FKBP12-rapamycin-associated protein (FRAP) autophosphorylates at serine 2481 under translationally repressive conditions. The Journal of Biological Chemistry. 2000;**275**(10):7416-7423

[64] Laplante M, Sabatini DM. mTOR signaling. Cold Spring Harbor Perspectives in Biology. 2012;**4**(2):a011593

[65] Pearson RB, Dennis PB, Han JW, Williamson NA, Kozma SC, Wettenhall RE, et al. The principal target of rapamycin-induced p70s6k inactivation is a novel phosphorylation site within a conserved hydrophobic domain. The EMBO Journal. 1995;**14**(21):5279-5287

[66] Han J-W, Pearson RB, Dennis PB, Thomas G. Rapamycin, Wortmannin, and the methylxanthine SQ20006 inactivate p70s6k by inducing dephosphorylation of the same subset of sites. The Journal of Biological Chemistry. 1995;**270**(36):21396-21403

[67] Ali SM, Sabatini DM. Structure of S6 kinase 1 determines whether raptor-mTOR or rictor-mTOR phosphorylates its hydrophobic motif site. The Journal of Biological Chemistry. 2005;**280**(20):19445-19448

[68] Düvel K, Broach JR. The role of phosphatases in TOR signaling in yeast. In: TOR. Springer; 2004. p. 19-38. ISBN 978-3-642-18930-2

[69] Peterson RT, Desai BN, Hardwick JS, Schreiber SL. Protein phosphatase 2A interacts with the 70-kDa S6 kinase and is activated by inhibition of FKBP12–rapamycin associated protein. Proceedings of the National Academy of Sciences. 1999;**96**(8):4438-4442

[70] Hahn K, Miranda M, Francis VA, Vendrell J, Zorzano A, Teleman AA. PP2A regulatory subunit PP2A-B′ counteracts S6K phosphorylation. Cell Metabolism. 2010;**11**(5):438-444

[71] Fenton TR, Gwalter J, Ericsson J, Gout IT. Histone acetyltransferases interact with and acetylate p70 ribosomal S6 kinases in vitro and in vivo. The International Journal of Biochemistry & Cell Biology. 2010;**42**(2):359-366

[72] Fenton TR, Gwalter J, Cramer R, Gout IT. S6K1 is acetylated at lysine 516 in response to growth factor stimulation. Biochemical and Biophysical Research Communications. 2010;**398**(3):400-405

[73] Gwalter J, Wang M-L, Gout I. The ubiquitination of ribosomal S6 kinases is independent from the mitogen-induced phosphorylation/activation of the kinase. The International Journal of Biochemistry & Cell Biology. 2009;**41**(4):828-833

[74] Wang M-L, Panasyuk G, Gwalter J, Nemazanyy I, Fenton T, Filonenko V, et al. Regulation of ribosomal protein S6 kinases by ubiquitination. Biochemical and Biophysical Research Communications. 2008;**369**(2):382-387

[75] Panasyuk G, Nemazanyy I, Filonenko V, Gout I. Ribosomal protein S6 kinase 1 interacts with and is ubiquitinated by ubiquitin ligase ROC1. Biochemical and Biophysical Research Communications. 2008;**369**(2):339-343

[76] Barilari M, Bonfils G, Treins C, Koka V, De Villeneuve D, Fabrega S, et al. ZRF1 is a novel S6 kinase substrate that drives the senescence programme. The EMBO Journal. 2017;**36**(6):736-750

[77] Pende M, Um SH, Mieulet V, Sticker M, Goss VL, Mestan J, et al. S6K1−/−/S6K2−/− mice exhibit perinatal lethality and rapamycin-sensitive 5′-terminal oligopyrimidine mRNA translation and reveal a mitogen-activated protein kinase-dependent S6 kinase pathway. Molecular and Cellular Biology. 2004;**24**(8):3112-3124

[78] Ruvinsky I, Sharon N, Lerer T, Cohen H, Stolovich-Rain M, Nir T, et al. Ribosomal protein S6 phosphorylation is a determinant of cell size and glucose homeostasis. Genes & Development. 2005;**19**(18):2199-2211

[79] Holz MK, Ballif BA, Gygi SP, Blenis J. mTOR and S6K1 mediate assembly of the translation preinitiation complex through dynamic protein interchange and ordered phosphorylation events. Cell. 2005;**123**(4):569-580

[80] Raught B, Peiretti F, Gingras A-C, Livingstone M, Shahbazian D, Mayeur GL, et al. Phosphorylation of eucaryotic translation initiation factor 4B Ser422 is modulated by S6 kinases. The EMBO Journal. 2004;**23**(8):1761-1769

[81] Dorrello NV, Peschiaroli A, Guardavaccaro D, Colburn NH, Sherman NE, Pagano M. S6K1-and \s sTRCP-mediated degradation of PDCD4 promotes protein translation and cell growth. Science. 2006;**314**(5798):467-471

[82] Wang X, Li W, Williams M, Terada N, Alessi DR, Proud CG. Regulation of elongation factor 2 kinase by p90RSK1 and p70 S6 kinase. The EMBO Journal. 2001;**20**(16):4370-4379

[83] Yamnik RL, Digilova A, Davis DC, Brodt ZN, Murphy CJ, Holz MK. S6 kinase 1 regulates estrogen receptor α in control of breast cancer cell proliferation. The Journal of Biological Chemistry. 2009;**284**(10):6361-6369

[84] Hannan KM, Thomas G, Pearson RB. Activation of S6K1 (p70 ribosomal protein S6 kinase 1) requires an initial calcium-dependent priming event involving formation of a high-molecular-mass signalling complex. The Biochemical Journal. 2003;**370**(2):469-477

[85] McMullen JR, Shioi T, Zhang L, Tarnavski O, Sherwood MC, Dorfman AL, et al. Deletion of ribosomal S6 kinases does not attenuate pathological, physiological, or insulin-like growth factor 1 receptor-phosphoinositide 3-kinase-induced cardiac hypertrophy. Molecular and Cellular Biology. 2004;**24**(14):6231-6240

[86] Nakamura JL, Garcia E, Pieper RO. S6K1 plays a key role in glial transformation. Cancer Research. 2008;**68**(16):6516-6523

[87] Hou G, Xue L, Lu Z, Fan T, Tian F, Xue Y. An activated mTOR/p70S6K signaling pathway in esophageal squamous cell carcinoma cell lines and inhibition of the pathway by rapamycin and siRNA against mTOR. Cancer Letters. 2007;**253**(2):236-248

[88] Wilkinson JE, Burmeister L, Brooks SV, Chan C-C, Friedline S, Harrison DE, et al. Rapamycin slows aging in mice. Aging Cell. 2012;**11**(4):675-682

[89] Carracedo A, Ma L, Teruya-Feldstein J, Rojo F, Salmena L, Alimonti A, et al. Inhibition of mTORC1 leads to MAPK pathway activation through a PI3K-dependent feedback loop in human cancer. The Journal of Clinical Investigation. 2008;**118**(9):3065-3074

[90] Wang X, Yue P, Chan C-B, Ye K, Ueda T, Watanabe-Fukunaga R, et al. Inhibition of mammalian target of rapamycin induces phosphatidylinositol 3-kinase-dependent and Mnk-mediated eukaryotic translation initiation factor 4E phosphorylation. Molecular and Cellular Biology. 2007;**27**(21):7405-7413

[91] Brown MC, Gromeier M. MNK controls mTORC1: Substrate association through regulation of TELO2 binding with mTORC1. Cell Reports. 2017;**18**(6):1444-1457

[92] Pyronnet S, Imataka H, Gingras A-C, Fukunaga R, Hunter T, Sonenberg N. Human eukaryotic translation initiation factor 4G (eIF4G) recruits mnk1 to phosphorylate eIF4E. The EMBO Journal. 1999;**18**(1):270-279